Sandra Tisdell-Clifford

DEVELOPMENTAL MATHEMATICS BOOK 2

Founding authors
Allan Thompson · Effie Wrightson

Series editor
Robert Yen

Australia • Brazil • Japan • Korea • Mexico • Singapore • Spain • United Kingdom • United States

Developmental Mathematics Book 2
5th Edition
Sandra Tisdell-Clifford

Publishing editor: Robert Yen
Project editor: Sarah Broomhall
Editor: Lisa Schmidt
Text design: Sarah Hazell
Cover design: Sarah Hazell
Cover image: iStockphoto/shuoshu; shutterstock.com/javarman
Permissions researcher: Wendy Duncan
Production controller: Erin Dowling
Typeset by Cenveo Publishing Services

For product information and technology assistance,
in Australia call **1300 790 853**;
in New Zealand call **0800 449 725**

For permission to use material from this text or product, please email
aust.permissions@cengage.com

National Library of Australia Cataloguing-in-Publication Data
Tisdell-Clifford, Sandra, author.
Developmental mathematics. Book 2 / Sandra Tisdell-Clifford.

1st edition.
9780170350990 (paperback)
For secondary school age.

Mathematics--Study and teaching (Secondary)--Australia.
Mathematics--Problems, exercises, etc.

510.76

Cengage Learning Australia
Level 7, 80 Dorcas Street
South Melbourne, Victoria Australia 3205

Cengage Learning New Zealand
Unit 4B Rosedale Office Park
331 Rosedale Road, Albany, North Shore 0632, NZ

For learning solutions, visit **cengage.com.au**

Printed in China by 1010 Printing International Limited
7 8 9 10 11 12 25 24 23 22 21

CONTENTS

#Pythagoras' theorem is a Year 9 topic in the Australian Curriculum and a Stage 4 (Year 8) topic in the NSW syllabus

ISBN 9780170350990

PREFACE

In schools for over four decades, *Developmental Mathematics* has been a unique, well-known and trusted Years 7–10 mathematics series with a strong focus on key numeracy and literacy skills. This 5th edition of the series has been revised for the new Australian curriculum as well as the NSW syllabus Stages 4 and 5.1. The four books of the series contain short chapters with worked examples, definitions of key words, graded exercises, a language activity and a practice test. Each chapter covers a topic that should require about two weeks of teaching time.

Developmental Mathematics supports students with mathematics learning, encouraging them to experience more confidence and success in the subject. This series presents examples and exercises in clear and concise language to help students master the basics and improve their understanding. We have endeavoured to equip students with the essential knowledge required for success in junior high school mathematics, with a focus on basic skills and numeracy.

Developmental Mathematics Book 2 is written for students in Years 8–9, covering the Australian curriculum (mostly Year 8 content) and NSW syllabus (see the curriculum grids on the following pages and the teaching program on the NelsonNet teacher website). This book presents concise and highly structured examples and exercises, with each new concept or skill on a double-page spread for convenient reading and referencing.

Students learning mathematics need to be taught by dynamic teachers who use a variety of resources. Our intention is that teachers and students use this book as their primary source or handbook, and supplement it with additional worksheets and resources, including those found on the NelsonNet teacher website (access conditions apply). We hope that teachers can use this book effectively to help students achieve success in secondary mathematics. Good luck!

ABOUT THE AUTHOR

Sandra Tisdell-Clifford teaches at Newcastle Grammar School and was the Mathematics coordinator at Our Lady of Mercy College (OLMC) in Parramatta for 10 years. Sandra is best known for updating *Developmental Mathematics* for the 21st century (4th edition, 2003) and writing its blackline masters books. She also co-wrote *Nelson Senior Maths 11 General for the Australian curriculum*, teaching resources for the NSW senior series *Maths in Focus* and the Years 7–8 homework sheets for *New Century Maths/NelsonNet.*

Sandra expresses her thanks and appreciation to the Headmaster and staff of Newcastle Grammar School and dedicates this book to her husband, Ray Clifford, for his support and encouragement. She also thanks series editor **Robert Yen** and editors **Lisa Schmidt, Sarah Broomhall** and **Alan Stewart** at Cengage Learning for their leadership on this project.

Original authors **Allan Thompson** and **Effie Wrightson** wrote the first three editions of *Developmental Mathematics* (published 1974, 1981 and 1988) and taught at Smith's Hill High School in Wollongong. Sandra thanks them for their innovative pioneering work, which has paved the way for this new edition for the Australian curriculum.

ISBN 9780170350990

FEATURES OF THIS BOOK

- Each chapter begins with a table of contents and list of chapter outcomes
- Each teaching section of a chapter is presented clearly on a double-page spread

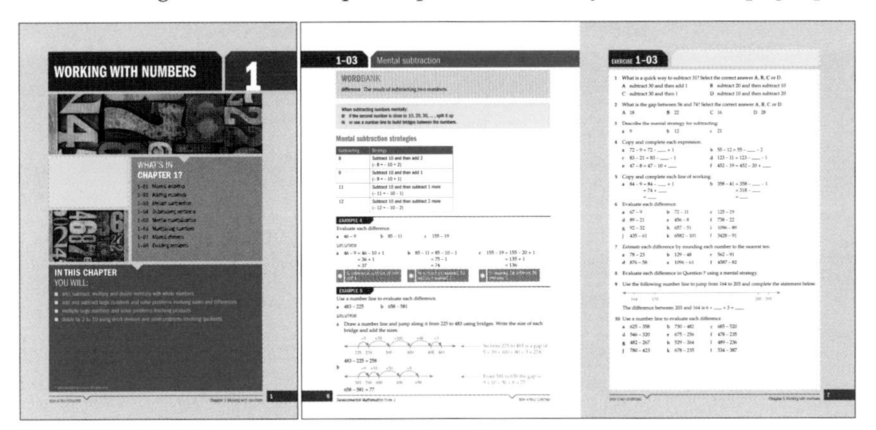

- The left page contains explanations, worked examples, and if appropriate, a Wordbank of mathematical terminology and a fact box
- The right page contains an exercise set, including multiple-choice questions, scaffolded solutions and realistic applications of mathematics
- Each chapter concludes with a **Language activity** (puzzle) that reinforces mathematical terminology in a fun way, and a **Practice test** containing non-calculator questions on general topics and topic questions grouped by chapter subheading

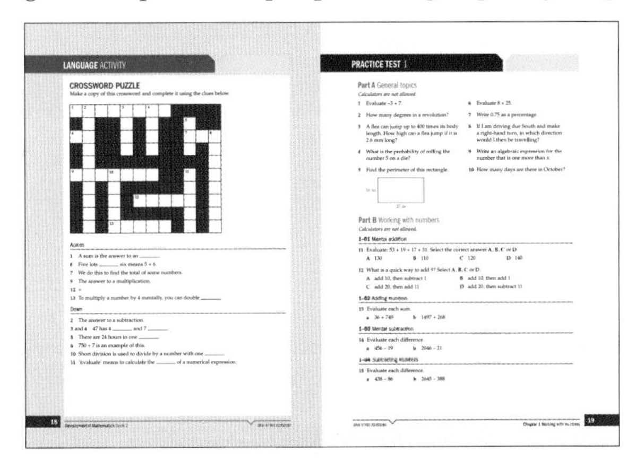

- **Answers** and **index** are at the back of the book
- Additional teaching resources can be downloaded from the NelsonNet teacher website at **www.nelsonnet.com.au**: worksheets, puzzle sheets, skillsheets, video tutorials, technology worksheets, teaching program, curriculum grids, chapter PDFs of this book
- Note: NelsonNet access is available to teachers who use *Developmental Mathematics* as a core educational resource in their classroom. Contact your sales representative for information about access codes and conditions.

CURRICULUM GRID
AUSTRALIAN CURRICULUM

STRAND AND SUBSTRAND	DEVELOPMENTAL MATHEMATICS BOOK 1 CHAPTER	DEVELOPMENTAL MATHEMATICS BOOK 2 CHAPTER
NUMBER AND ALGEBRA		
Number and place value	1 Integers and the number plane 3 Working with numbers 4 Factors and primes 5 Powers and decimals 6 Multiplying and dividing decimals 9 Algebra and equations	1 Working with numbers 2 Primes and powers 4 Integers
Real numbers	5 Powers and decimals 6 Multiplying and dividing decimals 7 Fractions 8 Multiplying and dividing fractions 17 Percentages and ratios	5 Decimals 11 Fractions 12 Percentages 16 Ratios and rates
Money and financial mathematics	6 Multiplying and dividing decimals	12 Percentages
Patterns and algebra	9 Algebra and equations	6 Algebra 15 Further algebra
Linear and non-linear relationships	1 Integers and the number plane 9 Algebra and equations	15 Further algebra 17 Graphing lines
MEASUREMENT AND GEOMETRY		
Using units of measurement	12 Length and time 13 Area and volume	9 Length and time 10 Area and volume
Shape	10 Shapes and symmetry	
Location and transformation	10 Shapes and symmetry	7 Angles and symmetry
Geometric reasoning	2 Angles 11 Geometry	7 Angles and symmetry 8 Triangles and quadrilaterals
Pythagoras and trigonometry		3 Pythagoras' theorem#
STATISTICS AND PROBABILITY		
Chance	16 Probability	14 Probability
Data representation and interpretation	14 Statistical graphs 15 Analysing data	13 Investigating data

#Pythagoras' theorem is a Year 9 topic in the Australian Curriculum and a Stage 4 (Year 8) topic in the NSW syllabus

ISBN 9780170350990

CURRICULUM GRID
AUSTRALIAN CURRICULUM

STRAND AND SUBSTRAND	DEVELOPMENTAL MATHEMATICS BOOK 3 CHAPTER	DEVELOPMENTAL MATHEMATICS BOOK 4 CHAPTER
NUMBER AND ALGEBRA		
Real numbers	**2** Whole numbers and decimals **3** Integers and fractions **6** Percentages **7** Indices **16** Ratios and rates	**1** Working with numbers **2** Percentages **7** Ratios and rates
Money and financial mathematics	**6** Percentages	**3** Earning and saving money
Patterns and algebra	**4** Algebra **7** Indices	**4** Algebra **10** Indices
Linear and non-linear relationships	**9** Equations **14** Graphing lines	**13** Equations and inequalities **15** Coordinate geometry **16** Graphing lines and curves
MEASUREMENT AND GEOMETRY		
Using units of measurement	**12** Length and time **13** Area and volume	**9** Length and time **11** Area and volume
Geometric reasoning	**8** Geometry	**8** Congruent and similar figures
Pythagoras and trigonometry	**1** Pythagoras' theorem **5** Trigonometry	**5** Pythagoras' theorem **6** Trigonometry
STATISTICS AND PROBABILITY		
Chance	**15** Probability	**14** Probability
Data representation and interpretation	**11** Investigating data	**12** Investigating data AC

SERIES OVERVIEW

BOOK 1

1 Integers and the number plane
2 Angles
3 Working with numbers
4 Factors and primes
5 Powers and decimals
6 Multiplying and dividing decimals
7 Fractions
8 Multiplying and dividing fractions
9 Algebra and equations
10 Shapes and symmetry
11 Geometry
12 Length and time
13 Area and volume
14 Statistical graphs
15 Analysing data
16 Probability
17 Percentages and ratios

BOOK 2

1 Working with numbers
2 Primes and powers
3 Pythagoras' theorem
4 Integers
5 Decimals
6 Algebra
7 Angles and symmetry
8 Triangles and quadrilaterals
9 Length and time
10 Area and volume
11 Fractions
12 Percentages
13 Investigating data
14 Probability
15 Further algebra
16 Ratios and rates
17 Graphing lines

BOOK 3

1 Pythagoras' theorem
2 Whole numbers and decimals
3 Integers and fractions
4 Algebra
5 Trigonometry
6 Percentages
7 Indices
8 Geometry
9 Equations
10 Earning money
11 Investigating data
12 Length and time
13 Area and volume
14 Graphing lines
15 Probability
16 Ratios and rates

BOOK 4

1 Working with numbers
2 Percentages
3 Earning and saving money
4 Algebra
5 Pythagoras' theorem
6 Trigonometry
7 Ratios and rates
8 Congruent and similar figures
9 Length and time
10 Indices
11 Area and volume
12 Investigating data
13 Equations and inequalities
14 Probability
15 Coordinate geometry
16 Graphing lines and curves

ISBN 9780170350990

WORKING WITH NUMBERS

WHAT'S IN CHAPTER 1?

IN THIS CHAPTER YOU WILL:

- add, subtract, multiply and divide mentally with whole numbers
- add and subtract large numbers and solve problems involving sums and differences
- multiply large numbers and solve problems involving products
- divide by 2 to 10 using short division and solve problems involving quotients

1-01 Mental addition

WORDBANK

sum The answer to an addition (+) of two or more numbers.

mental Using the mind, not a calculator.

evaluate To find the value or amount.

estimate To make a good (educated) guess of the answer to a problem in round figures.

Mental addition strategies

Adding	Strategy
8	Add 10 and then subtract 2 (+ 8 = + 10 – 2)
9	Add 10 and then subtract 1 (+ 9 = + 10 – 1)
11	Add 10 and then add 1 more (+ 11 = + 10 + 1)
12	Add 10 and then add 2 more (+ 12 = + 10 + 2)

EXAMPLE 1

Evaluate each sum.

a $128 + 9$ **b** $84 + 11$ **c** $4056 + 18$

SOLUTION

a $128 + 9 = 128 + 10 - 1$
$= 138 - 1$
$= 137$

b $84 + 11 = 84 + 10 + 1$
$= 94 + 1$
$= 95$

c $4056 + 18 = 4056 + 20 - 2$
$= 4076 - 2$
$= 4074$

When adding numbers mentally:

- remember that numbers can be added in any order
- look for unit digits that add to 10, such as 6 and 4.

EXAMPLE 2

Evaluate each sum.

a $36 + 68 + 24 + 12$ **b** $163 + 29 + 8 + 237$

SOLUTION

a $36 + 68 + 24 + 12 = (36 + 24) + (68 + 12)$
$= 60 + 80$
$= 140$
(Check by estimating: $36 + 68 + 24 + 12 \approx 40 + 70 + 20 + 10 = 140$)

b $163 + 29 + 8 + 237 = (163 + 237) + (29 + 8)$
$= 400 + 37$
$= 437$
(Check by estimating: $163 + 29 + 8 + 237 \approx 160 + 30 + 10 + 240 = 440$)

ISBN 9780170350990

EXERCISE 1–01

1 What is a quick way to add 12 mentally? Select the correct answer **A**, **B**, **C** or **D**.

A add 10, then subtract 2 **B** add 10, then add 2

C add 20, then add 2 **D** add 20, then subtract 2

2 Which expression gives the same answer as 52 + 45? Select **A**, **B**, **C** or **D**.

A 52 + 54 **B** 50 + 2 + 45 **C** 45 + 25 **D** 52 + 40 + 3

3 Which two numbers should be grouped together to add 36 + 72 + 64 mentally?

4 Describe the mental method for adding:

a 11 b 8 c 19

5 Copy and complete each equation.

a 48 + 11 = 48 + ___ + 1 b 72 + 8 = 72 + ___ − 2

c 125 + 19 = 125 + ___ − 1 d 464 + 12 = 464 + ___ + 2

e 1537 + 9 = 1537 + 10 − ___ f 6852 + 21 = 6852 + 20 + ___

6 Evaluate each sum using a mental strategy.

a 54 + 8 b 27 + 11 c 256 + 19

d 312 + 21 e 56 + 9 f 487 + 12

g 68 + 22 h 246 + 31 i 652 + 99

j 1095 + 41 k 296 + 82 l 4582 + 101

7 Estimate each sum by rounding each number to the nearest ten.

a 68 + 12 b 231 + 49 c 98 + 32

d 125 + 61 e 435 + 89 f 3854 + 21

g 678 + 48 h 8424 + 71 i 20 964 + 52

8 Evaluate each sum in Question 7 using a mental strategy.

9 Evaluate each sum mentally by pairing numbers that have units digits adding to ten.

a 48 + 21 + 19 + 120 b 64 + 230 + 9 + 111

c 74 + 32 + 109 + 28 d 231 + 45 + 119

e 48 + 29 + 112 + 51 f 452 + 61 + 49 + 108

g 118 + 7 + 32 + 253 h 432 + 56 + 18 + 104

10 A magic square has all rows, columns and diagonals adding to the same number. Complete each magic square.

a

3		6
	5	
		7

b

4		7
		3
		8

1-02 Adding numbers

To add large numbers:
- write them underneath each other in their place value columns: units, tens, hundreds, and so on
- add the digits in columns: units first, then tens, and so on
- some additions will involve carrying from one column to its left column.

EXAMPLE 3

Evaluate each sum.

a 962 + 76

b 1355 + 483

SOLUTION

✱ Set out units under units, tens under tens, and so on.

a

$$\begin{array}{r} {}^{1}9\ 6\ 2\ + \\ 7\ 6 \\ \hline 1\ 0\ 3\ 8 \\ \hline \end{array}$$

962 + 76 = 1038

(Estimating: 962 + 76 ≈ 960 + 80 = 1040)

✱ rounding to the nearest ten

b

$$\begin{array}{r} 1\ {}^{1}3\ 5\ 5\ + \\ 4\ 8\ 3 \\ \hline 1\ 8\ 3\ 8 \\ \hline \end{array}$$

1355 + 483 = 1838

(Estimating: 1355 + 483 ≈ 1400 + 500 = 1900)

✱ rounding to the nearest hundred

Shutterstock.com/kurhan

ISBN 9780170350990

EXERCISE 1-02

1 What is the sum of 19, 36, 8 and 15? Select the correct answer **A**, **B**, **C** or **D**.

A 76 **B** 74 **C** 75 **D** 78

2 Increase $365 by $23. Select **A**, **B**, **C** or **D**.

A $368 **B** $388 **C** $378 **D** $366

3 Evaluate each sum.

a 28 + 785 **b** 497 + 56 **c** 54 + 49
d 93 + 56 **e** 2786 + 428 **f** 98 + 5632
g 348 + 74 **h** 525 + 376 **i** 540 + 3779
j 649 + 2876 **k** 642 + 297 **l** 652 + 78

4 What is 34 more than 799?

5 An ice-cream stand sells 48 ice-creams on Monday, 52 on Tuesday, 47 on Wednesday, 56 on Thursday and 148 on Friday. Find the total number of ice-creams sold.

6 Dilani saves $75 one week, $88 the next and $115 in the third week. How much did she save altogether?

7 **a** Increase 984 cm by 78 cm.
b Increase $3276 by $596.

8 Connor went to the supermarket and bought:

a jar of Vegemite for $4.50
a punnet of strawberries for $3.80
3 kg of apples for $9.60
a box of chocolates for $12.25.

What was the total cost of his purchases?

Shutterstock.com/nulinukas

Shutterstock.com/graja

9 Find the sum of 4682, 466, 1327, 18 750 and 848.

10 **a** Estimate the total of $52.60, $129.20, 78c, $2.24 and $368.50.
b Find the exact total of the amounts in part **a**.

1-03 Mental subtraction

WORDBANK

difference The result of subtracting two numbers.

When subtracting numbers mentally:

- if the second number is close to 10, 20, 30, ... , split it up
- or use a number line to build bridges between the numbers.

Mental subtraction strategies

Subtracting	Strategy
8	Subtract 10 and then add 2 (– 8 = – 10 + 2)
9	Subtract 10 and then add 1 (– 9 = – 10 + 1)
11	Subtract 10 and then subtract 1 more (– 11 = – 10 – 1)
12	Subtract 10 and then subtract 2 more (– 12 = – 10 – 2)

EXAMPLE 4

Evaluate each difference.

a $46 - 9$ **b** $85 - 11$ **c** $155 - 19$

SOLUTION

a $46 - 9 = 46 - 10 + 1$
$= 36 + 1$
$= 37$

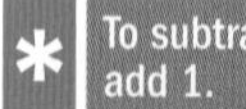
To subtract 9, subtract 10 and add 1.

b $85 - 11 = 85 - 10 - 1$
$= 75 - 1$
$= 74$

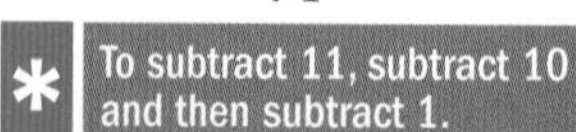
To subtract 11, subtract 10 and then subtract 1.

c $155 - 19 = 155 - 20 + 1$
$= 135 + 1$
$= 136$

* To subtract 19, subtract 20 and add 1.

EXAMPLE 5

Use a number line to evaluate each difference.

a $483 - 225$ **b** $658 - 581$

SOLUTION

a Draw a number line and jump along it from 225 to 483 using bridges. Write the size of each bridge and add the sizes.

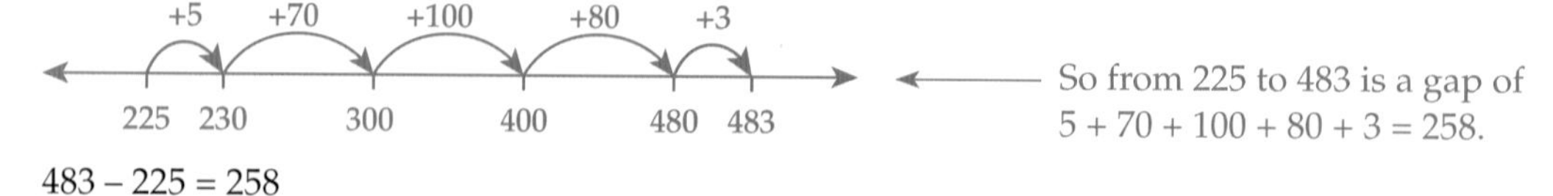

$483 - 225 = 258$

b

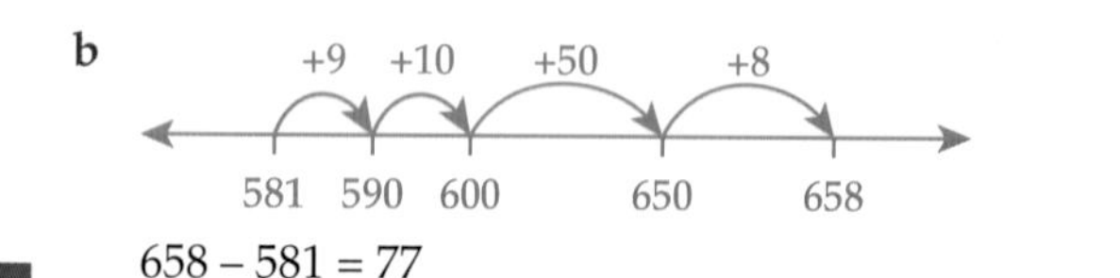

From 581 to 658 the gap is $9 + 10 + 50 + 8 = 77$.

$658 - 581 = 77$

ISBN 9780170350990

EXERCISE 1–03

1 What is a quick way to subtract 31? Select the correct answer **A**, **B**, **C** or **D**.

A subtract 30 and then add 1 **B** subtract 20 and then subtract 10

C subtract 30 and then 1 **D** subtract 10 and then subtract 20

2 What is the gap between 56 and 74? Select **A**, **B**, **C** or **D**.

A 18 **B** 22 **C** 16 **D** 28

3 Describe the mental strategy for subtracting:

a 9 **b** 12 **c** 21

4 Copy and complete each expression.

a 72 – 9 = 72 – ___ + 1 **b** 55 – 12 = 55 – ___ – 2

c 83 – 21 = 83 – ___ – 1 **d** 123 – 11 = 123 – ___ – 1

e 47 – 8 = 47 – 10 + ___ **f** 452 – 19 = 452 – 20 + ___

5 Copy and complete each line of working.

a 84 – 9 = 84 – ___ + 1
= 74 + ___
= ___

b 358 – 41 = 358 – ___ – 1
= 318 – ___
= ___

6 Evaluate each difference.

a 67 – 9 **b** 72 – 11 **c** 125 – 19

d 89 – 21 **e** 456 – 8 **f** 738 – 22

g 92 – 32 **h** 657 – 51 **i** 1096 – 89

j 435 – 61 **k** 6582 – 101 **l** 3428 – 91

7 *Estimate* each difference by rounding each number to the nearest ten.

a 78 – 23 **b** 129 – 48 **c** 562 – 91

d 876 – 58 **e** 1096 – 61 **f** 4587 – 82

8 Evaluate each difference in Question 7 using a mental strategy.

9 Use the following number line to jump from 164 to 203 and complete the statement below.

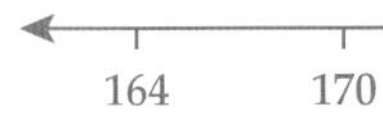

The difference between 203 and 164 is 6 + ___ + 3 = ___.

10 Use a number line to evaluate each difference.

a 625 – 358 **b** 730 – 482 **c** 685 – 520

d 546 – 320 **e** 675 – 256 **f** 478 – 235

g 482 – 267 **h** 529 – 264 **i** 489 – 236

j 780 – 423 **k** 678 – 235 **l** 534 – 387

1–04 Subtracting numbers

To subtract large numbers:

- write them underneath each other in their place value columns: units, tens, hundreds, and so on
- subtract the digits in columns: units first, then tens, and so on
- some subtractions will involve trading from one column to its left column.

EXAMPLE 6

Evaluate each difference.

a 92 – 37 **b** 764 – 328

SOLUTION

a

$$\begin{array}{r} {}^{8}\not{9}\ {}^{1}2\ - \\ 3\ \ 7 \\ \hline 5\ \ 5 \\ \hline \end{array}$$

In the units column, change 2 – 7 to 12 – 7 by taking 10 from the tens column. So 9 in the tens column becomes 8.

92 – 37 = 55 (Estimating: 92 – 37 ≈ 90 – 40 = 50)

b

$$\begin{array}{r} 7\ {}^{5}\not{6}\ {}^{1}4\ - \\ 3\ \ 2\ \ 8 \\ \hline 4\ \ 3\ \ 6 \\ \hline \end{array}$$

In the units column, change 4 – 8 to 14 – 8 by taking 10 from the tens column. So 6 in the tens column becomes 5.

764 – 328 = 436 (Estimating: 764 – 328 ≈ 800 – 300 = 500)

iStockphoto/kershawj

ISBN 9780170350990

EXERCISE 1–04

1 What is the difference between 529 and 76? Select the correct answer **A**, **B**, **C** or **D**.

A 443	**B** 453	**C** 473	**D** 463

2 Decrease $858 by $33. Select **A**, **B**, **C** or **D**.

A $835	**B** $815	**C** $845	**D** $825

3 Evaluate each difference.

a 78 – 34	**b** 210 – 65	**c** 94 – 38
d 73 – 48	**e** 2536 – 478	**f** 7564 – 383
g 428 – 86	**h** 753 – 186	**i** 7208 – 2165
j 6829 – 278	**k** 3652 – 294	**l** 15 682 – 228

4 What is the difference between 8253 and 6089?

5 Natalie saved $465 and then spent $88 on a present. How much did she have left?

6 Jude set out on a road trip from Melbourne to Warrnambool, a distance of 348 km. He travelled 164 km in the first 2 hours before taking a break.

a Estimate how far he still had to travel.

b Calculate the exact number of kilometres he still had to travel.

7 Decrease 12 000 by 288.

8 A train was carrying 82 passengers. At Fortitude Valley, 15 passengers got off the train. At Bowen Hills, 6 passengers got off but 9 got on the train. At Eagle Junction, another 8 passengers got off the train. How many passengers are now on the train?

9 Bianca had a $50 note when she went to the supermarket. She bought:

a box of chocolates for $12.65	2 kg of apples for $5.80
a watermelon for $4.20	a packet of biscuits for $2.75

a What was the total cost of her purchases?

b How much change would Bianca receive from $50?

Shutterstock.com/topseller

10 Year 8 students are asked to set up the hall for assembly with 868 chairs. They already have 296 chairs set up.

a Estimate how many chairs still need to be set up.

b Calculate exactly how many chairs need to be set up.

ISBN 9780170350990

1-05 Mental multiplication

WORDBANK

product The answer to a multiplication (×) of two or more numbers.

Multiplication table

×	1	2	3	4	5	6	7	8	9	10
1	1	2	3	4	5	6	7	8	9	10
2	2	4	6	8	10	12	14	16	18	20
3	3	6	9	12	15	18	21	24	27	30
4	4	8	12	16	20	24	28	32	36	40
5	5	10	15	20	25	30	35	40	45	50
6	6	12	18	24	30	36	42	48	54	60
7	7	14	21	28	35	42	49	56	63	70
8	8	16	24	32	40	48	56	64	72	80
9	9	18	27	36	45	54	63	72	81	90
10	10	20	30	40	50	60	70	80	90	100

Mental multiplication strategies

Multiplying by	Strategy
2	Double
4	Double twice
5	Multiply by 10, then halve
8	Double 3 times
9	Multiply by 10, then subtract the number
10	Add a 0 to the end
100	Add 00 to the end

EXAMPLE 7

Evaluate each product.

a 42×9

b 34×4

c 76×5

SOLUTION

a
$$\begin{aligned} 42 \times 9 &= 42 \times (10 - 1) \\ &= 42 \times 10 - 42 \times 1 \\ &= 420 - 42 \\ &= 378 \end{aligned}$$

b
$$\begin{aligned} 34 \times 4 &= 34 \times 2 \times 2 \\ &= 68 \times 2 \\ &= 136 \end{aligned}$$

* double twice

c
$$\begin{aligned} 76 \times 5 &= 76 \times 10 \times \frac{1}{2} \\ &= 760 \div 2 \\ &= 380 \end{aligned}$$

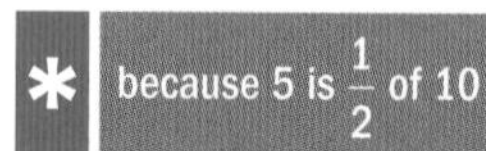

EXAMPLE 8

Evaluate each product by changing the order.

a $3 \times 4 \times 25$

b $7 \times 5 \times 6 \times 2$

SOLUTION

a
$$\begin{aligned} 3 \times 4 \times 25 &= 4 \times 25 \times 3 \\ &= 100 \times 3 \\ &= 300 \end{aligned}$$

← group convenient numbers together

b
$$\begin{aligned} 7 \times 5 \times 6 \times 2 &= 5 \times 2 \times 7 \times 6 \\ &= 10 \times 42 \\ &= 420 \end{aligned}$$

ISBN 9780170350990

EXERCISE 1–05

1 What is a quick method for multiplying by 5 mentally? Select the correct answer **A**, **B**, **C** or **D**.

A multiply by 10 and then double **B** multiply by 2 and then multiply by 3
C multiply by 10 and then halve **D** multiply by 20 and then halve

2 Evaluate 28×4. Select **A**, **B**, **C** or **D**.

A 112 **B** 102 **C** 122 **D** 56

3 Describe the mental strategy for multiplying by:

a 4 **b** 1000 **c** 9 **d** 8

4 Evaluate each product.

a 13×4 **b** 45×1000 **c** 18×5 **d** 22×8
e 47×10 **f** 54×2 **g** 362×100 **h** 19×9

5 Evaluate each product.

a 4×5 **b** 6×7 **c** 8×3 **d** 5×6
e 7×4 **f** 9×5 **g** 6×8 **h** 10×6
i 7×8 **j** 5×8 **k** 7×7 **l** 9×8

6 Evaluate each product.

a 28×2 **b** 52×5 **c** 65×100 **d** 14×8
e 132×1000 **f** 38×9 **g** 5608×10 **h** 24×5
i 22×9 **j** 73×4 **k** 18×5 **l** 15×9

7 Copy and complete each line of working.

a $2 \times 18 \times 5 = 2 \times$ ___ $\times 18$
$=$ ___ $\times 18$
$=$ ___

b $4 \times 31 \times 25 = 4 \times$ ___ $\times 31$
$=$ ___ $\times 31$
$=$ ___

8 Evaluate each product by changing the order.

a $5 \times 4 \times 25$ **b** $13 \times 5 \times 6$ **c** $24 \times 2 \times 50$
d $10 \times 5 \times 20$ **e** $5 \times 20 \times 7$ **f** $4 \times 3 \times 15$
g $4 \times 14 \times 25$ **h** $5 \times 11 \times 200$ **i** $10 \times 8 \times 5$
j $2 \times 7 \times 5$ **k** $6 \times 25 \times 4$ **l** $30 \times 9 \times 10$

ISBN 9780170350990

1–06 Multiplying numbers

WORDBANK

short multiplication A method of multiplying by a number with one digit (1 to 9).

long multiplication A method of multiplying by a number with two or more digits.

EXAMPLE 9

Evaluate each product.

a 428 × 6 **b** 384 × 74

SOLUTION

a Use short multiplication.

$$\begin{array}{r} {}^{1}4\,{}^{4}2\,8 \times \\ 6 \\ \hline 2\,5\,6\,8 \end{array}$$

In the units column, 8 × 6 = 48: write down 8 and carry 4.
In the tens column, 2 × 6 = 12, + 4 = 16: write down 6 and carry 1.
In the hundreds column, 4 × 6 = 24, + 1 = 25.

428 × 6 = 2568 (Estimating: 428 × 6 ≈ 400 × 6 = 2400)

b Use long multiplication.

$$\begin{array}{r} 3\,8\,4 \times \\ 7\,4 \\ \hline 1\,5\,3\,6 \\ 2\,6\,8\,8\,0 \\ \hline 2\,8\,4\,1\,6 \end{array}$$

1536 ← 384 × 4 = 1536
26880 ← Place a 0 in the units column, then 384 × 7 = 2688.
28416 ← 1536 + 26 880 = 28 416

384 × 74 = 28 416 (Estimating: 384 × 74 ≈ 400 × 70 = 28 000)

Shutterstock.com/Giorgio Rossi

ISBN 9780170350990

EXERCISE 1-06

1 What is the product of 15 and 11? Select the correct answer **A**, **B**, **C** or **D**.

A 155 B 165 C 156 D 175

2 What is the product of 150 and 110? Select **A**, **B**, **C** or **D**.

A 1550 B 15 500 C 1650 D 16 500

3 Evaluate each product by short multiplication.

a 83×4 **b** 56×7 **c** 79×8 **d** 218×6

e 328×8 **f** 583×5 **g** 1349×4 **h** 5268×9

i 456×7 **j** 1268×6 **k** 2056×9 **l** 1854×8

4 Evaluate each product by long multiplication.

a 484×16 **b** 963×25 **c** 125×38 **d** 497×18

e 625×26 **f** 1248×36 **g** 346×80 **h** 4800×49

5 If Jack sleeps 11 hours each night, how many hours sleep does he get in a fortnight?

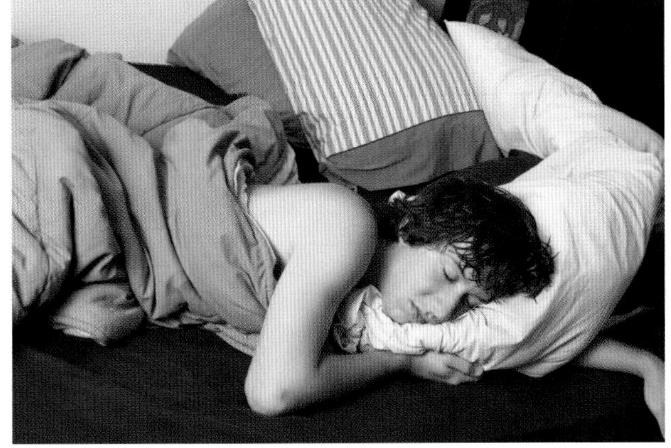

iStockphoto/ktaylorg

6 Tia plants 64 seeds in each row of her garden. How many seeds does she plant altogether if there are 17 rows?

7 How many hours are there in:

a 1 day? **b** 18 days? **c** 1 year?

8 True or false?

a $27 \times 4 = 27 \times 2 \times 2$ **b** $54 \times 9 = 54 \times (10 + 1)$ **c** $42 \times 11 = 42 \times (10 + 1)$

d $36 \times 12 = 36 \times (10 + 2)$ **e** $73 \times 5 = 73 \times 10 \div 2$ **f** $93 \times 8 = 93 \times (10 - 1)$

9 Evaluate each product.

a 28×5 **b** 46×11 **c** 63×9 **d** 82×4

e 35×10 **f** 52×8 **g** 75×2 **h** 97×12

i 17×20 **j** 57×10 **k** 74×40 **l** 68×90

m 27×300 **n** 48×600 **o** 82×500 **p** 91×4000

1-07 Mental division

WORDBANK

quotient The result of dividing (÷) a number by another number. For example, if $12 \div 4 = 3$, the quotient is 3.

Division is the **opposite** of multiplication. Division can be written as: $12 \div 4$ or $\frac{12}{4}$ or $4\overline{)12}$.

Mental division strategies

Dividing by	Strategy
2	Halve
4	Halve twice
5	Divide by 10, then double
8	Halve three times
10	Move the decimal point one place left, or for a whole number ending in 0, drop a 0 from the end of the number
20	Divide by 10, then halve
100	Move the decimal point two places left, or for a whole number ending in 0s, drop two 0s from the end of the number

EXAMPLE 10

Evaluate each quotient.

a $2940 \div 10$ b $716 \div 4$ c $368 \div 100$

d $600 \div 5$ e $568 \div 8$ f $420 \div 20$

SOLUTION

a $2940 \div 10 = 294$

b $716 \div 4 = 716 \div 2 \div 2$

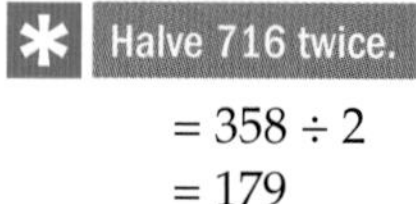

$= 358 \div 2$
$= 179$

c $368 \div 100 = 3.68$

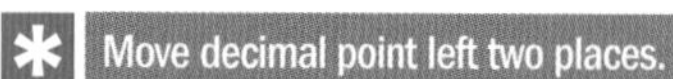

d $600 \div 5 = 600 \div 10 \times 2$

✱ Divide by 10 and double.

$= 60 \times 2$
$= 120$

e $568 \div 8 = 568 \div 2 \div 2 \div 2$

✱ Divide by 2 three times.

$= 284 \div 2 \div 2$
$= 142 \div 2$
$= 71$

f $420 \div 20 = 420 \div 10 \div 2$

✱ Divide by 10, then halve.

$= 42 \div 2$
$= 21$

ISBN 9780170350990

EXERCISE 1-07

1 What is a quick way to divide by 5 mentally? Select the correct answer **A**, **B**, **C** or **D**.

A divide by 10 and then double

B divide by 10 and then multiply by 5

C divide by 10 and then halve

D divide by 25 and then double

2 Evaluate 268 ÷ 4. Select **A**, **B**, **C** or **D**.

A 57 **B** 77 **C** 134 **D** 67

3 Evaluate each quotient.

a 15 ÷ 3 **b** 24 ÷ 6 **c** 32 ÷ 8 **d** 25 ÷ 5

e 42 ÷ 7 **f** 56 ÷ 8 **g** 45 ÷ 9 **h** 20 ÷ 4

i 22 ÷ 2 **j** 90 ÷ 10 **k** 63 ÷ 9 **l** 64 ÷ 8

m 30 ÷ 6 **n** 27 ÷ 3 **o** 35 ÷ 7 **p** 72 ÷ 8

4 Describe the mental method for dividing by:

a 4 **b** 5 **c** 20

5 Evaluate each quotient.

a 568 ÷ 2 **b** 1096 ÷ 4 **c** 7800 ÷ 20

d 850 ÷ 5 **e** 348 ÷ 4 **f** 1620 ÷ 20

g 236 ÷ 2 **h** 576 ÷ 8 **i** 420 ÷ 5

j 184 ÷ 8 **k** 260 ÷ 5 **l** 340 ÷ 20

6 Evaluate each quotient.

a 6500 ÷ 100 **b** 5430 ÷ 10

c 68 000 ÷ 1000 **d** 458 ÷ 10

e 1256 ÷ 1000 **f** 2340 ÷ 100

g 678 ÷ 100 **h** 2450 ÷ 100

i 4050 ÷ 10 **j** 218 ÷ 100

k 49 ÷ 10 **l** 321 ÷ 1000

7 Is each equation true or false?

a 128 ÷ 2 = 64 ÷ 8 **b** 96 ÷ 4 = 48 ÷ 2

c 96 ÷ 8 = 24 ÷ 2 **d** 150 ÷ 10 = 15

e 2800 ÷ 100 = 2.8 **f** 54 ÷ 10 = 0.54

g 236 ÷ 100 = 2.36

ISBN 9780170350990

1-08 Dividing numbers

WORDBANK

short division A method of dividing by a one-digit number (2 to 9).

remainder An amount or number left over from a division.

EXAMPLE 11

Use short division to evaluate each quotient.

a $1624 \div 8$ **b** $387 \div 3$

SOLUTION

a The steps of short division are shown below.

$$\begin{array}{r} 2\_\_ \\ 8\overline{)1624} \end{array} \longleftarrow 16 \div 8 = 2$$

$$\begin{array}{r} 20\_ \\ 8\overline{)162^{2}4} \end{array} \longleftarrow 2 \div 8 = 0\text{, remainder } 2$$

$$\begin{array}{r} 203 \\ 8\overline{)162^{2}4} \end{array} \longleftarrow 24 \div 8 = 3\text{, no remainder}$$

So $1624 \div 8 = 203$.

b
$$\begin{array}{r} 129 \\ 3\overline{)38^{2}7} \end{array} \longleftarrow 3 \div 3 = 1,\ 8 \div 3 = 2 \text{ remainder } 2,\ 27 \div 3 = 9 \text{ no remainder}$$

So $387 \div 3 = 129$.

EXAMPLE 12

Evaluate $5248 \div 9$.

SOLUTION

Sometimes when dividing, there is a **remainder** at the end.

When this happens, we can write the remainder as a fraction of the number we are dividing by.

$$\begin{array}{r} 5\_\_ \\ 9\overline{)52^{7}48} \end{array} \longleftarrow 52 \div 9 = 5\text{, remainder } 7$$

$$\begin{array}{r} 58\_ \\ 9\overline{)52^{7}4^{2}8} \end{array} \longleftarrow 74 \div 9 = 8\text{, remainder } 2$$

$$\begin{array}{r} 583\text{r}1 \\ 9\overline{)52^{7}4^{2}8} \end{array} \longleftarrow 28 \div 9 = 3\text{, remainder } 1$$

So $5248 \div 9 = 583\frac{1}{9}$.

ISBN 9780170350990

EXERCISE 1–08

1 Evaluate 685 ÷ 5. Select the correct answer **A**, **B**, **C** or **D**.

A 137	**B** 127	**C** 147	**D** 117

2 Evaluate 3264 ÷ 4. Select **A**, **B**, **C** or **D**.

A 826	**B** 816	**C** 806	**D** 818

3 Evaluate each quotient.

a 84 ÷ 4	**b** 612 ÷ 6	**c** 4914 ÷ 7	**d** 896 ÷ 8
e 315 ÷ 5	**f** 1854 ÷ 9	**g** 6285 ÷ 3	**h** 4165 ÷ 5
i 1830 ÷ 6	**j** 20 571 ÷ 3	**k** 5624 ÷ 8	**l** 27 189 ÷ 9

4 Sally worked for 7 hours at the local market and was paid $147. How much did she earn per hour?

iStockphoto/Steve Debenport

5 Dinner at a restaurant costs $354 for a table of 6. How much should each person pay if they divide the bill equally between them?

6 A bag of 126 chocolates was shared equally between 9 friends. How many chocolates did each friend receive?

7 At Northbridge Catholic College, there are 125 students in Year 8. If they are placed evenly into five classes, how many students are in each class?

8 Evaluate each quotient, showing the remainder as a fraction.

a 89 ÷ 3	**b** 724 ÷ 6	**c** 957 ÷ 7	**d** 618 ÷ 8
e 3246 ÷ 5	**f** 7259 ÷ 9	**g** 9185 ÷ 3	**h** 1659 ÷ 5

9 **a** Find the length of timber to the nearest cm if 3268 mm is divided into 8 equal pieces.
  b How much is left over?

10 Profits for a business were $24 780. Nikitha wants to share this profit with her two business partners. How much will they each receive if Nikitha rounds each share to the nearest hundred dollars?

CROSSWORD PUZZLE

Make a copy of this crossword and complete it using the clues below.

Across

1 A sum is the answer to an ________.

6 Five lots ________ six means 5×6.

7 We do this to find the total of some numbers.

9 The answer to a multiplication.

12 =

13 To multiply a number by 4 mentally, you can double ________.

Down

2 The answer to a subtraction.

3 and 4 47 has 4 ________ and 7 ________.

5 There are 24 hours in one ________.

8 $750 \div 7$ is an example of this.

10 Short division is used to divide by a number with one ________.

11 'Evaluate' means to calculate the ________ of a numerical expression.

ISBN 9780170350990

PRACTICE TEST 1

Part A General topics

Calculators are not allowed.

1 Evaluate –3 + 7.

2 How many degrees in a revolution?

3 A flea can jump up to 400 times its body length. How high can a flea jump if it is 2.6 mm long?

4 What is the probability of rolling the number 5 on a die?

5 Find the perimeter of this rectangle.

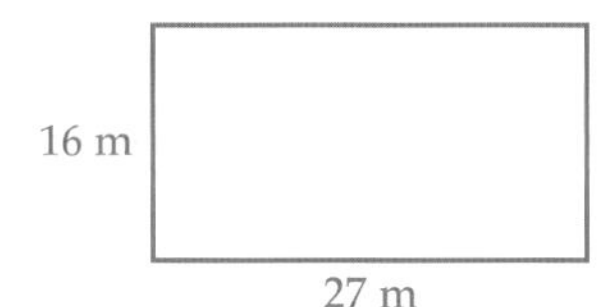

6 Evaluate 8 × 25.

7 Write 0.75 as a percentage.

8 If I am driving due South and make a right-hand turn, in which direction would I then be travelling?

9 Write an algebraic expression for the number that is one more than x.

10 How many days are there in October?

Part B Working with numbers

Calculators are not allowed.

1-01 Mental addition

11 Evaluate: 53 + 19 + 17 + 31. Select the correct answer **A**, **B**, **C** or **D**.

A 130 **B** 110 **C** 120 **D** 140

12 What is a quick way to add 9? Select **A**, **B**, **C** or **D**.

A add 10, then subtract 1

B add 10, then add 1

C add 20, then add 11

D add 20, then subtract 11

1-02 Adding numbers

13 Evaluate each sum.

a 36 + 749 **b** 1497 + 268

1-03 Mental subtraction

14 Evaluate each difference.

a 456 – 19 **b** 2046 – 21

1-04 Subtracting numbers

15 Evaluate each difference.

a 438 – 86 **b** 2645 – 388

1-05 Mental multiplication

16 Evaluate each product.

a 28×5 b 54×20

1-06 Multiplying numbers

17 Evaluate each product.

a 48×6 b 120×38

1-07 Mental division

18 Evaluate each quotient.

a $1428 \div 4$ b $27\,500 \div 1000$

1-08 Dividing numbers

19 Evaluate each quotient.

a $4671 \div 3$ b $5684 \div 8$

ISBN 9780170350990

PRIMES AND POWERS

2

WHAT'S IN CHAPTER 2?

IN THIS CHAPTER YOU WILL:

- test whether a number is divisible by 2, 3, 4, 5, 6, 8 or 9
- identify prime and composite numbers
- evaluate expressions involving powers, square root and cube root
- multiply and divide terms with the same base
- find a power of a power
- use the power of zero

* Shutterstock.com/ChameleonsEye

2-01 Divisibility tests

WORDBANK

divisible A number is divisible by another number if you divide by it and there is no remainder. For example, 10 is divisible by 5 as 10 ÷ 5 = 2, no remainder.

Is 4716 divisible by 3? Does 3 go into 4716 evenly, with no remainder? How do you know? **Divisibility tests** are rules for deciding whether a number is divisible by any number from 2 to 10.

Divisible by:	Divisibility test
2	The number ends with 0, 2, 4, 6, or 8.
3	The sum of the digits in the number is divisible by 3.
4	The last two digits form a number divisible by 4.
5	The number ends with 0 or 5.
6	The number is divisible by *both* 2 and 3.
7	There is no simple divisibility test for 7.
8	The last three digits form a number divisible by 8.
9	The sum of the digits in the number is divisible by 9.
10	The number ends with 0.

EXAMPLE 1

Test whether 4716 is divisible by:

a 3 **b** 4 **c** 6 **d** 8 **e** 9

SOLUTION

a Sum of digits = 4 + 7 + 1 + 6 = 18, which is divisible by 3. ← 18 ÷ 3 = 6
So 4716 is divisible by 3. ← (4716 ÷ 3 = 1572)

b Last two digits = 16, which is divisible by 4. ← 16 ÷ 4 = 4
So 4716 is divisible by 4. ← (4716 ÷ 4 = 1179)

c The number ends in 6, so it is divisible by 2 (even).
The number is divisible by 3 (from **a**)
So 4716 is divisible by 6. ← (4716 ÷ 6 = 786)

d Last three digits = 716, which is *not* divisible by 8.
So 4716 is *not* divisible by 8.

e Sum of digits = 4 + 7 + 1 + 6 = 18, which is divisible by 9. ← 18 ÷ 9 = 2
So 4716 is divisible by 9. ← (4716 ÷ 9 = 524)

ISBN 9780170350990

EXERCISE 2–01

1 What is the test for divisibility by 5? Select the correct answer **A**, **B**, **C** or **D**.

A the number ends in 0 or 5

B the last two digits are divisible by 5

C the number is divisible by 2 and by 3

D the number ends in 5

iStockphoto/selensergen

2 Is 558 divisible by 6? Select **A**, **B**, **C** or **D**.

A Yes, as its digits add to 18, which is divisible by 6

B No, as it does not end in 0 or 6

C No, as the last two digits are not divisible by 6

D Yes, as the number is divisible by 2 and by 3

3 Test whether each number is divisible by 2, 5 and 10.

a 452 **b** 1065 **c** 4580

d 868 **e** 12 550 **f** 6824

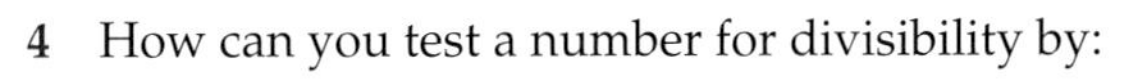

4 How can you test a number for divisibility by:

a 3? **b** 9?

5 Test whether each number is divisible by 3 and by 9.

a 189 **b** 235 **c** 4095

d 12 786 **e** 32 745 **f** 108 963

6 Test whether each number is divisible by 4, by 6 and by 8.

a 456 **b** 826 **c** 468

d 1024 **e** 2598 **f** 18 453

7 There are 168 students in Year 8. Test whether they can be placed into equal groups of:

a 3 **b** 4 **c** 8 **d** 9

8 A cash prize of $1203 is to be divided evenly between a group of winners. If each winner is to receive a share in whole dollars, can this prize money be shared evenly between:

a 3 people **b** 6 people **c** 9 people?

9 Write all the numbers between 1 and 50 that are divisible by both 3 and 5.

2-02 Prime and composite numbers

WORDBANK

factor A value that divides evenly into a given number. For example, 4 is a factor of 28 as $28 \div 4 = 7$.

prime number A number with only two factors, 1 and itself. For example, 5 is a prime number because it has exactly two factors: 1 and 5.

composite number A number with more than two factors. For example, 10 is a composite number because it has four factors: 1, 2, 5 and 10.

- 1 is neither prime nor composite because it has only one factor, 1.
- 2 is the first prime number and the only even prime number (all other even numbers are composite).
- The first five prime numbers are 2, 3, 5, 7 and 11.

EXAMPLE 2

State whether each number is prime or composite.

a 23 **b** 39 **c** 47 **d** 21

SOLUTION

a 23 is prime as its only factors are 1 and 23. ← $1 \times 23 = 23$

b 39 is composite as it has more than two factors, including 3 and 13. ← $3 \times 13 = 39$

c 47 is prime as its only factors are 1 and 47. ← $1 \times 47 = 47$

d 21 is composite as it has more than two factors, including 3 and 7. ← $3 \times 7 = 21$

EXERCISE 2-02

1 Is 1 prime or composite? Select the correct answer **A**, **B**, **C** or **D**.

A prime **B** composite

C both prime and composite **D** neither prime nor composite

2 Is 2 prime or composite? Select **A**, **B**, **C** or **D**.

A prime **B** composite

C both prime and composite **D** neither prime nor composite

3 Write the numbers between 10 and 30 that are prime.

4 Write the numbers between 30 and 50 that are composite.

ISBN 9780170350990

EXERCISE 2-02

5 The prime numbers from 1 to 120 can be found by listing the numbers and using the Sieve of Eratosthenes (pronounced Siv of Era-tos-the-nees). Eratosthenes was an ancient Greek mathematician who 'sifted' out the prime numbers by crossing out all of the multiples of numbers (the composite numbers).

a Copy the grid below for 1 to 120, or print out the worksheet 'Sieve of Eratosthenes' from Book 1's NelsonNet website.

1	2	3	4	5	6
7	8	9	10	11	12
13	14	15	16	17	18
19	20	21	22	23	24
25	26	27	28	29	30
31	32	33	34	35	36
37	38	39	40	41	42
43	44	45	46	47	48
49	50	51	52	53	54
:	:	:	:	:	:

b Cross out 1. It is neither prime nor composite.

c Except for 2, cross out every multiple of 2: 4, 6, 8, 10, … and notice the pattern.

d Except for 3, cross out every multiple of 3: 6, 9, 12, 15, … and notice the pattern.

e Except for 5, cross out every multiple of 5: 10, 15, 20, 25, … and notice the pattern.

f Except for 7, cross out every multiple of 7: 14, 21, 28, 35, … and notice the pattern.

g Write the remaining 30 prime numbers between 1 and 120 (the last one is 113).

6 Look at the prime numbers from Question **5**. There are pairs of numbers called **twin primes** that are only two apart. List all 10 pairs of twin primes between 1 and 120.

7 State whether each number is prime (P) or composite (C).

a 54 **b** 78 **c** 91 **d** 101

e 15 **f** 63 **g** 51 **h** 99

i 281 **j** 643 **k** 225 **l** 357

8 Use the divisibility tests to help test whether each number is prime or composite.

a 491 **b** 279 **c** 1065 **d** 3131

9 What number am I?

a I am prime.
I have two digits.
I am less than 50.
The sum of my digits is 7.

b I am composite.
I am divisible by 6.
I am between 150 and 200.
The sum of my digits is 12.

2-03 Powers and roots

- $5^2 = 5 \times 5$, where 5 is called the **base** and 2 is the **power** or **index**.
- For 5^2, we say '5 squared' or '5 to the power of 2'.
- For $6^3 = 6 \times 6 \times 6$, we say '6 cubed' or '6 to the power of 3'.
- The power shows how many times the base appears in the repeated multiplication.

EXAMPLE 3

Write each expression using index notation.

a $5 \times 5 \times 5 \times 5 \times 5 \times 5$

b $8 \times 8 \times 8 \times 8$

c $12 \times 12 \times 12$

SOLUTION

a $5 \times 5 \times 5 \times 5 \times 5 \times 5 = 5^6$

b $8 \times 8 \times 8 \times 8 = 8^4$

c $12 \times 12 \times 12 = 12^3$

Your calculator has keys for calculating squares x^2, cubes x^3, powers x^y, square roots $\sqrt{\ }$ and cube roots $\sqrt[3]{\ }$.

EXAMPLE 4

Evaluate each expression.

a 7^3 b 2^5 c $\sqrt{16}$

d $\sqrt[3]{8}$ e $\sqrt[3]{-125}$

SOLUTION

a $7^3 = 343$ ← On calculator, enter 7 x^3 =

b $2^5 = 32$ ← On calculator, enter 2 x^y 5 =

c $\sqrt{16} = 4$ ← On calculator, enter $\sqrt{\ }$ 16 =

d $\sqrt[3]{8} = 2$ ← On calculator, enter $\sqrt[3]{\ }$ 8 =

e $\sqrt[3]{-125} = -5$ ← On calculator, enter $\sqrt[3]{\ }$ (−) 125 =

* You need to press the SHIFT or 2ndF key to use $\sqrt[3]{\ }$

EXAMPLE 5

Evaluate each root correct to two decimal places.

a $\sqrt{15}$ b $\sqrt{124}$

c $\sqrt[3]{-29}$ d $\sqrt[3]{238}$

SOLUTION

a $\sqrt{15} = 3.8729$
$= 3.87$

b $\sqrt{124} = 11.1355$
≈ 11.14

c $\sqrt[3]{-29} = -3.0723$
≈ -3.07

d $\sqrt[3]{238} = 6.1971$
≈ 6.20

ISBN 9780170350990

EXERCISE 2-03

1 What is the base number for 5^3? Select the correct answer **A**, **B**, **C** or **D**.

A 3 **B** 5 **C** 25 **D** 125

2 Evaluate 4^5. Select **A**, **B**, **C** or **D**.

A 4 **B** 5 **C** 45 **D** 1024

3 For each expression, write the base and the power.

a 5^8 **b** 7^4 **c** 3^9 **d** 4^2

e 8^5 **f** 3^1 **g** 15^1 **h** 20^4

4 Use the x^y key on your calculator to evaluate each expression in Question **3**.

5 Write each expression in index notation.

a $2 \times 2 \times 2 \times 2 \times 2$ **b** $5 \times 5 \times 5 \times 5$

c $8 \times 8 \times 8 \times 8 \times 8 \times 8$ **d** $9 \times 9 \times 9 \times 9 \times 9$

e $4 \times 4 \times 4 \times 4$ **f** $11 \times 11 \times 11 \times 11 \times 11 \times 11$

g 7 **h** $21 \times 21 \times 21 \times 21$

6 Use the x^y key on your calculator to evaluate each expression in Question **5**.

7 Copy and complete each statement.

a $\sqrt{36} =$ ___ as $6^2 = 36$ **b** $\sqrt[3]{64} =$ ___ as $4^3 = 64$

c $\sqrt{16} = 4$ as ___ $= 16$ **d** $\sqrt[3]{8} = 2$ as ___ $= 8$

8 Evaluate each root.

a $\sqrt{81}$ **b** $\sqrt{100}$ **c** $\sqrt[3]{125}$

d $\sqrt[3]{216}$ **e** $\sqrt{49}$ **f** $\sqrt[3]{1000}$

g $\sqrt{4}$ **h** $\sqrt[3]{1}$ **i** $\sqrt{144}$

j $\sqrt[3]{512}$ **k** $\sqrt{196}$ **l** $\sqrt[3]{729}$

9 Evaluate each root correct to two decimal places.

a $\sqrt{115}$ **b** $\sqrt{38}$ **c** $\sqrt[3]{75}$

d $\sqrt[3]{118}$ **e** $\sqrt{556}$ **f** $\sqrt[3]{107}$

10 **a** Evaluate 30^2, 70^2 and 300^2.

b Hence find $\sqrt{900}$, $\sqrt{4900}$ and $\sqrt{90\,000}$.

11 **a** Evaluate 30^3, 50^3 and 200^3.

b Hence find $\sqrt[3]{27\,000}$, $\sqrt[3]{125\,000}$ and $\sqrt[3]{8\,000\,000}$.

2–04 Multiplying terms with the same base

WORDBANK

index Another word for power.

indices The plural of index, pronounced 'in-der-sees'.

When multiplying terms with the same base, such as $2^5 \times 2^4$, we can add the powers because:

$2^5 \times 2^4 = \underbrace{(2 \times 2 \times 2 \times 2 \times 2)}_{\text{5 times}} \times \underbrace{(2 \times 2 \times 2 \times 2)}_{\text{4 times}} = 2^9$ ← $(5 + 4 = 9)$

> To multiply terms with the same base, add the indices.
> $a^m \times a^n = a^{m+n}$

EXAMPLE 6

a Write $4^2 \times 4^3$ in expanded form.

b Simplify $4^2 \times 4^3$, writing the answer in index notation.

SOLUTION

a $4^2 \times 4^3 = (4 \times 4) \times (4 \times 4 \times 4)$

b $4^2 \times 4^3 = 4^{2+3} = 4^5$

EXAMPLE 7

Simplify each expression using index notation.

a $3^4 \times 3^8$ **b** $7^3 \times 7^5$ **c** $9^5 \times 9$

SOLUTION

a $3^4 \times 3^8 = 3^{4+8}$
$= 3^{12}$

b $7^3 \times 7^5 = 7^{3+5}$
$= 7^8$

c $9^5 \times 9 = 9^5 \times 9^1$
$= 9^{5+1}$
$= 9^6$

✱ $9 = 9^1$

Shutterstock.com/Cherryson

ISBN 9780170350990

EXERCISE 2–04

1 Simplify $2^5 \times 2^3$. Select the correct answer **A**, **B**, **C** or **D**.

A 2^{15} **B** 2^{35} **C** 2^8 **D** 4^8

2 Simplify $2^3 \times 3^2$. Select **A**, **B**, **C** or **D**.

A 2^6 **B** 2^{32} **C** 2^5 **D** cannot be simplified

3 **a** What is the base number of 5^4?

b What is the index of 5^4?

c Write 5^4 in expanded form.

4 Write each expression in expanded form.

a $7^8 \times 7^5$ **b** $9^6 \times 9^3$ **c** $10^3 \times 10^2$

5 Simplify each expression in index notation.

a $3^5 \times 3^3$ **b** $6^4 \times 6^3$ **c** $5^3 \times 5^6$

d $7^8 \times 7^4$ **e** $9^5 \times 9^3$ **f** $10^6 \times 10^2$

g $8^6 \times 8^4$ **h** $12^5 \times 12^7$ **i** $20^3 \times 20^5$

j $5^3 \times 5^1 \times 5^2$ **k** $3^8 \times 3^2 \times 3$ **l** $2^5 \times 2^0 \times 2^2$

6 Write $6^4 \times 6 \times 6^5$ in expanded form and hence evaluate it in index notation.

7 Is each equation true or false?

a $3^6 \times 3^3 = 3^{18}$ **b** $5^4 \times 5^2 = 5^6$

c $7^3 \times 7^4 \times 7 = 7^7$ **d** $4^5 \times 4^0 \times 4^2 = 4^7$

8 Can $3^3 \times 4^4$ be simplified by adding indices? Use your calculator to evaluate it.

9 Use a calculator to evaluate each expression.

a $2^6 \times 4^2$ **b** $6^4 \times 10^2$ **c** $5^3 \times (-3)^2$ **d** $(-2)^7 \times 4^2$

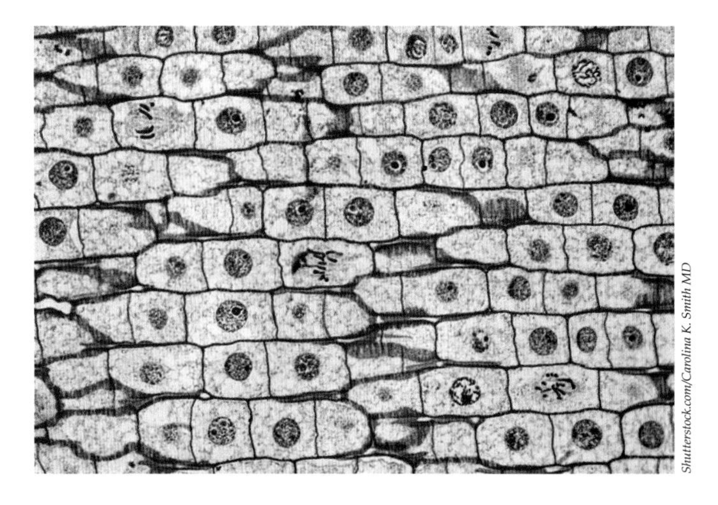

Shutterstock.com/Carolina K. Smith MD

ISBN 9780170350990

2-05 Dividing terms with the same base

When dividing terms with the same base, such as $3^8 \div 3^6$, we can subtract the powers because:

$$3^8 \div 3^6 = \frac{3\times3\times\cancel{3}\times\cancel{3}\times\cancel{3}\times\cancel{3}\times\cancel{3}\times\cancel{3}}{\cancel{3}\times\cancel{3}\times\cancel{3}\times\cancel{3}\times\cancel{3}\times\cancel{3}} = 3\times3 = 3^2 \leftarrow 8 - 6 = 2$$

To divide terms with the same base, subtract the indices.

$$a^m \div a^n = \frac{a^m}{a^n} = a^{m-n}$$

EXAMPLE 8

a Write $4^5 \div 4^2$ in expanded form.

b Simplify $4^5 \div 4^2$, writing the answer in index notation.

SOLUTION

a $4^5 \div 4^2 = \frac{4\times4\times4\times4\times4}{4\times4}$

b $4^5 \div 4^2 = 4^{5-2} = 4^3$

EXAMPLE 9

Simplify each expression, writing the answer in index notation.

a $4^8 \div 4^5$ **b** $8^9 \div 8^3$ **c** $10^7 \div 10^6$

SOLUTION

a $4^8 \div 4^5 = 4^{8-5} = 4^3$

b $8^9 \div 8^3 = 8^{9-3} = 8^6$

c $10^7 \div 10^6 = 10^{7-6} = 10^1 = 10$

$10^1 = 10$

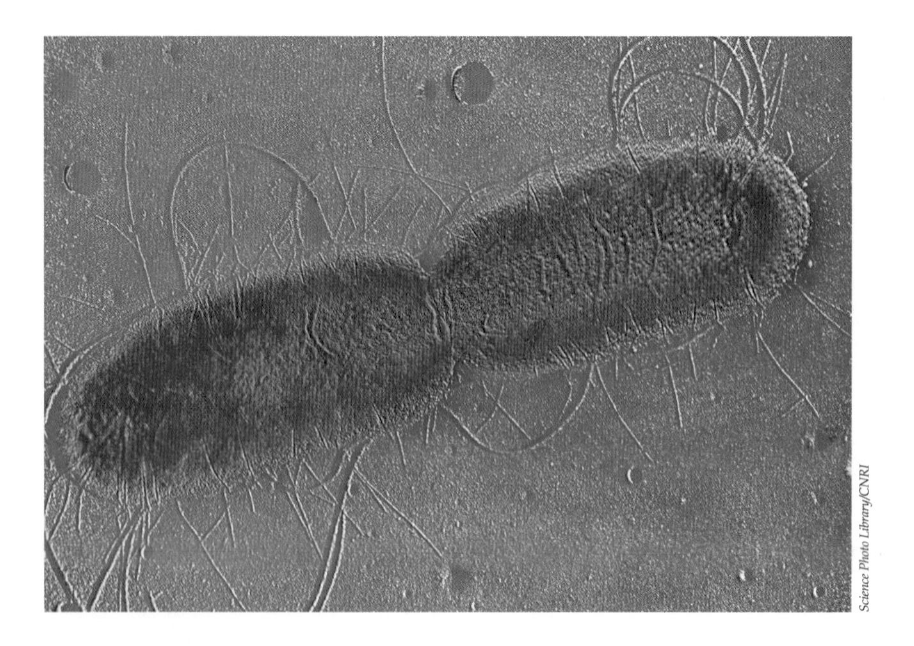

Science Photo Library/CNRI

ISBN 9780170350990

EXERCISE 2-05

1 Simplify $2^8 \div 2^4$. Select the correct answer **A**, **B**, **C** or **D**.

A 2^2 **B** 2^4 **C** 2^8 **D** 1^4

2 Simplify $2^8 \div 3^2$. Select **A**, **B**, **C** or **D**.

A 2^6 **B** 2^4 **C** 3^4 **D** cannot be simplified

3 **a** What is the base number of 8^3?

b What is the index of 8^3?

c Write 8^3 in expanded form.

4 Write each expression in expanded form.

a $7^{12} \div 7^5$ **b** $9^6 \div 9^3$ **c** $\frac{10^9}{10^2}$

5 Simplify each expression using index notation.

a $3^5 \div 3^3$ **b** $6^8 \div 6^3$ **c** $5^9 \div 5^6$

d $\frac{7^8}{7^4}$ **e** $9^5 \div 9^2$ **f** $10^6 \div 10^2$

g $8^6 \div 8^4$ **h** $\frac{12^8}{12^7}$ **i** $20^9 \div 20^5$

j $5^8 \div 5 \div 5^2$ **k** $3^8 \times 3^2 \div 3^0$ **l** $\frac{6^5 \times 6^0}{6}$

6 Is each equation true or false?

a $3^6 \div 3^3 = 3^3$ **b** $5^8 \div 5^4 = 5^2$

c $7^3 \times 7^4 \div 7 = 7^7$ **d** $4^8 \times 4^0 \div 4^2 = 4^7$

7 Can $3^3 \div 4^3$ be simplified by subtracting indices? Use your calculator to evaluate it as a fraction.

iStockphoto/Eduard Andras

2–06 Power of a power

When finding a power of a power with the same base, such as $(3^2)^4$, we can multiply powers because:

$(3^2)^4 = 3^2 \times 3^2 \times 3^2 \times 3^2 = (3 \times 3) \times (3 \times 3) \times (3 \times 3) \times (3 \times 3) = 3^8$ ← $(4 \times 2 = 8)$

> To find a power of a power, multiply the indices.
> $(a^m)^n = a^{m \times n} = a^{mn}$

EXAMPLE 10

Simplify each expression in index notation.

a $(2^3)^5$ **b** $(6^4)^3$ **c** $(10^4)^8$

SOLUTION

a $(2^3)^5 = 2^{3\times5}$
$= 2^{15}$

b $(6^4)^3 = 6^{4\times3}$
$= 6^{12}$

c $(10^4)^8 = 10^{4\times8}$
$= 10^{32}$

Shutterstock.com/Leonardo Gonzalez

 ISBN 9780170350990

EXERCISE 2-06

1 Simplify $(3^5)^3$. Select the correct answer **A**, **B**, **C** or **D**.

A 3^8 **B** 3^2 **C** 3^{15} **D** 3^{53}

2 Simplify $(2^3)^5$. Select **A**, **B**, **C** or **D**.

A 2^2 **B** 2^{15} **C** 2^8 **D** 2^{53}

3 Copy and complete each equation.

a $(2^{10})^2 = 2^{10 \times \_} = 2^{\_}$ **b** $(8^4)^7 = 8^{\_ \times 7} = \_^{28}$

4 Simplify each expression in index notation.

a $(3^4)^3$ **b** $(5^6)^2$ **c** $(9^2)^5$ **d** $(11^3)^4$

e $(2^7)^3$ **f** $(5^3)^8$ **g** $(8^3)^4$ **h** $(3^5)^6$

i $(4^3)^6$ **j** $(7^6)^4$ **k** $(10^5)^3$ **l** $(6^3)^7$

m $(5^5)^4$ **n** $(7^3)^3$ **o** $(4^3)^6$ **p** $(8^2)^5$

5 Is each equation true or false?

a $(5^8)^3 = 5^{11}$ **b** $(3^6)^4 = 3^{24}$ **c** $(2^2)^3 = 4^6$

iStockphoto/Andrew Rich

ISBN 9780170350990

2-07 The zero index

What does $9^6 \div 9^6$ equal?

$9^6 \div 9^6 = \frac{9\times9\times9\times9\times9\times9}{9\times9\times9\times9\times9\times9} = 1$ because any number divided by itself equals 1.

However, when dividing terms with the same base we subtract the powers:

$9^6 \div 9^6 = 9^{6-6} = 9^0$

So $9^0 = 1$.

This rule works for any number as a base (not just 9).

> Any term raised to the **power of 0** is 1.
> $a^0 = 1$

EXAMPLE 11

Evaluate each expression.

a 5^0 **b** $(-3)^0$ **c** 7×6^0 **d** $(3 \times 4)^0$

SOLUTION

a $5^0 = 1$

b $(-3)^0 = 1$

c $7 \times 6^0 = 7 \times 1$
$= 7$

d $(3 \times 4)^0 = 12^0$
$= 1$

Dreamstime/Vlue

ISBN 9780170350990

EXERCISE 2-07

1 Evaluate 5×2^0. Select the correct answer **A**, **B**, **C** or **D**.

A 5 **B** 2 **C** 10 **D** 1

2 Evaluate $(5 \times 2)^0$. Select **A**, **B**, **C** or **D**.

A 5 **B** 2 **C** 10 **D** 1

3 Evaluate each expression.

a 7^0 **b** $(-5)^0$ **c** 3×6^0

d $(2 \times 8)^0$ **e** $3 + 5^0$ **f** $(-7 \times 2)^0$

g $6 - 4^0$ **h** $2 \times (-8)^0$ **i** $\left(\frac{1}{2}\right)^0$

j $24 \div 3^0$ **k** $(4^5)^0$ **l** $(1 - 9)^0$

4 **a** Simplify $7^3 \div 7^3$ in index notation, then evaluate the answer.

b Describe the rule shown in part **a**.

5 Is each equation true or false?

a $2^0 = 1$ **b** $7^0 = 7$ **c** $10^0 = 1$

d $6^0 = 6$ **e** $5 \times 3^0 = 5$ **f** $(-4)^0 = -1$

g $(2 \times 3)^0 = 1$ **h** $-2 \times 9^0 = -18$ **i** $3 \times (-6)^0 = 3$

j $(-2)^0 = -2^0$ **k** $3 \times 4^0 = (3 \times 4)^0$

6 Evaluate $(6 \times 2)^0 + 6 \times 2^0 - (-2)^0$.

7 Use index laws to simplify each expression using index notation.

a $4^6 \times 4^5$ **b** $6^8 \div 6^4$ **c** $(2^4)^3$

d $3^{12} \div 3^6$ **e** $(5^4)^3$ **f** $7^5 \times 7^3$

g $3^4 \times 3^6 \div 3^{10}$ **h** $7^8 \div 7^5 \times 7^3 \div 7^6$

i $(4^3)^4 \div 4^5 \times 4^2$ **j** $5^{12} \div 5^6 \times (5^2)^3 \div 5^8$

iStockphoto/jerry2313

CODE PUZZLE

Use this table to decode the words and phrases used in this chapter.

1	2	3	4	5	6	7	8	9	10	11	12	13
A	B	C	D	E	F	G	H	I	J	K	L	M

14	15	16	17	18	19	20	21	22	23	24	25	26
N	O	P	Q	R	S	T	U	V	W	X	Y	Z

1 9 – 14 – 4 – 5 – 24

2 4 – 9 – 21 – 9 – 19 – 9 – 2 – 12 – 5

3 19 – 17 – 21 – 1 – 18 – 5   18 – 15 – 15 – 20

4 2 – 1 – 19 – 5

5 3 – 21 – 2 – 5   18 – 15 – 15 – 20

6 16 – 18 – 9 – 13 – 5

7 13 – 21 – 12 – 20 – 9 – 16 – 12 – 25

8 3 – 15 – 13 – 16 – 15 – 19 – 9 – 20 – 5

9 16 – 15 – 23 – 5 – 18

10 26 – 5 – 18 – 15

11 4 – 9 – 22 – 9 – 4 – 5

12 19 – 21 – 2 – 20 – 18 – 1 – 3 – 20

13 20 – 5 – 18 – 13

14 14 – 21 – 13 – 2 – 5 – 18

15 4 – 9 – 22 – 9 – 19 – 9 – 2 – 9 – 12 – 9 – 20 – 25

ISBN 9780170350990

PRACTICE TEST 2

Part A General topics

Calculators are not allowed.

1 Evaluate 300×8.

2 Copy and complete: 18 km = _____ m.

3 List all the factors of 12.

4 Simplify $x + y - y + x$.

5 Find the mode of the scores:
12, 8, 7, 6, 8, 7, 5, 7.

6 Given that $19 \times 6 = 114$, evaluate 1.9×6.

7 Find 25% of $640.

8 Find the value of d if the area of this rectangle is 45 m^2.

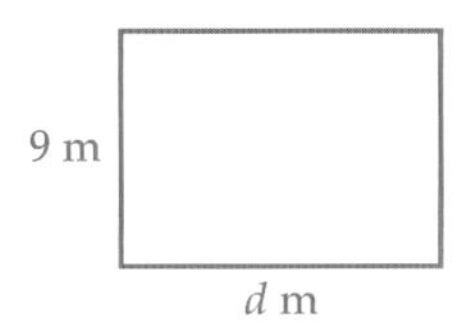

9 Simplify $3 \times r \times 7$.

10 Copy and complete: $\frac{3}{8} = \frac{}{32}$.

Part B Primes and powers

Calculators are not allowed.

2-01 Divisibility tests

11 Which of these numbers is divisible by 4? Select the correct answer **A**, **B**, **C** or **D**.

A 365 **B** 812 **C** 950 **D** 729

12 Which of these numbers is divisible by 3? Select **A**, **B**, **C** or **D**.

A 365 **B** 812 **C** 950 **D** 729

2-02 Prime and composite numbers

13 Which of these numbers is composite? Select **A**, **B**, **C** or **D**.

A 11 **B** 51 **C** 23 **D** 41

14 Write all the prime numbers between 12 and 25.

2-03 Powers and roots

15 Write $6 \times 6 \times 6 \times 6 \times 6$ using index notation.

16 Evaluate each expression.

a 5^4 **b** $\sqrt{169}$ **c** $\sqrt[3]{-64}$

ISBN 9780170350990

PRACTICE TEST 2

2-04 Multiplying terms with the same base

17 Simplify each expression using index notation.

a $4^3 \times 4^5$ **b** $8^4 \times 8$ **c** $2^3 \times 2^3$

2-05 Dividing terms with the same base

18 Simplify each expression using index notation.

a $5^8 \div 5^3$ **b** $7^{12} \div 7^4$ **c** $9^7 \div 9$

2-06 Power of a power

19 Simplify each expression using index notation.

a $(3^2)^4$ **b** $(6^3)^5$ **c** $(4^2)^6$

2-07 The zero index

20 Evaluate each expression.

a 5^0 **b** $3 + 4^0$ **c** $(5 - 3)^0$

ISBN 9780170350990

PYTHAGORAS' THEOREM

3

WHAT'S IN CHAPTER 3?

IN THIS CHAPTER YOU WILL:

- understand and write Pythagoras' theorem for right-angled triangles
- use Pythagoras' theorem to find the length of the hypotenuse in a right-angled triangle, giving the answer as a surd or a decimal approximation
- use Pythagoras' theorem to find the length of a shorter side in a right-angled triangle, giving the answer as a surd or a decimal approximation
- solve problems involving Pythagoras' theorem

Note: Pythagoras' theorem is a Year 9 topic in the Australian Curriculum and appears in Book 3 of this series but it has been included here because it is a Stage 4 topic in the NSW syllabus.

* Shutterstock.com/Moving Moment

3-01 Pythagoras' theorem

WORDBANK

right-angled triangle A triangle with one angle exactly 90°. This angle is called the right angle.

hypotenuse The longest side of a right-angled triangle, the side opposite the right angle.

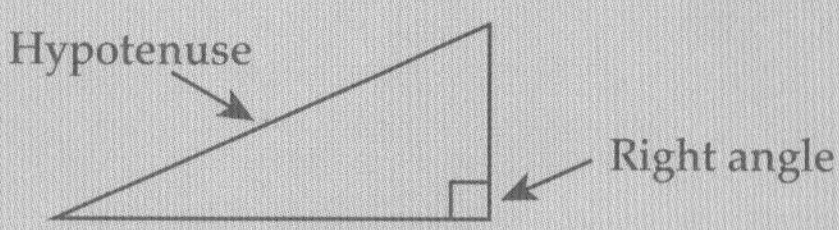

Pythagoras' theorem The rule or formula $c^2 = a^2 + b^2$ that relates the lengths of the sides of a right-angled triangle. Pythagoras was an ancient Greek mathematician who discovered this rule ('theorem' means rule).

PYTHAGORAS' THEOREM

- In any right-angled triangle, the square of the hypotenuse is equal to the sum of the squares of the other two sides.
- If the lengths of the sides of a right-angled triangle are a, b and c, where c is the length of the hypotenuse (longest side), then $c^2 = a^2 + b^2$.

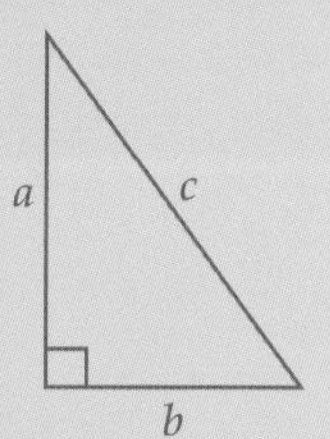

Pythagoras' theorem can be demonstrated by the diagram below.

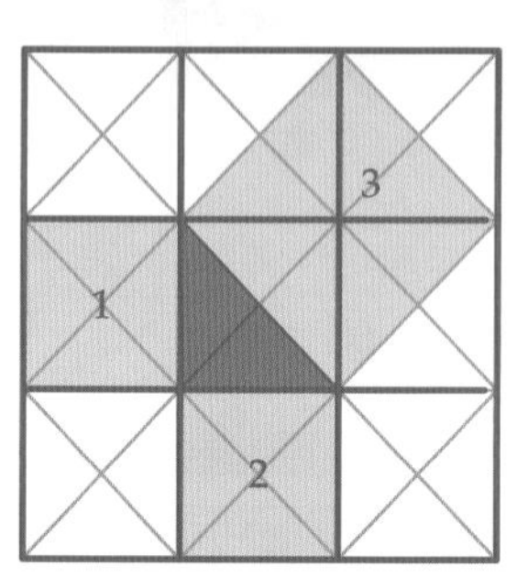

- The dark triangle here is a right-angled triangle. A square has been drawn on each side of the triangle in a lighter colour. Each square has been divided into smaller identical triangles.
- The square of the hypotenuse is the area of square 3, which is 8 triangular units.
- The squares of the other two sides are the areas of square 1 and square 2, which are 4 triangular units each.
- $4 + 4 = 8$, so the square of the hypotenuse equals the sum of the squares of the other two sides.

EXAMPLE 1

Name the hypotenuse and state Pythagoras' theorem for each right-angled triangle below.

a

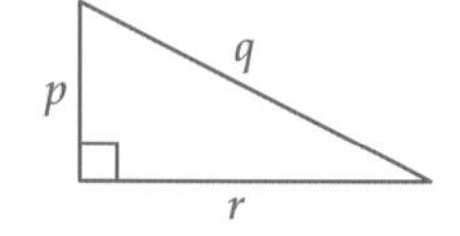

b

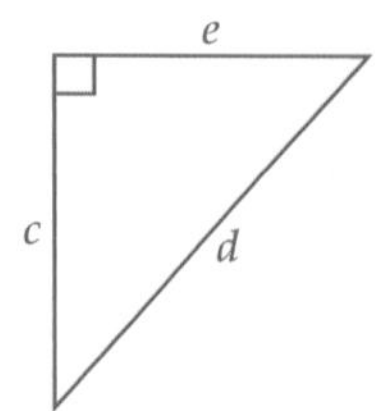

SOLUTION

a The hypotenuse is q. ← the longest side, opposite the right angle

Pythagoras' theorem is $q^2 = p^2 + r^2$.

✱ $(\text{hypotenuse})^2$ = the sum of the squares of the other two sides

b The hypotenuse is d.

Pythagoras' theorem is $d^2 = c^2 + e^2$.

ISBN 9780170350990

EXAMPLE 2

Show whether Pythagoras' theorem is true for each triangle below.

a

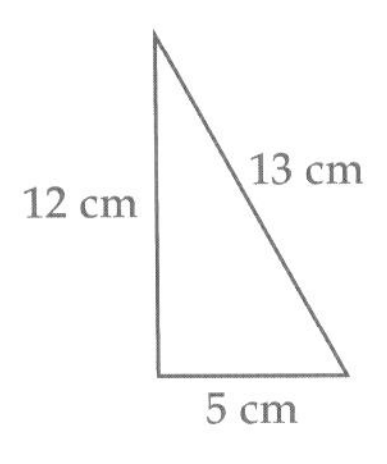

b

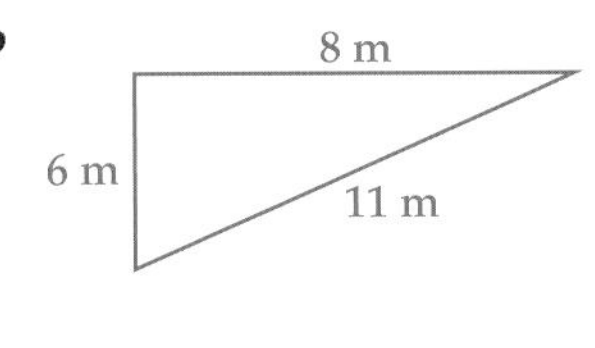

SOLUTION

a For Pythagoras' theorem to be true, hypotenuse^2 = sum of squares of the other two sides.
Does $13^2 = 5^2 + 12^2$?
$169 = 25 + 144$ ⟵ True, so Pythagoras' theorem is true.
This means the triangle is right-angled.

b Does $11^2 = 6^2 + 8^2$?
$121 = 36 + 64$ ⟵ False, so Pythagoras' theorem is not true.
This means the triangle is not right-angled.

EXERCISE 3–01

1 Which side of this right-angled triangle is the hypotenuse? Select the correct answer **A**, **B** or **C**.

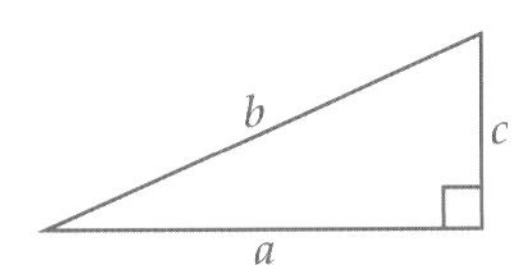

A a **B** b **C** c

2 What is Pythagoras' theorem for the triangle in Question **1**? Select **A**, **B** or **C**.

A $b^2 = a^2 + c^2$ **B** $c^2 = a^2 + b^2$ **C** $a^2 = b^2 + c^2$

3 Name the hypotenuse for each right-angled triangle.

a

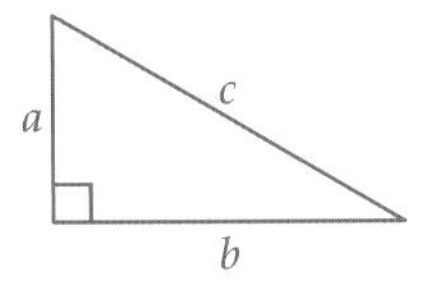

b

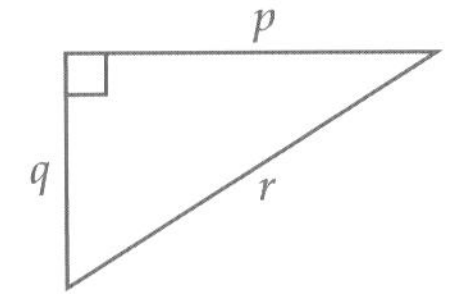

c

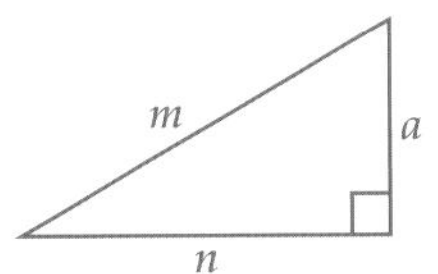

d

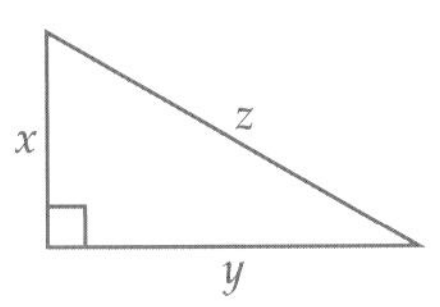

e

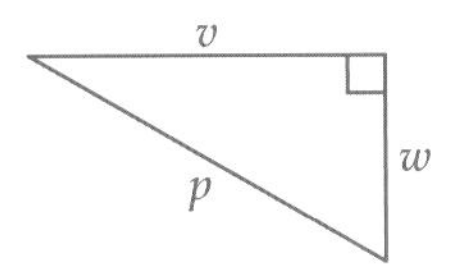

f

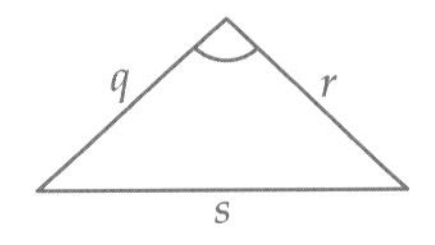

4 State Pythagoras' theorem for each right-angled triangle in Question **3**.

EXERCISE 3–01

5 Show whether Pythagoras' theorem is true for each triangle below.

a

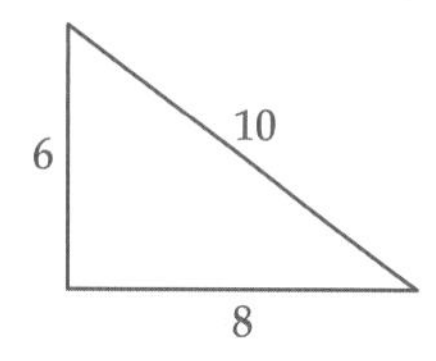

b

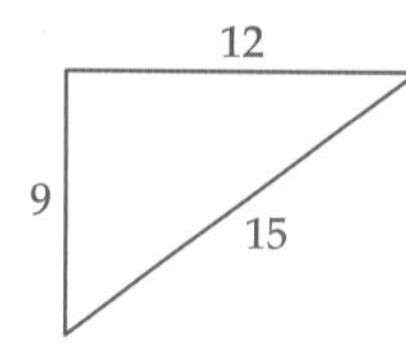

c

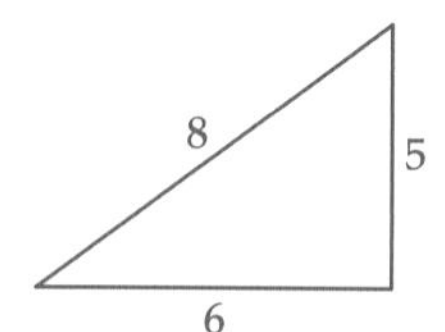

d

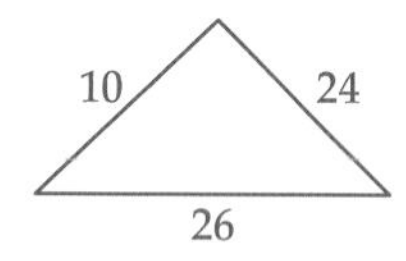

e

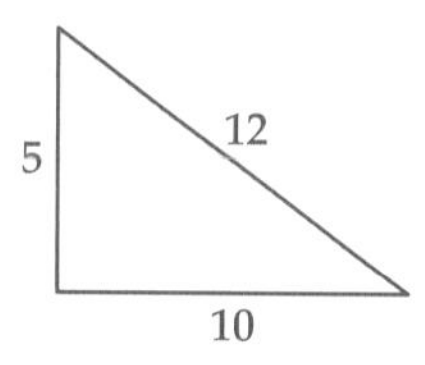

f

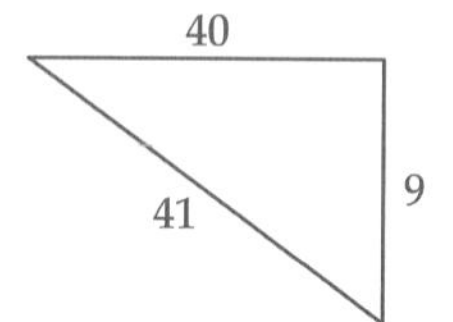

6 Which triangles in Question **5** are right-angled?

123RF/Syda Productions

ISBN 9780170350990

3–02 Finding the hypotenuse

WORDBANK

surd A square root where the answer is not an exact number. For example, $\sqrt{8} = 2.8284...$ is a surd because there isn't an exact number squared that is equal to 8. As a decimal, the digits of $\sqrt{8}$ run endlessly without any repeating pattern.

exact form When an answer is written as an exact number, such as a whole number, decimal or surd, and is not rounded.

To find the length of the hypotenuse in a right-angled triangle:

- write Pythagoras' theorem in the form $c^2 = a^2 + b^2$ where c is the length of the hypotenuse
- solve the equation
- check that your answer is the longest side.

EXAMPLE 3

Find the length of the hypotenuse in each triangle, writing your answer in exact form.

a

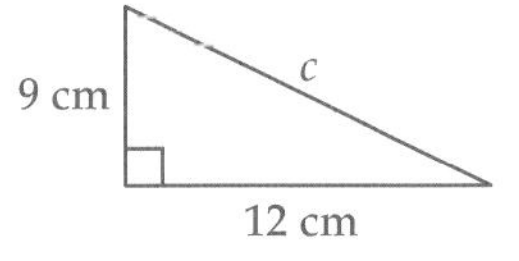

b

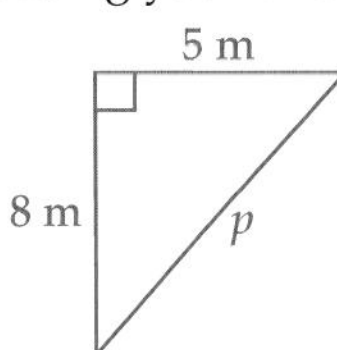

SOLUTION

a $c^2 = a^2 + b^2$

$= 9^2 + 12^2$

$= 225$

$c = \sqrt{225}$

$= 15$ cm ⟵ This is in exact form.

From the diagram, a hypotenuse of length 15 cm looks reasonable. It is also the longest side.

b $c^2 = a^2 + b^2$

$p^2 = 5^2 + 8^2$ ⟵ p is the hypotenuse.

$= 89$

$p = \sqrt{89}$ m ⟵ This is in exact surd form.

EXAMPLE 4

Find d correct to one decimal place.

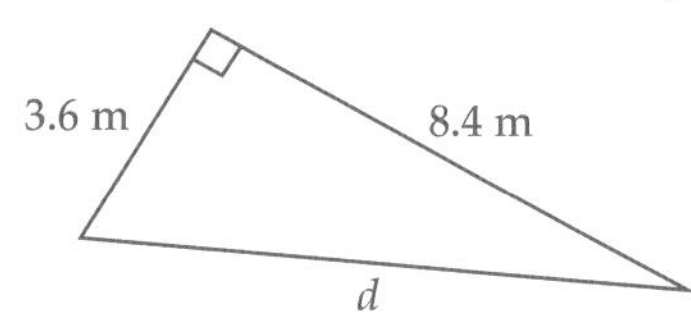

SOLUTION

$c^2 = a^2 + b^2$

$d^2 = 3.6^2 + 8.4^2$

$= 83.52$

$d = \sqrt{83.52}$

$= 9.1389...$

≈ 9.1 m ⟵ rounded to one decimal place

ISBN 9780170350990

EXERCISE 3–02

1 What is Pythagoras' theorem for this right-angled triangle? Select the correct answer **A, B, C** or **D**.

A $12^2 = 5^2 + c^2$ **B** $c^2 = 12^2 - 5^2$ **C** $c^2 = 5^2 + 12^2$ **D** $5^2 = c^2 + 12^2$

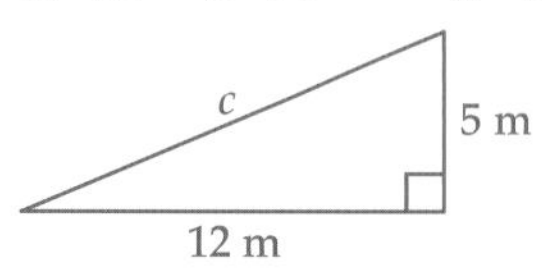

2 What is the length of side c in Question **1**? Select **A, B, C** or **D**.

A 7 m **B** 13 m **C** 15 m **D** $\sqrt{119}$ m

3 Copy and complete for this triangle.

$c^2 = a^2 + b^2$

$c^2 = 18^2 + \_\_^2$

$c^2 = \_\_\_$

$c = \sqrt{\_\_}$

$= \_\_\_$

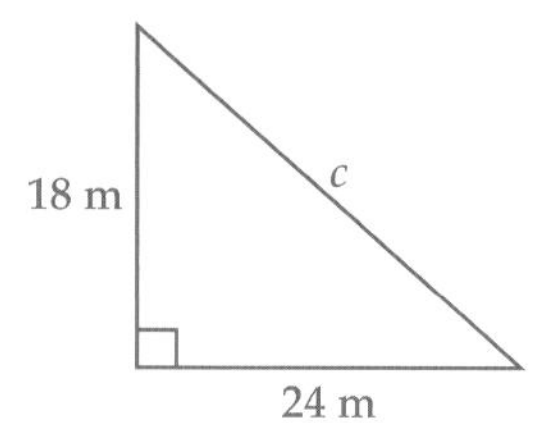

4 Find the value of the pronumeral in each triangle below. Answer in exact form.

a

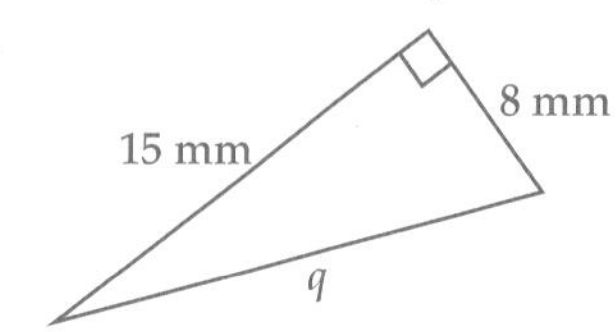

b

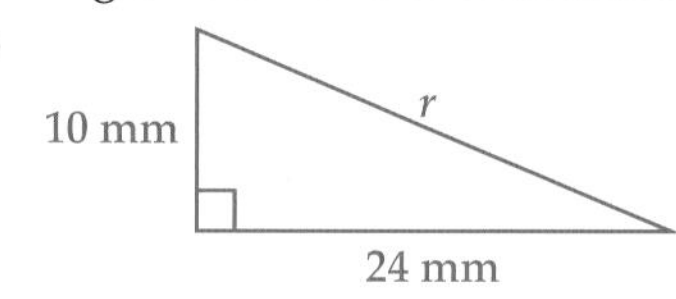

c

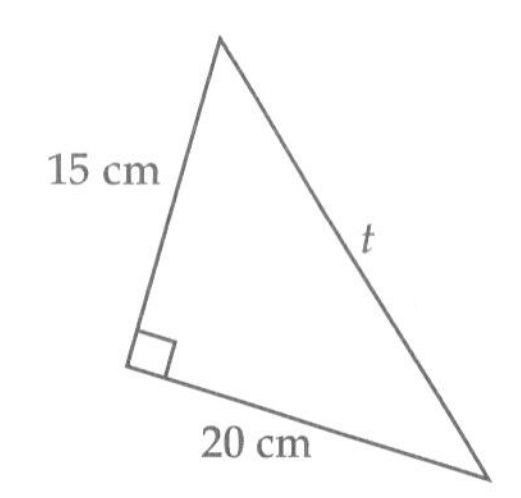

d

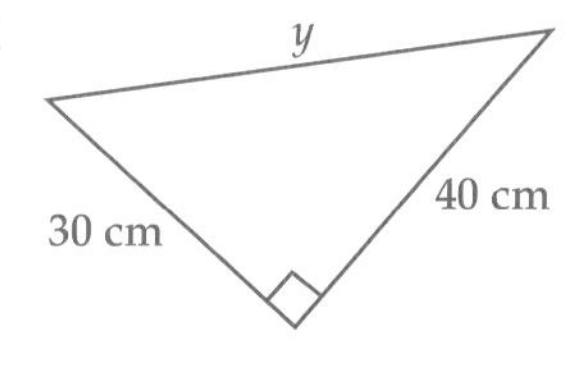

5 Find the length of the hypotenuse in each triangle below. Answer correct to one decimal place.

a

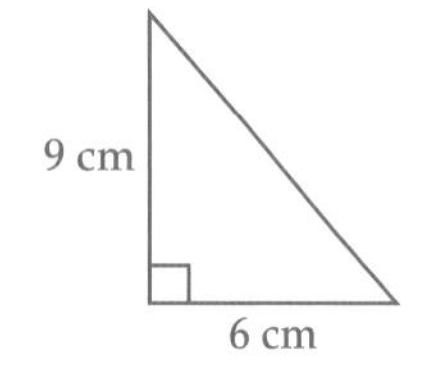

b

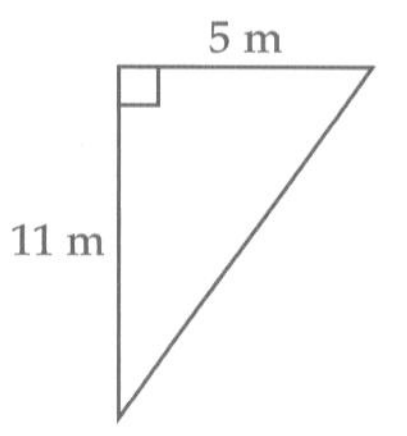

c

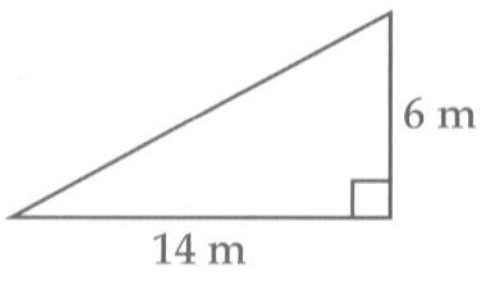

d

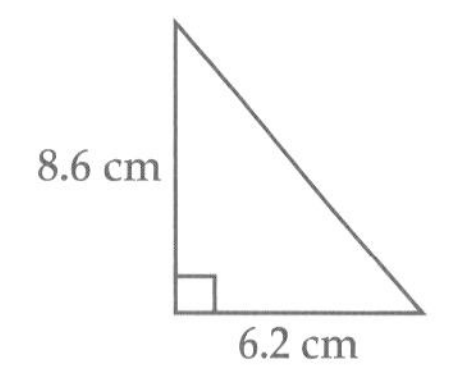

e

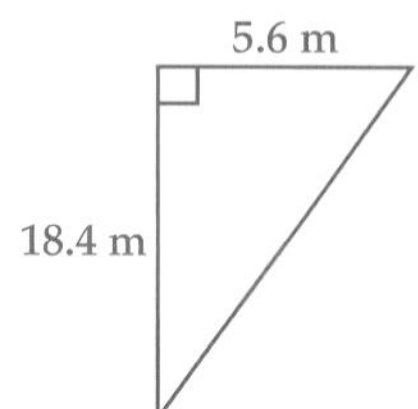

f

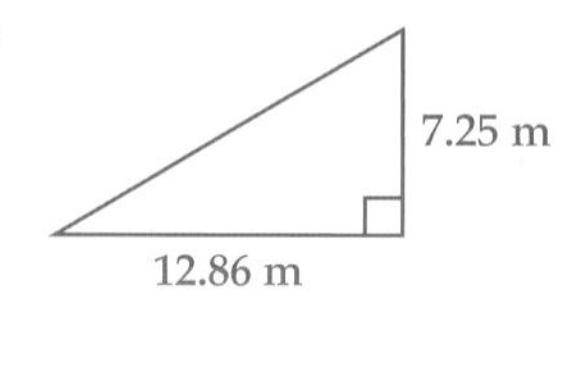

ISBN 9780170350990

3-03 Finding a shorter side

To find the length of a shorter side in a right-angled triangle:

- write Pythagoras' theorem in the form $c^2 = a^2 + b^2$ where c is the length of the hypotenuse
- rearrange the equation so that the shorter side is on the LHS (left-hand-side)
- solve the equation
- check that your answer is shorter than the hypotenuse.

EXAMPLE 5

Find the length of the unknown side in each triangle. Answer in exact form.

a

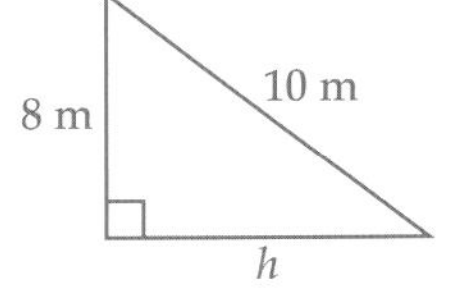

b

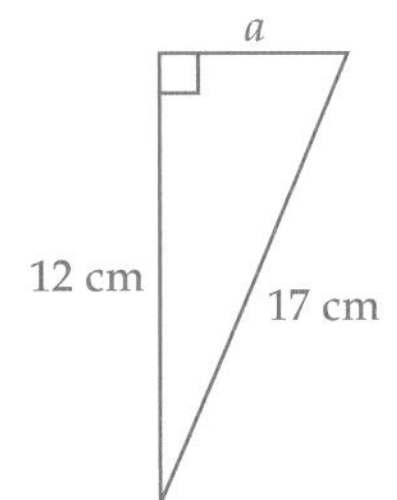

SOLUTION

a

$$10^2 = h^2 + 8^2$$ ← 10 is the hypotenuse.

$$h^2 + 8^2 = 10^2$$ ← Rearranging equation so that h is on the LHS.

$$h^2 = 10^2 - 8^2$$
$$= 36$$
$$h = \sqrt{36}$$
$$= 6 \text{ m}$$

From the diagram, a length of 6 m looks reasonable. It is also shorter than the hypotenuse, 10 m.

b

$$17^2 = a^2 + 12^2$$
$$a^2 + 12^2 = 17^2$$
$$a^2 = 17^2 - 12^2$$
$$= 145$$
$$a = \sqrt{145} \text{ cm}$$ ← in exact form

EXAMPLE 6

Find the value of the pronumeral in each triangle. Answer correct to one decimal place.

a

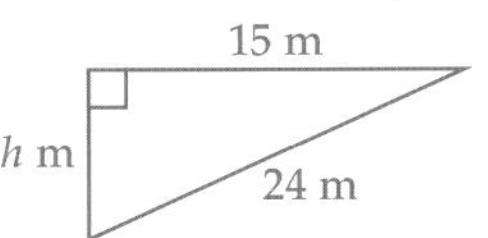

b

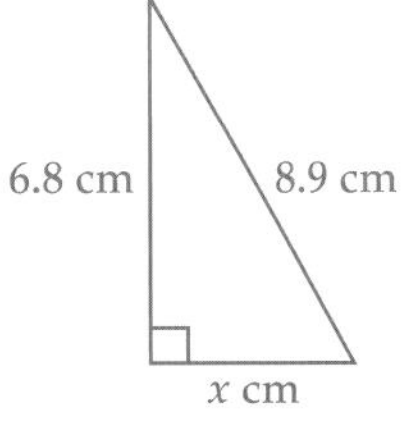

a

$$24^2 = h^2 + 15^2$$
$$h^2 + 15^2 = 24^2$$
$$h^2 = 24^2 - 15^2$$
$$= 351$$
$$h = \sqrt{351}$$
$$= 18.7349\ldots$$
$$\approx 18.7$$ ← rounded to one decimal place

b

$$8.9^2 = x^2 + 6.8^2$$
$$x^2 + 6.8^2 = 8.9^2$$
$$x^2 = 8.9^2 - 6.8^2$$
$$= 32.97$$
$$x = \sqrt{32.97}$$
$$= 5.7419\ldots$$
$$\approx 5.7$$

ISBN 9780170350990

EXERCISE 3–03

1 To find the length of the shorter side, a, in the triangle below, which rule is correct? Select the correct answer **A**, **B**, **C** or **D**.

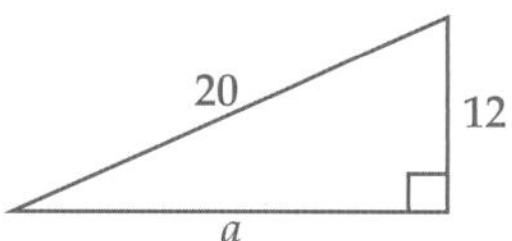

A $20^2 = a^2 + 12^2$ **B** $20^2 = a^2 - 12^2$ **C** $a^2 = 12^2 - 20^2$ **D** $a^2 = 12^2 + 20^2$

2 What is the value of a in the triangle above? Select **A**, **B**, **C** or **D**.

A 15 **B** 18 **C** 16 **D** 14

3 Copy and complete this solution to find b.

$13^2 = b^2 + 5^2$

$b^2 + \_\_^2 = 13^2$

$b^2 = 13^2 - \_\_^2$

$= 169 - \_\_$

$b = \sqrt{\_\_}$

$= \_\_$

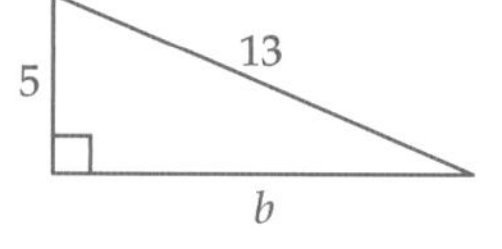

4 Find the value of b for each triangle below. Leave your answers in exact form.

a

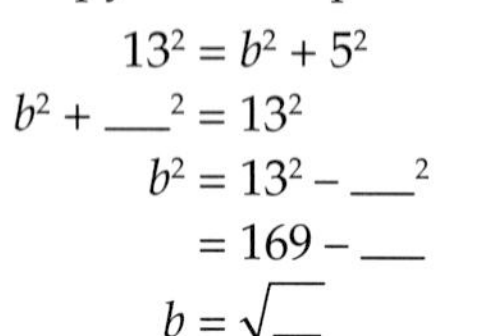

b

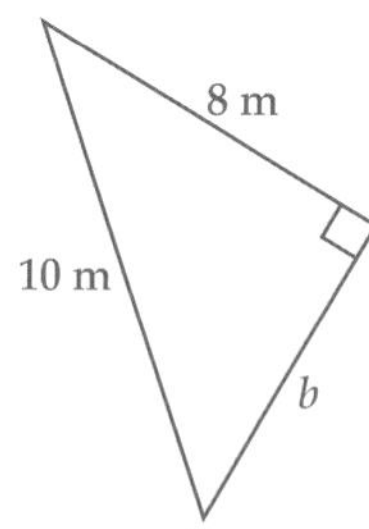

c

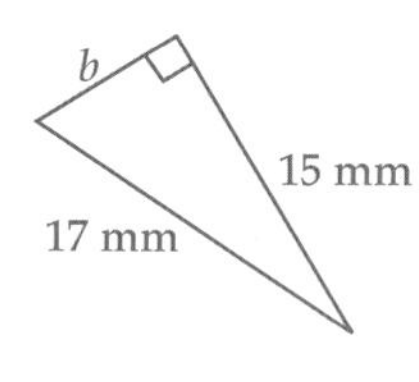

d

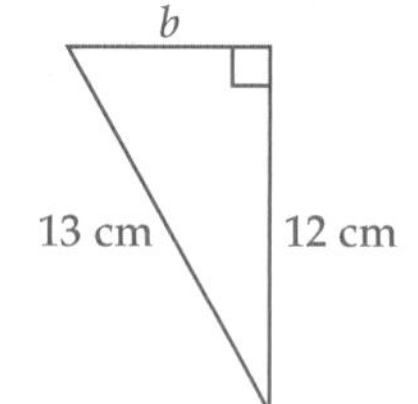

e

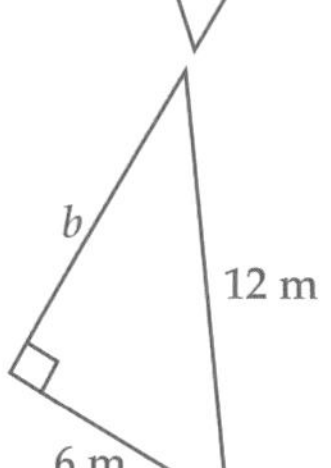

f

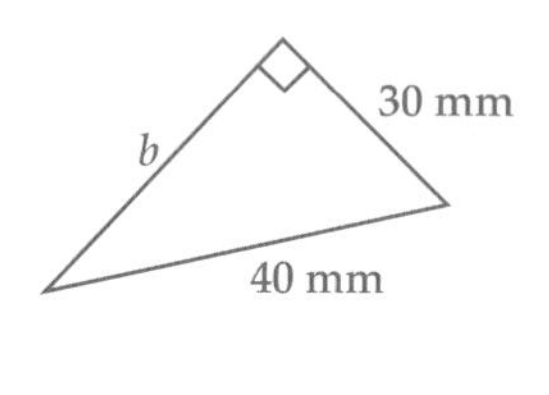

5 Find correct to one decimal place the value of each pronumeral.

a

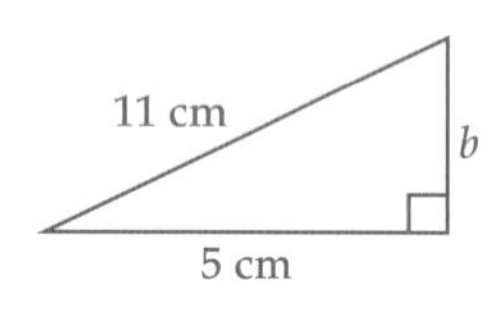

b

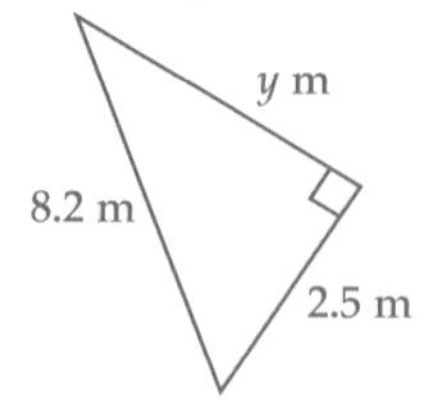

c

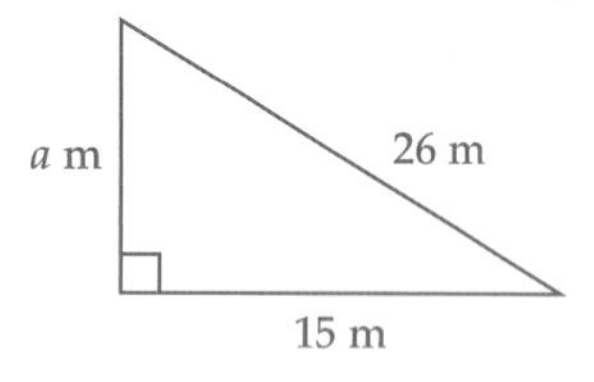

d

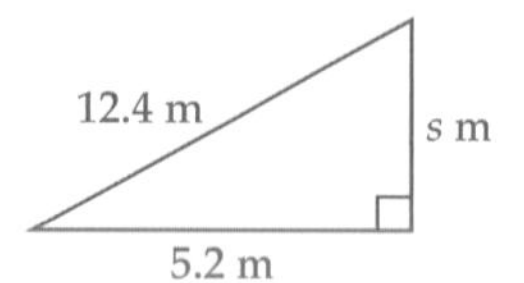

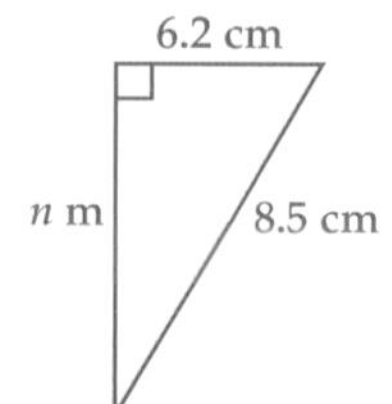

ISBN 9780170350990

3–04 Mixed problems

For this right-angled triangle:

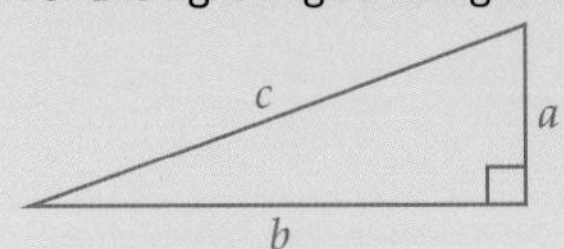

- **to find the length of the hypotenuse**, use $c^2 = a^2 + b^2$ where c is the hypotenuse
- **to find the length of one of the shorter sides**, use $b^2 = c^2 - a^2$ to find side b or $a^2 = c^2 - b^2$ to find side a.

EXAMPLE 7

Find the length of the unknown side in each triangle below. Leave your answer in exact form.

a

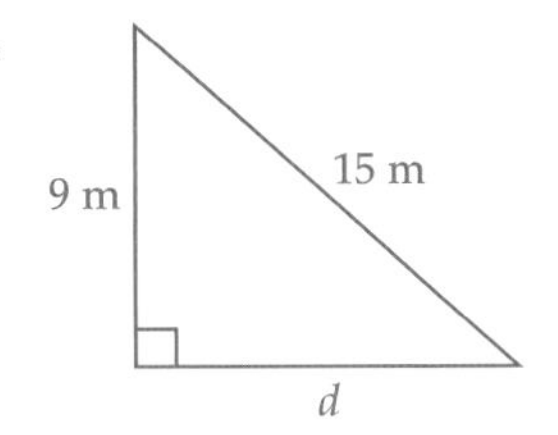

b

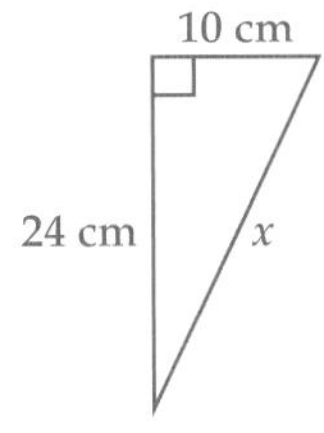

c

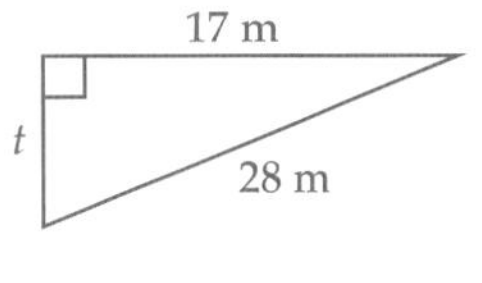

SOLUTION

a d is a shorter side
square and subtract

$$d^2 = 15^2 - 9^2$$
$$= 225 - 81$$
$$= 144$$
$$d = \sqrt{144}$$
$$= 12 \text{ m}$$

b x is the hypotenuse
square and add

$$x^2 = 10^2 + 24^2$$
$$= 100 + 576$$
$$= 676$$
$$x = \sqrt{676}$$
$$= 26 \text{ cm}$$

c t is a shorter side
square and subtract

$$t^2 = 28^2 - 17^2$$
$$= 784 - 289$$
$$= 495$$
$$t = \sqrt{495} \text{ m}$$

EXAMPLE 8

A rectangle has a diagonal of length 34 cm and width of 18 cm. What is the length of the rectangle correct to two decimal places?

SOLUTION

Draw a diagram. Let the length of the rectangle be x cm.

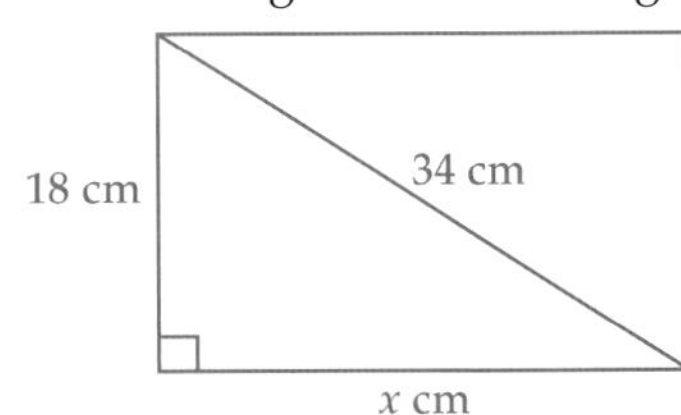

$$34^2 = x^2 + 18^2$$
$$x^2 = 34^2 - 18^2$$
$$= 832$$
$$x = \sqrt{832}$$
$$= 28.8444\ldots$$
$$\approx 28.84$$

The length of the rectangle is 28.84 cm.

ISBN 9780170350990

EXERCISE 3–04

1 To find the length of the hypotenuse, j, in the triangle below, which rule is correct? Select the correct answer **A**, **B**, **C** or **D**.

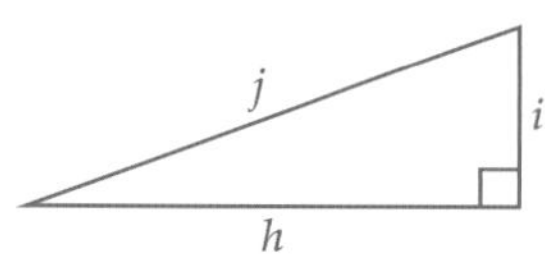

A $j^2 = i^2 + h^2$ **B** $h^2 = i^2 - j^2$ **C** $j^2 = h^2 - i^2$ **D** $i^2 = h^2 + j^2$

2 To find the length of the shorter side, r, in the triangle below, which rule is correct? Select **A**, **B**, **C** or **D**.

A $r^2 = q^2 + p^2$ **B** $p^2 = r^2 - q^2$

C $q^2 = r^2 - p^2$ **D** $r^2 = q^2 - p^2$

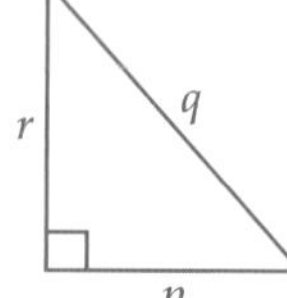
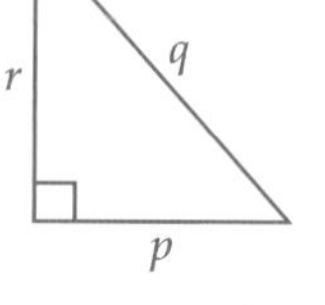

3 Find the length of the unknown side in each triangle. Leave your answers in exact form.

a

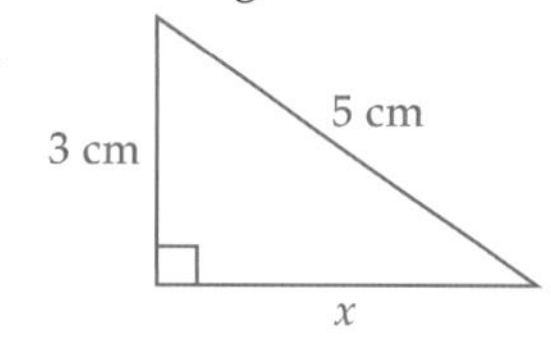

b

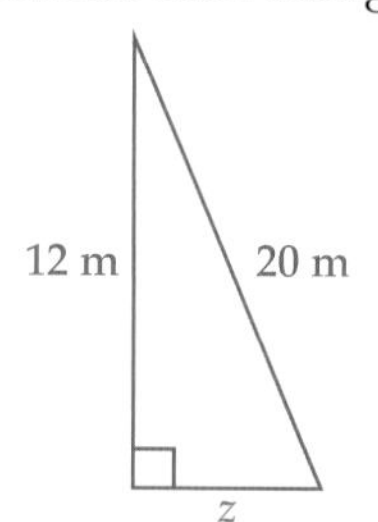

c

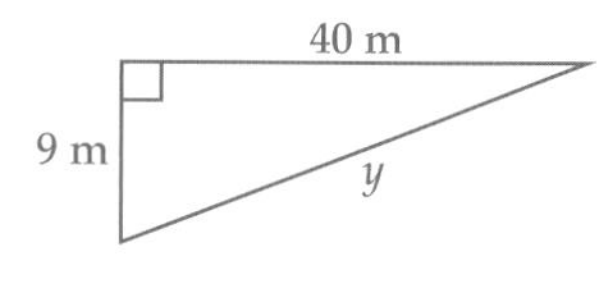

d

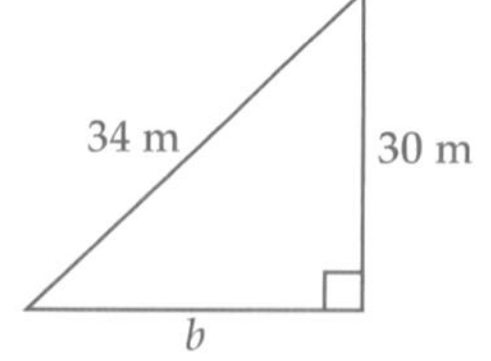

e

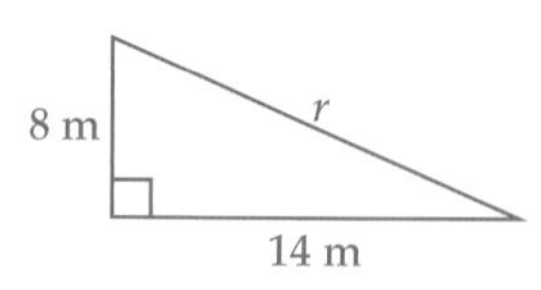

f

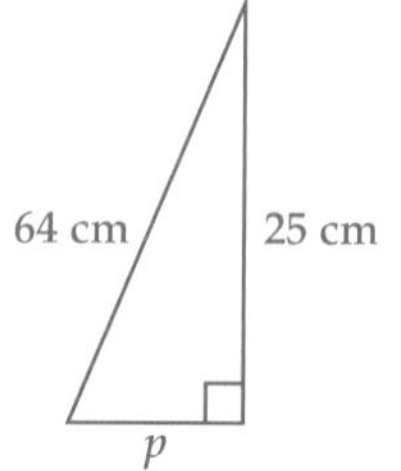

4 Find correct to one decimal place the value of the pronumeral in each triangle.

a

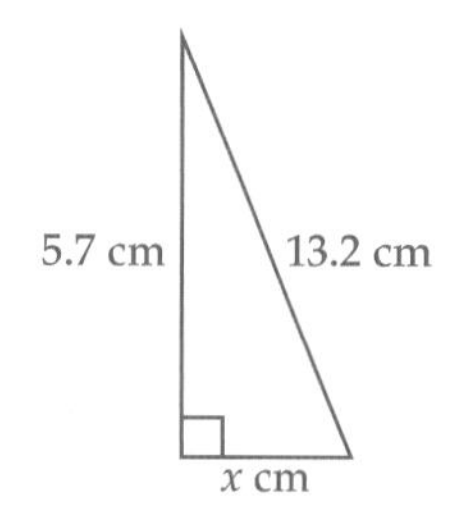

b

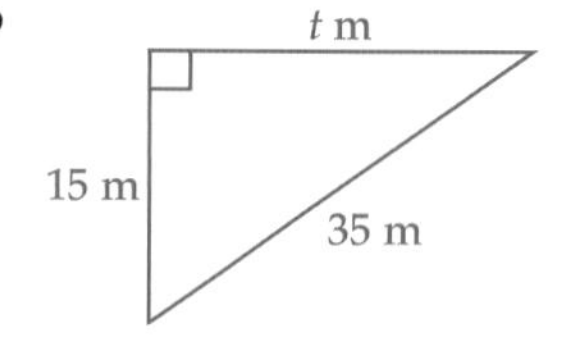

c

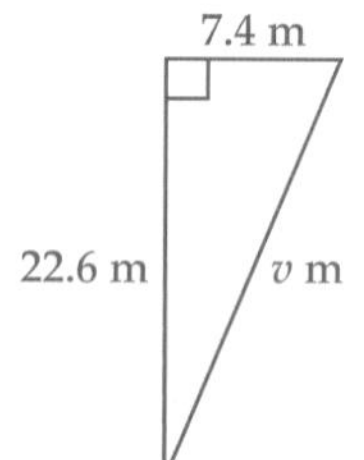

ISBN 9780170350990

5 A ladder is placed against a building to reach a window on the second floor 6 m high. Find correct to two decimal places the length of the ladder.

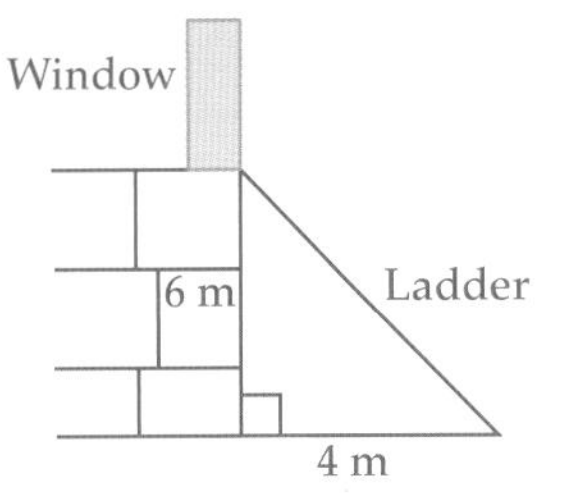

6 For this triangle, find correct to one decimal place the length of BC.

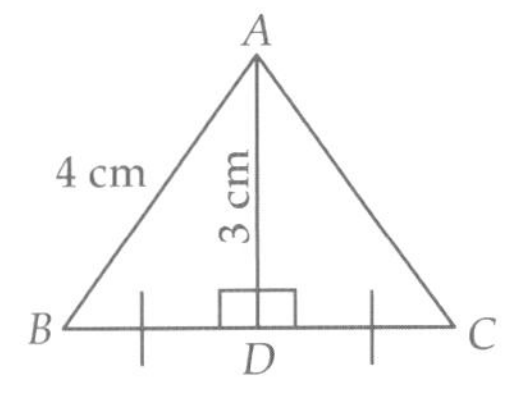

Alamy/Tetra Images

CROSSWORD PUZZLE

Make a copy of this crossword, then complete it using the clues below.

Across

1 The ancient Greek mathematician who discovered a rule for right-angled triangles.

3 You must do this to numbers when you use Pythagoras' theorem.

7 The opposite to squaring a number (two words).

9 To find the hypotenuse, you do this to the squares of the two shorter sides.

10 A shape with three straight sides.

Down

2 The longest side in a right-angled triangle.

4 Another word for a rule.

5 The number of sides in a triangle.

6 To find a shorter side, you do this to the squares of the hypotenuse and to the other short side.

8 Pythagoras' theorem is true for this type of triangle only.

ISBN 9780170350990

PRACTICE TEST 3

Part A General topics

Calculators are not allowed.

1 Write 1456 in 12-hour time.

2 Complete: 45.8 cm = ________ mm.

3 Evaluate 44 × 5.

4 Find the perimeter of this shape.

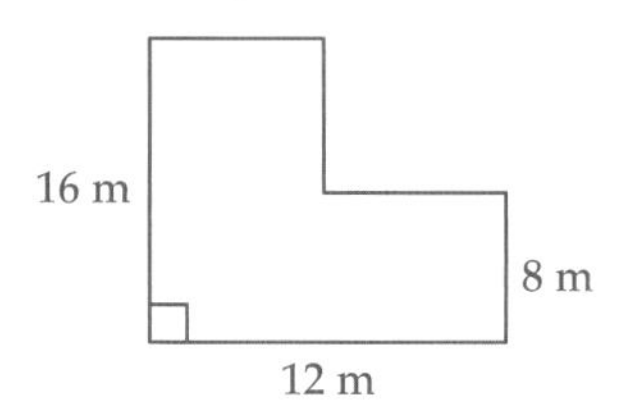

5 Evaluate $\frac{3}{8}+\frac{2}{5}$.

6 Evaluate $(-3)^3$.

7 Find $\frac{3}{4}$ of $56.

8 Find the range of 8, 2, 5, 3, 7.

9 How many faces has a rectangular prism?

10 Ben pays a grocery bill of $86.35 with a $100 note. Calculate the change.

Part B Pythagoras' theorem

Calculators are allowed.

3-01 Pythagoras' theorem

11 Name the hypotenuse in this right-angled triangle.

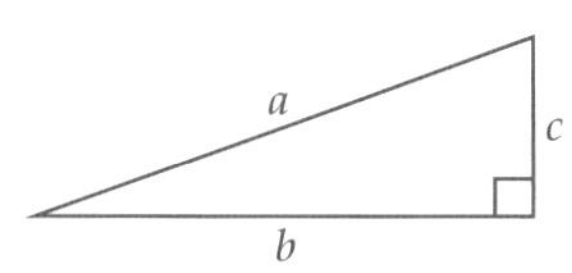

12 What is Pythagoras' theorem for the triangle in Question 1? Select the correct answer **A, B, C** or **D**.

A $b^2 = a^2 + c^2$ **B** $c^2 = a^2 + b^2$ **C** $a^2 = b^2 + c^2$ **D** $b^2 = a^2 - c^2$

3-02 Finding the hypotenuse

13 Find the value of each pronumeral, giving your answer in exact form.

a

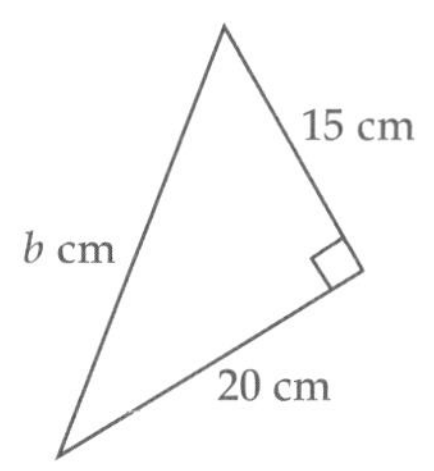

b

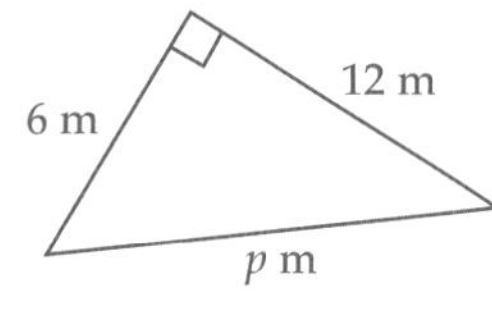

3-03 Finding a shorter side

14 Find the value of each pronumeral, giving your answer correct to two decimal places.

a

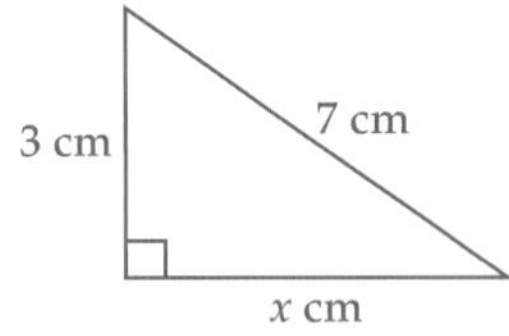

b

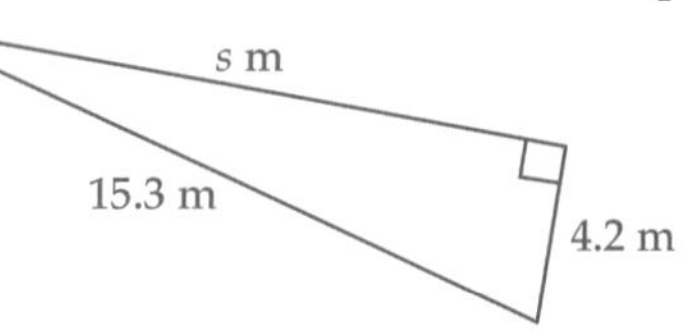

3-04 Mixed problems

15 Find the value of each pronumeral, correct to one decimal place.

a

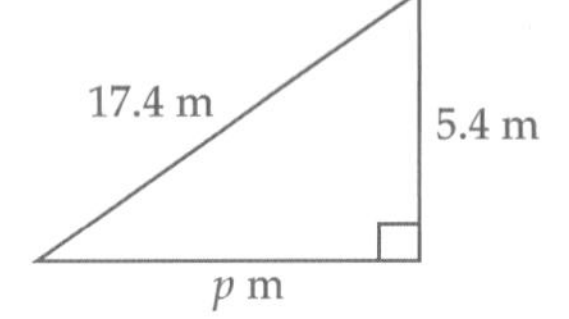

b

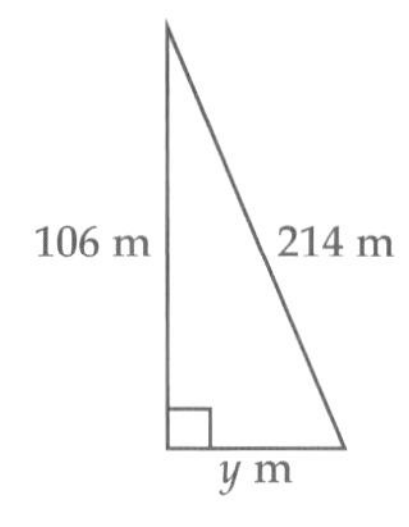

16 Find the length of the diagonal in a square of side length 3 cm.

17 Find the height that a 15 m ladder will reach if it is placed 9 m from the base of a wall.

ISBN 9780170350990

INTEGERS

4

WHAT'S IN CHAPTER 4?

IN THIS CHAPTER YOU WILL:

- arrange integers in order, including on a number line
- add and subtract integers
- multiply and divide integers
- use the order of operations for evaluating mixed expressions

* Shutterstock.com/Sychugina

4-01 Ordering integers

WORDBANK

negative number A number that is less than 0; for example, –5.

integer A positive or negative whole number or zero.

ascending Going up, increasing from smallest to largest.

descending Going down, decreasing from largest to smallest.

Positive and negative numbers can be used to describe everyday situations such as:
- withdrawing \$20 from a bank: – \$20
- 5 degrees above zero: +5°C or just 5°C

Integers can be ordered on a **number line**.

The numbers on the right are larger than the numbers on the left.
- The **positive integers** are on the right because they are greater than 0.
- The **negative integers** are on the left because they are less than 0.
- The number line extends in both directions forever, so we place arrows on both ends.
- We can delete the + sign for the positive integers, so 3 is the same as +3.

EXAMPLE 1

Plot these integers on a number line: –2, 3, 0, –1, 1.

SOLUTION

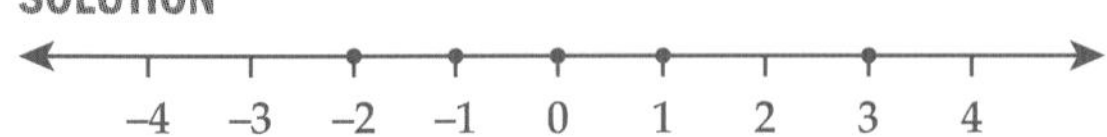

$<$ means is less than; for example, $4 < 9$

$>$ means is greater than; for example, $6 > 2$

- Think of the symbol pointing to the smaller number with the other side opening up to the larger number.

EXAMPLE 2

Is each statement true or false?

a $-9 < 8$ **b** $7 < -2$ **c** $-4 < -3$ **d** $0 > -6$

SOLUTION

a True, –9 is less than 8.

b False, 7 is not less than –2.

* Use the number line to work out the position of each number.

c True, –4 is less than –3.

d True, 0 is greater than –6.

EXAMPLE 3

Write these integers in ascending order: 5, –6, 9, –3, 0, –2, 8, –4.

SOLUTION

From smallest to largest: –6, –4, –3, –2, 0, 5, 8, 9.

 ISBN 9780170350990

EXERCISE 4-01

1 Which of these integers 18, –4, 9, –2, –16, 21 is the smallest? Select the correct answer **A**, **B**, **C** or **D**.

A –2 **B** –16 **C** –4 **D** 21

2 Which statement is true? Select **A**, **B**, **C** or **D**.

A $-2 < -3$ **B** $0 > -4$ **C** $9 < -9$ **D** $-4 > -3$

3 What is an integer? Select **A**, **B**, **C** or **D**.

A a negative number **B** a positive number
C zero **D** any positive or negative whole number or zero

4 Write an integer to represent each situation.

a depositing $22 in a bank account **b** going down 5 steps
c twelve degrees below zero **d** going 28 m above sea level
e losing $50 **f** winning $100

5 Is each statement true or false?

a $6 > -8$ **b** $0 < -4$ **c** $9 > -2$
d $-12 < -9$ **e** $-3 < -6$ **f** $-5 > 5$
g $-15 < -13$ **h** $0 > -8$ **i** $-18 < -12$

6 Plot these integers on a number line:
–8, 5, 6, –3, 4, –7, 9, 0, –5, 3, 2, –1

7 Write these integers in ascending order:
18, –4, 12, 0, –9, 1, –6, 21, –15, 8, –2, 9

8 Write these integers in descending order:
23, –14, –6, 8, –12, 0, 6, –17, 25, 18, –3, –4

9 Use the number line below to decode each message.

a –1 –2 0 –3 2 –3 1 4
b –4 1 –3
c –1 –2 0 1 –1 2 3 –1 –2 2

Each number on the number line has a letter above it to use in the code.

A	E	N	I	T	R	G	U	S
–4	–3	–2	–1	0	1	2	3	4

10 Copy and complete each statement using a $>$ or $<$ symbol.

a 3 ___ 2 **b** 0 ___ 5 **c** 6 ___ –6 **d** 0 ___ –2
e 8 ___ –3 **f** –2 ___ –4 **g** –8 ___ 3 **h** –9 ___ –3
i –12 ___ 3 **j** –10 ___ –4 **k** –17 ___ 7 **l** 26 ___ –6

4–02 Adding integers

Integers can be **added** using a number line.

- Move **right** if adding a **positive** integer.
- Move **left** if adding a **negative** integer.
- Adding a **negative integer** is the same as **subtracting its opposite**; for example, $10 + (-1) = 10 - 1 = 9$

EXAMPLE 4

Use a number line to evaluate each sum.

a $-5 + 4$ **b** $5 + (-3)$

SOLUTION

a $-5 + 4$

Start at –5 and move 4 to the right.

$-5 + 4 = -1$

b $5 + (-3)$

Start at 5 and move 3 to the left.

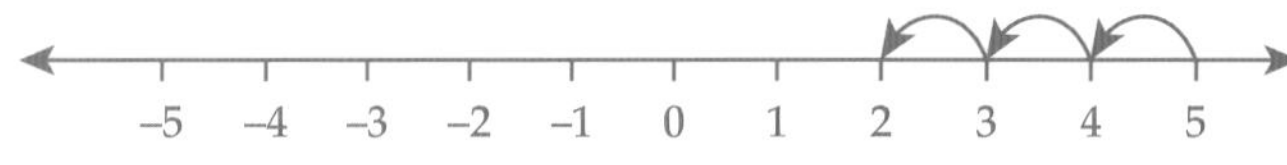

$5 + (-3) = 5 - 3 = 2$

Negative numbers can be entered into a calculator using the (–) or +/– key.

EXAMPLE 5

Use a calculator to evaluate each sum.

a $28 + (-32)$ **b** $-45 + (-23)$ **c** $-128 + 115$

SOLUTION

a $28 + (-32) = -4$ ← On calculator, enter: 28 + (–) 32 =

b $-45 + (-23) = -68$ ← On calculator, enter: (–) 45 + (–) 23 =

c $-128 + 115 = -13$ ← On calculator, enter: (–) 128 + 115 =

 ISBN 9780170350990

EXERCISE 4–02

1 What is the sum of 12 and –7? Select the correct answer **A**, **B**, **C** or **D**.

A –5 **B** –19 **C** 19 **D** 5

2 Evaluate –6 + 11. Select **A**, **B**, **C** or **D**.

A 5 **B** –17 **C** 17 **D** –5

3 Is each equation true or false?

a $-5 + 8 = -3$ **b** $9 + (-3) = 6$ **c** $-12 + 8 = -4$ **d** $-7 + (-2) = 9$

4 Evaluate each sum.

a $-8 + 4$	**b** $8 + (-4)$	**c** $-8 + (-4)$
d $-7 + 5$	**e** $7 + (-5)$	**f** $-7 + (-5)$
g $-9 + 6$	**h** $9 + (-6)$	**i** $-9 + (-6)$
j $-11 + 7$	**k** $11 + (-7)$	**l** $-11 + (-7)$
m $-12 + 5$	**n** $-8 + (-3)$	**o** $19 + (-4)$
p $10 + (-6)$	**q** $-14 + (-2)$	**r** $25 + (-6)$
s $-16 + (-2)$	**t** $18 + (-8)$	**u** $-7 + (-9)$
v $-18 + (-9)$	**w** $-22 + 4$	**x** $54 + (-32)$

5 A lift was malfunctioning so that each time the button was pressed it would go up 3 levels and then slip back down 1 level. Ben got in the lift and pressed the button four times. On which level did he end up if he was on the second level when he got into the lift?

6 Baby Kyle walked forward 5 steps, back 3, forward 2, back 3, forward 6 steps and then fell over. If his steps follow the same pattern, how many attempts like this would it take him to reach a toy 20 steps away?

Shutterstock.com/Oksana Kuzmina

4-03 Subtracting integers

Integers can be **subtracted** using a number line.

- Move **left** if subtracting a **positive** integer.
- Move **right** if subtracting a **negative** integer.
- Subtracting a **negative integer** is the same as **adding its opposite**; for example, $3 - (-4) = 3 + 4 = 7$

Think of how we speak:
A single negative means negative: 'I am *not* going to the movies'.
A double negative reverses the meaning to positive: 'I am *not not* going to the movies' means 'I am going to the movies'.

iStockphoto/monkeybusinessimages

EXAMPLE 6

Evaluate each difference.

a $-1 - 3$ **b** $4 - (-2)$

SOLUTION

a $-1 - 3$

Start at –1 and move 3 left.

$-1 - 3 = -4$

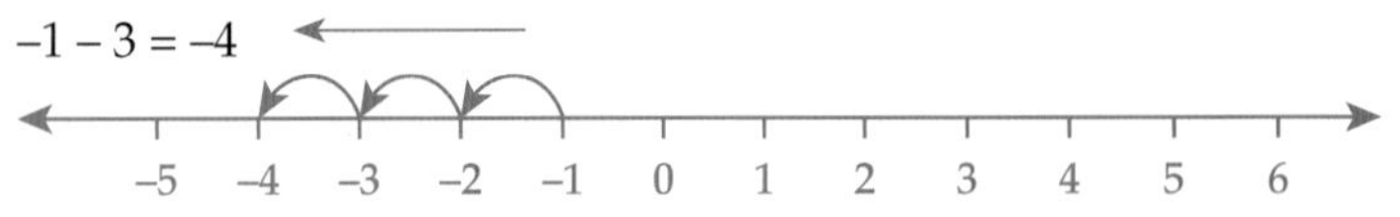

b $4 - -2$

Start at 4 and move 2 right.

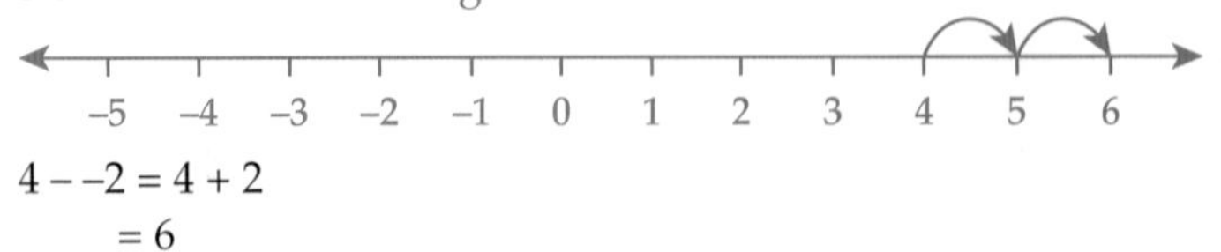

$4 - -2 = 4 + 2$
$= 6$

EXAMPLE 7

Use a calculator to evaluate $29 - (-24)$.

SOLUTION

$29 - (-24) = 53$ On calculator, enter: 29 [–] [(–)] 24 [=]

ISBN 9780170350990

EXERCISE 4–03

1 Find the difference between –8 and 5. Select the correct answer **A**, **B**, **C** or **D**.

A 3 B 13 C –3 D –13

2 Evaluate –7 – 4. Select **A**, **B**, **C** or **D**.

A 11 B –3 C 3 D –11

3 Evaluate each difference.

a –4 – 5	**b** 4 – (–5)	**c** –4 – (–5)
d –6 – 3	**e** 6 – (–3)	**f** –6 – (–3)
g –7 – 2	**h** 7 – (–2)	**i** –7 – (–2)
j –8 – 1	**k** 8 – (–1)	**l** –8 – (–1)
m –6 – 8	**n** –7 – (–4)	**o** 11 – (–5)
p –12 – 3	**q** –4 – (–18)	**r** 23 – (–12)
s –19 – 11	**t** –7 – 8	**u** 32 – (–21)
v –24 – 9	**w** –15 – 6	**x** 44 – (–21)

4 Use a calculator to evaluate each difference.

a –48 – 23	**b** –56 – (–31)	**c** 22 – (–19)
d –82 – 64	**e** 35 – (–92)	**f** –56 – 89
g 100 – (–32)	**h** –54 – 96	**i** –86 – (–114)

5 Jackson opened a bank account and deposited $250. A week later, he withdrew $28 but then deposited $56 the next day. How much money does he have in his bank account now?

6 Gian was lost, but tried to find her way by walking 25 steps forward, 3 steps back, 18 steps forward again and then 12 steps back. How many steps is she now from her starting point?

Alamy/Steven May

ISBN 9780170350990

4–04 Multiplying integers

Look at this pattern:

$3 \times 7 = 7 + 7 + 7 = 21$ — **positive × positive = positive number**

$3 \times (-7) = -7 + (-7) + (-7) = -21$ — **positive × negative = negative number**

$-3 \times (-7) = -[(-7) + (-7) + (-7)] = -[-21] = 21$ — **negative × negative = positive number**

When **multiplying integers**:

- **two positive** integers give a **positive** answer (+ × + = +)
- a **positive** and a **negative** integer give a **negative** answer (+ × − = −)
- **two negative** integers give a **positive** answer (− × − = +).

EXAMPLE 8

Find each product.

a -6×5 **b** $4 \times (-7)$ **c** $-4 \times (-9)$ **d** $(-5)^2$

SOLUTION

a $-6 \times 5 = -30$ ← negative × positive = negative

b $4 \times (-7) = -28$ ← positive × negative = negative

c $-4 \times (-9) = 36$ ← negative × negative = positive

d $(-5)^2 = -5 \times (-5) = 25$ ← negative × negative = positive

On calculator, enter: ( 5   

iStockphoto/HS3RUS

 ISBN 9780170350990

EXERCISE 4-04

1 Find the product of –6 and 8. Select the correct answer **A**, **B**, **C** or **D**.

A 48 **B** –54 **C** –48 **D** –56

2 Evaluate –12 × (–11). Select **A**, **B**, **C** or **D**.

A 132 **B** 144 **C** –132 **D** –144

3 Evaluate each product.

a –3 × 4 **b** 8 × (–5) **c** –4 × (–9)
d 5 × (–7) **e** –6 × (–4) **f** 8 × (–6)
g –11 × 5 **h** 12 × (–7) **i** –7 × (–9)
j 8 × (–9) **k** –7 × (–6) **l** –12 × 8
m 11 × (–8) **n** –6 × 9 **o** –4 × (–12)

4 Copy and complete this table.

×	-2	5	-6	8	-4
3					
-7					
-9					
10					

5 Evaluate each product using a calculator.

a 27 × (–3) **b** –38 × (–6) **c** 52 × (–8)
d 37 × (–9) **e** –62 × (–4) **f** –88 × 7
g –78 × (–5) **h** 129 × (–2) **i** 389 × (–8)
j 185 × (–23) **k** –246 × (–45) **l** –682 × 22
m –458 × 12 **n** 923 × (–54) **o** –1024 × (–7)

iStockphoto/MikeCherim

4-05 Dividing integers

For dividing integers, we can use the same rules as for multiplying integers because division and multiplication are inverse operations.
Look at this pattern:

$3 \times 7 = 21$, so $21 \div 7 = 3$	**positive ÷ positive = positive number**
$3 \times (-7) = -21$, so $-21 \div (-7) = 3$	**negative ÷ negative = positive number**
$-3 \times (-7) = 21$, so $21 \div (-7) = (-3)$	**positive ÷ negative = negative number**

When **dividing integers**:
- **two positive** integers give a **positive** answer $(+ \div + = +)$
- a **positive** and a **negative** integer give a **negative** answer $(+ \div - = -)$
- **two negative** integers give a **positive** answer $(- \div - = +)$

EXAMPLE 9

Evaluate each quotient.

a $-24 \div 8$ b $45 \div (-9)$ c $-42 \div (-6)$

SOLUTION

a $-24 \div 8 = -3$ ← negative ÷ positive = negative

b $45 \div (-9) = -5$ ← positive ÷ negative = negative

c $-42 \div (-6) = 7$ ← negative ÷ negative = positive

Shutterstock.com/Wichai Sittipan

ISBN 9780170350990

EXERCISE 4–05

1 What is the quotient of –54 and 9? Select the correct answer **A**, **B**, **C** or **D**.

A 7	**B** 6	**C** –6	**D** –8

2 Evaluate 72 ÷ (–8). Select **A**, **B**, **C** or **D**.

A –9	**B** –7	**C** 9	**D** –8

3 Evaluate each quotient.

a –8 ÷ 4	**b** 25 ÷ (–5)	**c** –48 ÷ (–8)
d 56 ÷ (–7)	**e** –27 ÷ (–3)	**f** 63 ÷ (–9)
g –108 ÷ (–12)	**h** 72 ÷ 8	**i** –40 ÷ (–5)
j –66 ÷ 11	**k** –30 ÷ (–5)	**l** 49 ÷ (–7)
m –60 ÷ (–5)	**n** –44 ÷ 4	**o** –96 ÷ (–12)

4 Use your calculator to evaluate each quotient.

a –24 ÷ 3	**b** 16 ÷ (–8)	**c** –27 ÷ (–3)	**d** 48 ÷ (–6)
e –49 ÷ (–7)	**f** 36 ÷ (–9)	**g** –64 ÷ 8	**h** –35 ÷ 5
i 81 ÷ (–9)	**j** –300 ÷ (–10)	**k** 280 ÷ (–7)	**l** –560 ÷ 80
m $72 \div (-3)^2$	**n** –120 ÷ 60	**o** $64 \div (-2)^2$	**p** –2400 ÷ (–120)

Shutterstock.com/Volt Collection

ISBN 9780170350990

4–06 Order of operations

WORDBANK

operations The four operations in mathematics are addition (+), subtraction (–), multiplication (×) and division (÷).

order of operations The correct order to evaluate a mixed expression with more than one operation, such as $16 \times 2 - (20 + 4)$.

brackets Grouping symbols around expressions such as round brackets ( ) or square brackets [ ].

When evaluating **mixed expressions**, calculate using this **order of operations**:

- brackets ( ) first,
- then powers (x^y) and square roots ($\sqrt{\ }$)
- then multiplication (×) and division (÷) from left to right
- then addition (+) and subtraction (−) from left to right.

EXAMPLE 10

Evaluate each expression.

a $18 - 12 \div (-4)$

b $160 + (-4) \times (-8)$

c $[-6 + (-3)] \times (-28 \div 7)$

d $(-8)^2 \div (-4) + \sqrt{16} \times (-2)$

e $(-48) \div 8 + (-60) \times (-2) - 18$

SOLUTION

a $18 - \underbrace{12 \div (-4)}_{-3} = 18 - (-3)$ ← ÷ first

$= 21$

b $160 + \underbrace{(-4) \times (-8)}_{32} = 160 + 32$ ← × first

$= 192$

c $\underbrace{[-6 + (-3)]}_{-9} \times \underbrace{(-28 \div 7)}_{-4} = -9 \times (-4)$ ← Work left to right: do [ ] first.

$= 36$

d $(-8)^2 \div (-4) + \sqrt{16} \times (-2) = 64 \div (-4) + 4 \times (-2)$ ← Powers and square roots first.

$= -16 + (-8)$ ← Then work left to right: ÷ and ×.

$= -24$

e $\underbrace{(-48) \div 8}_{-6} + \underbrace{(-60) \times (-2)}_{120} - 18 = -6 + 120 - 18$ ← ÷ and × first

$= 96$

ISBN 9780170350990

EXERCISE 4–06

1 Evaluate $24 - 12 \times (-3)$. Select the correct answer **A**, **B**, **C** or **D**.

A -12 **B** 60 **C** 12 **D** -60

2 Evaluate $48 + (-56) \div (-8)$. Select **A**, **B**, **C** or **D**.

A 41 **B** -41 **C** 56 **D** 55

3 Which operation do you do first if evaluating a mixed expression that involves:

a $+$ and $\times$? **b** $+$ and $-$? **c** $-$ and $\div$? **d** $\times$ and $\div$?

4 What operation would you do first in each expression?

a $28 - 5 \times (-3)$ **b** $24 + (-8) \div (-2)$ **c** $62 - (-63) \div 9$

d $38 + 81 \div (-9)$ **e** $-12 + (-7) \times 4$ **f** $56 - 120 \div (-12)$

g $19 - 35 \div (-7)$ **h** $48 + (-8) \times (-7)$ **i** $-72 + (-6) \times 8$

5 Evaluate each expression in Question **4**.

6 Evaluate each expression using a calculator.

a $(-4)^2 \times (-3) + 12$ **b** $-54 \div (-9) \times 26$

c $23 - (-6)^2 + 8$ **d** $29 - (-4) \times 16$

e $\sqrt{25} \times (-8) + (-3)^2$ **f** $238 - 123 \div (-3)$

g $\sqrt{49} + (-9)^2 \times (-3)$ **h** $\dfrac{42 - 16}{39 \div 3}$

i $\dfrac{28 - 12}{48 \div (-12)}$ **j** $(-4 \times 3) - [12 \div (-6)]$

k $-32 + (-9 + 6) \times (-3)$ **l** $65 - (-8 + 2) \times (-5)$

m $120 - (-8) \times (-9 - 2)$ **n** $[-7 \times (-6)] \div (-2 - 4)$

o $[-55 \div (-11)] \times (-7 + 15)$ **p** $-200 + [-4 \times (-8)] \div (-16)$

q $[-9 \times (-2)] \div [-5 + (-4)]$ **r** $\{500 - [-25 + (-35)]\} \div 80$

s $-240 \div [-8 + (-12 \times 3) + 4]$

7 At the supermarket, Georgia bought 3 cartons of milk for \$2.75 each, 5 blocks of chocolate for \$3.50 each and 4 packets of biscuits for \$1.95 each.

a How much did this cost altogether?

b How much change did she receive from a \$50 note?

8 Jake joined a tennis club and was charged \$25 per year to join, \$12.50 each visit during Monday to Friday and \$15 per visit on the weekend. How much will it cost him over a year if he plays twice during the week and once each weekend?

ISBN 9780170350990

CROSSWORD PUZZLE

Make a copy of this puzzle, then complete the crossword using the clues below.

Across

5 Word beginning with the letter 'E' meaning an educated guess of an answer.
8 The answer to a multiplication.
10 The opposite of multiply.
11 A positive number divided by a negative number gives this type of number.
12 Integers are positive or negative _____numbers or zero.

Down

1 Positive and negative whole numbers.
2 A number that is neither positive nor negative.
3 The operation of finding the difference.
4 The answer to an addition.
6 Word beginning with 'I' meaning going on forever, as with integers.
7 The operation of finding the sum.
9 Multiplying before adding is an example of _____ of operations.

ISBN 9780170350990

PRACTICE TEST 4

Part A General topics

Calculators are not allowed.

1 Evaluate 19 × 9.

2 Use index notation to simplify $(3^5)^3$

3 Complete: 56.4 km = _____ m.

4 Find $\frac{2}{5}$ of $300.

5 Find the perimeter of this triangle.

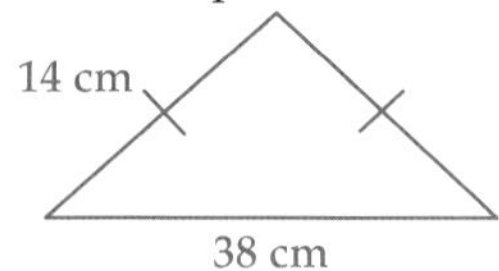

6 Evaluate 40 × 5.

7 Write 0.125 as a simple fraction.

8 List all the factors of 18.

9 How many axes of symmetry has a square?

10 How many days are there in May?

Part B Integers

Calculators are allowed.

4-01 Ordering integers

11 Arrange 6, –3, –5, 0 in ascending order. Select the correct answer **A**, **B**, **C** or **D**.

A 6, 0, –3, –5 **B** –3, –5, 0, 6 **C** –5, –3, 0, 6 **D** 6, 0 , –5, –3

12 Which statement is true? Select **A**, **B**, **C** or **D**.

A $-5 < -6$ **B** $0 < -2$ **C** $-5 < 5$ **D** $-8 > -7$

4-02 Adding integers

13 Evaluate –38 + 27. Select **A**, **B**, **C** or **D**.

A –65 **B** –11 **C** 11 **D** –9

14 Evaluate each sum.

a 56 + (–45) **b** –68 + 120

4-03 Subtracting integers

15 Evaluate each difference.

a –18 – 26 **b** 56 – (–24)

4-04 Multiplying integers

16 Evaluate each product.

a –6 × 9 **b** –8 × (–9)

4-05 Dividing integers

17 Evaluate each quotient.

a $-63 \div (-9)$ b $54 \div (-6)$

4-06 Order of operations

18 Evaluate $210 - (-6) \times 2 + (-72) \div (-9)$.

ISBN 9780170350990

DECIMALS

5

WHAT'S IN CHAPTER 5?

IN THIS CHAPTER YOU WILL:

- understand place value in decimals
- convert decimals to simple fractions
- compare and order decimals
- round decimals and money amounts
- add and subtract decimals
- multiply and divide decimals
- compare the unit cost of items of different sizes and brands to find the 'best buy' (value for money)
- understand and use terminating and recurring decimals
- convert fractions to decimals

* Shutterstock.com/patpitchaya

5-01 Decimals

WORDBANK

decimal A number that uses a decimal point and place value to show tenths $\left(\frac{1}{10}\right)$, hundredths $\left(\frac{1}{100}\right)$, thousandths $\left(\frac{1}{1000}\right)$ and so on.

decimal places The number of digits after the decimal point. For example, 6.24 has 2 decimal places as there are 2 digits after the decimal point.

The size of a decimal such as 6345.284 is shown by its **place value**.

Thousands (1000s)	Hundreds (100s)	Tens (10s)	Units (1s)	Decimal point	Tenths $\left(\frac{1}{10}\right)$ths	Hundredths $\left(\frac{1}{100}\right)$ths	Thousandths $\left(\frac{1}{1000}\right)$ths
6	3	4	5	.	2	8	4
6000	300	40	5		$\frac{2}{10}$	$\frac{8}{100}$	$\frac{4}{1000}$

So $6345.284 = (6 \times 1000) + (3 \times 100) + (4 \times 10) + (5 \times 1) + \left(2 \times \frac{1}{10}\right) + \left(8 \times \frac{1}{100}\right) + \left(4 \times \frac{1}{1000}\right)$

To write a decimal as a fraction, remember this:

One decimal place is 0.__ $= \frac{\square}{10}$ **1** number after the point, **1** zero in the denominator.

Two decimal places is 0.__ __ $= \frac{\square}{100}$ **2** numbers after the point, **2** zeros in the denominator.

Three decimal places is 0.__ __ __ $= \frac{\square}{1000}$ **3** numbers after the point, **3** zeros in the denominator.

EXAMPLE 1

Convert each decimal to a simple fraction.

a 0.4 **b** 0.07 **c** 0.32 **d** 1.016

SOLUTION

a $0.4 = \frac{4}{10} = \frac{2}{5}$

b $0.07 = \frac{7}{100}$

c $0.32 = \frac{32}{100} = \frac{8}{25}$

d $1.016 = 1\frac{16}{1000} = 1\frac{2}{125}$

✱ Always simplify the fraction if possible.

EXAMPLE 2

Convert each fraction to a decimal.

a $\frac{3}{10}$ **b** $\frac{37}{100}$ **c** $\frac{52}{10\,000}$

SOLUTION

a $\frac{3}{10} = 0.3$

1 zero, 1 decimal place

b $\frac{37}{100} = 0.37$

2 zeros, 2 decimal places

c $\frac{52}{10\,000} = 0.0052$

4 zeros, 4 decimal places

ISBN 9780170350990

EXERCISE 5-01

1 Which fraction is the same as 0.47? Select the correct answer **A**, **B**, **C** or **D**.

A $\frac{47}{10}$ B $\frac{47}{100}$ C $\frac{47}{1000}$ D $\frac{47}{10\,000}$

2 Convert 0.008 to a simplified fraction. Select **A**, **B**, **C** or **D**.

A $\frac{8}{10}$ B $\frac{4}{500}$ C $\frac{1}{125}$ D $\frac{8}{1000}$

3 For the decimal 5.178, write which digit has the place value of:

a tenth $\left(\frac{1}{10}\right)$ b thousandth $\left(\frac{1}{1000}\right)$

c unit d hundredth $\left(\frac{1}{100}\right)$

4 Convert each decimal to a fraction.

a 0.9 b 0.09 c 0.009 d 0.0009

5 Convert each decimal to a fraction with a denominator of 10, 100, 1000 or 10 000.

a 0.5 b 0.06 c 0.0007 d 0.004

e 0.06 f 0.54 g 0.075 h 0.386

i 0.0024 j 1.8 k 2.36 l 6.082

m 5.25 n 3.04 o 7.006 p 2.0186

6 Write the answers to Question **5** as simple fractions.

7 Convert each fraction to a decimal.

a $\frac{6}{10}$ b $\frac{3}{100}$ c $\frac{4}{1000}$ d $\frac{9}{10}$

e $\frac{21}{100}$ f $\frac{45}{1000}$ g $\frac{451}{1000}$ h $\frac{79}{10\,000}$

i $1\frac{2}{10}$ j $2\frac{3}{100}$ k $3\frac{25}{1000}$ l $\frac{652}{100}$

m $\frac{78}{10}$ n $\frac{524}{1000}$ o $\frac{178}{1000}$ p $6\frac{45}{10\,000}$

8 Convert each fraction to a denominator of 10 or 100 and then write as a decimal.

a $\frac{4}{5}$ b $\frac{3}{20}$ c $\frac{3}{4}$ d $\frac{9}{25}$

e $\frac{1}{5}$ f $\frac{7}{50}$ g $\frac{4}{20}$ h $\frac{11}{25}$

5-02 Ordering decimals

Adding 0s to the end of a decimal does not change the size of the decimal.

For example, 3.4 = 3.40 = 3.400 because $3\frac{4}{10} = 3\frac{40}{100} = 3\frac{400}{1000}$.

> To **compare and order decimals**, first add 0s to each decimal where required so that all decimals have the same number of decimal places. This makes them easier to compare.

EXAMPLE 3

Write these decimals in *ascending* order:

Ascending means from small to large

0.4, 0.45, 0.5, 0.49, 0.421, 0.405, 0.04

SOLUTION

First, write every decimal with 3 decimal places by adding 0s where required:

0.400, 0.450, 0.500, 0.490, 0.421, 0.405, 0.040

Now order them from smallest to largest, ignoring the decimal points:

0.040, 0.400, 0.405, 0.421, 0.450, 0.490, 0.500

Now write the decimals as they were in the question:

0.04, 0.4, 0.405, 0.421, 0.45, 0.49, 0.5

If asked to write numbers in *descending* order we would write them from *large to small.*

EXAMPLE 4

True or False?

a $0.6 > 0.65$ **b** $0.912 < 0.92$

SOLUTION

a $0.60 > 0.65$ ← First, write each decimal with the same number of decimal places. 0.60 is less than 0.65.

So $0.60 > 0.65$ is false.

b $0.912 < 0.920$ ← First, write each decimal with the same number of decimal places. 0.912 is less than 0.920.

So $0.912 < 0.920$ is true.

ISBN 9780170350990

EXERCISE 5–02

1 Which statement is true? Select the correct answer **A**, **B**, **C** or **D**.

A $0.8 > 0.81$ B $0.6 < 0.06$ C $0.45 > 0.405$ D $0.07 > 0.7$

2 Which statement is false? Select **A**, **B**, **C** or **D**.

A $0.4 < 0.402$ B $0.6 > 0.61$ C $0.12 > 0.01$ D $0.93 < 0.932$

3 Write each decimal below with 3 decimal places.

0.3 0.03 0.35 0.003

4 Write the decimals in Question **3** in ascending order.

5 Write the decimals below in descending order by first writing them with 3 decimal places.

0.06 0.62 0.6 0.061

6 Write each set of decimals in ascending order.

a 0.3, 0.03, 0.003, 0.31, 0.38, 0.312

b 0.56, 0.502, 0.05, 0.006, 0.516, 0.555

c 0.009, 0.92, 0.9, 0.09, 0.119, 0.911

d 1.4, 1.004, 1.04, 1.114, 1.41, 1.014

7 Write each set of decimals in descending order.

a 0.6, 0.666, 0.006, 0.06, 0.601, 0.61

b 0.24, 0.242, 0.002, 0.024, 0.244, 0.02

c 0.835, 0.08, 0.8, 0.853, 0.083, 0.008

d 4.5, 4.05, 4.005, 4.515, 4.55, 4.555

8 Is each statement true or false?

a $0.9 > 0.09$ **b** $0.08 < 0.8$ **c** $2.6 < 2.66$

d $4.02 > 4.2$ **e** $6.85 > 6.085$ **f** $12.34 < 12.4$

9 Copy and complete each statement with a $<$ or $>$ sign.

a 4.25 ____ 4.5 **b** 6.8 ____ 6.18

c 7.29 ____ 7.229 **d** 11.6 ____ 11.006

5–03 Rounding decimals and money

WORDBANK

rounding To write a number with approximately the same value using fewer digits.

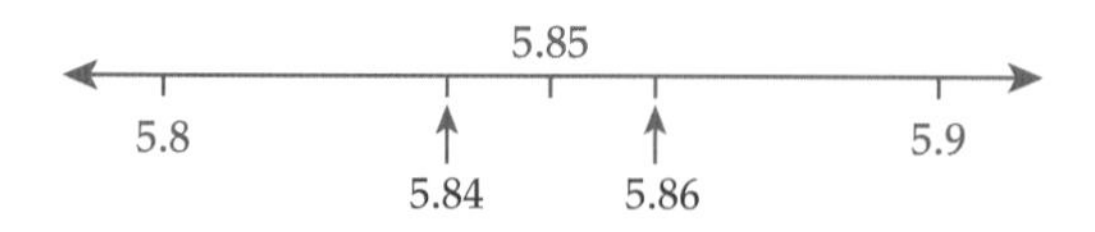

When rounding to 1 decimal place, 5.84 is closer to 5.8, whereas 5.86 is closer to 5.9. The halfway mark is 5.85.

To **round decimals** to a certain number of decimal places, look at the next digit. If it is 5 or more, round the decimal **up**. If it is less than 5, round the decimal **down**.

EXAMPLE 5

Round each decimal to 2 decimal places.

a 6.328 **b** 1.7631 **c** 28.415

SOLUTION

a Count 2 decimal places, then look at the next digit 8. It is more than 5, so round the digit 2 up to 3.

$\underline{6.32}8 \approx 6.33$

b Look at the 3. It is less than 5, so leave the digit 6 as is.

$\underline{1.76}31 \approx 1.76$

c Look at the 5. It is 5 or more, so round the digit 1 up to 2.

$\underline{28.41}5 \approx 28.42$

EXAMPLE 6

Round each amount to the nearest cent.

a \$6.238 **b** \$0.842 **c** \$59.455

SOLUTION

Rounding to the nearest cent means rounding to 2 decimal places.

a \$6.238 ≈ \$6.24 **b** \$0.842 ≈ \$0.84 **c** \$59.455 ≈ \$59.46

ISBN 9780170350990

EXERCISE 5-03

1 Round 72.4538 correct to 2 decimal places. Select the correct answer **A**, **B**, **C** or **D**.

A 72.4 **B** 72.453 **C** 72.46 **D** 72.45

2 Round $56.3817 to the nearest cent. Select **A**, **B**, **C** or **D**.

A $56.38 **B** $56.381 **C** $56.382 **D** $56.39

3 Round 128.6849 correct to 3 decimal places.

4 Write each number correct to 1 decimal place.

a 6.82 **b** 3.86 **c** 4.22 **d** 12.45

e 17.52 **f** 21.64 **g** 123.76 **h** 38.28

5 Round each decimal to 2 decimal places.

a 4.566 **b** 9.123 **c** 8.485 **d** 11.381

e 183.652 **f** 34.528 **g** 78.888 **h** 982.476

6 Write 45.829 451 6 correct to:

a 1 decimal place **b** 3 decimal places

c 4 decimal places **d** 5 decimal places.

7 Round each amount to the nearest cent.

a $4.568 **b** $23.623 **c** $0.258 × 233 **d** $6.28 ÷ 3

Shutterstock.com/

5–04 Adding and subtracting decimals

To add or subtract decimals:
- write the decimals underneath each other in columns
- ensure the decimal points line up underneath each other
- fill in the gaps with 0s
- add or subtract the digits in columns
- place the decimal point directly underneath.

Check your answer by estimating.

EXAMPLE 7

Evaluate each expression.

a 0.7 + 4.86 **b** 5.28 + 1.068 + 0.4 **c** 524.6 − 67.75

SOLUTION

a

${}^{1}0\ .\ 7\ 0\ +$	⟵	gaps filled in with 0s
$4\ .\ 8\ 6$	⟵	points under points
$5\ .\ 5\ 6$		

Check by estimating:

$0.7 + 4.86 \approx 1 + 5 = 6$

(5.56 is close to 6)

b

$5\ .\ {}^{1}2\ 8\ 0\ +$	⟵	gaps filled in with 0s
$1\ .\ 0\ 6\ 8$	⟵	points under points
$0\ .\ 4\ 0\ 0$		
$6\ .\ 7\ 4\ 8$		

Check by estimating:

$5.28 + 1.068 + 0.4 \approx 5 + 1 + 1 = 7$

(6.748 is close to 7)

c

${}^{4}\not{5}\ {}^{11}\not{2}\ {}^{13}\not{4}\ .\ {}^{15}\not{6}\ {}^{1}0\ -$	⟵	use trading to subtract
$6\ 7\ .\ 7\ 5$	⟵	points under points
$4\ 5\ 6\ .\ 8\ 5$		

EXAMPLE 8

Evaluate 123.78 − 68.9 using a calculator.

SOLUTION

123.78 − 68.9 = 54.88 ⟵ On calculator: 123.78 [−] 68.9 [=]

ISBN 9780170350990

EXERCISE 5-04

1 Find the sum of 45.6 and 228.59. Select the correct answer **A**, **B**, **C** or **D**.

A 264.19 B 273.19 C 274.19 D 263.19

2 Find the difference between 228.59 and 45.6. Select **A**, **B**, **C** or **D**.

A 182.99 B 182.89 C 172.99 D 182.19

3 Evaluate each sum.

a 0.8 + 21.7 b 2.17 + 3.8 c 23.7 + 8.95
d 28.76 + 1.6 e 12.4 + 0.986 f 128.4 + 3.76
g 54.79 + 2.543 h 0.95 + 43.6 i 75.8 + 106.452

4 Evaluate each difference.

a 28.4 – 0.3 b 32.5 – 0.6 c 47.3 – 0.05
d 15.6 – 2.9 e 28.3 – 4.7 f 98.45 – 3.62
g 128.5 – 45.8 h 342.7 – 23.8 i 763.25 – 18.6

5 Use a calculator to evaluate each expression.

a 245.6 + 23.94 b 129.65 – 65.8
c 1209.8 + 3.416 d 22.98 – 6.735
e 123.4 + 89.66 f 0.986 – 0.29
g 43.6 + 82.45 – 23.657 h 128.4 – 67 + 12.8
i 456.2 – 98.5 + 0.117

6 Nick went to the supermarket with a $50 note and bought a roast chicken for $12.45, some vegetables for $17.60 and dessert for $8.40. What change will he receive from a $50 note?

7 Imogen went shopping and bought the following items.

a dress for $125.95
a belt for $39.95
a pair of boots for $215.00
a necklace for $45.50
a handbag for $52.80

a How much did her shopping cost her?
b How much change would she have from $500?

8 Evaluate $62.688 + $125.752 – $149.678 + $521.674 correct to the nearest cent.

5-05 Multiplying decimals

To multiply decimals:

- multiply them as whole numbers without the decimal points
- count the total number of decimal places in the question
- write the answer using this number of decimal places.

Check your answer by estimating if possible.

EXAMPLE 9

Evaluate each product.

a 0.8×0.03 **b** 0.05×0.09

SOLUTION

a $8 \times 3 = 24$

0.8 has 1 decimal place and 0.03 has 2 decimal places. Total = 3 decimal places.

Write 24 with 3 decimal places: 0.<u>024</u>

$0.8 \times 0.03 = 0.024$

Check by estimating: $0.8 \times 0.03 \approx 1 \times 0.03 = 0.03$ (0.024 is close to 0.03)

b $5 \times 9 = 45$

0.05 has 2 decimal places and 0.09 also has 2 decimal places. Total = 4 decimal places.

Write 45 with 4 decimal places: 0.<u>0045</u>

$0.05 \times 0.09 = 0.0045$

Check by estimating: $0.05 \times 0.09 \approx 0.05 \times 0.1 = 0.005$ (0.0045 is close to 0.005)

EXAMPLE 10

Use a calculator to evaluate 58.92×1.56.

SOLUTION

$58.92 \times 1.56 = 91.9152$ ← On calculator: 58.92 [×] 1.56 [=]

* Note that the answer has 2 + 2 = 4 decimal places.

ISBN 9780170350990

EXERCISE 5–05

1 How many decimal places will the answer to 6.2×0.04 have? Select the correct answer **A**, **B**, **C** or **D**.

A 1 **B** 2 **C** 3 **D** 4

2 Evaluate 6.2×0.04. Select **A**, **B**, **C** or **D**.

A 2.48 **B** 0.248 **C** 24.8 **D** 0.0248

3 Write the number of decimal places in each product.

a 0.8×0.2 **b** 0.09×0.7 **c** 0.6×0.04

d 2.1×0.3 **e** 3.06×0.4 **f** 0.06×3.5

g 4.2×0.05 **h** 6.12×0.009 **i** 7.8×0.003

4 Evaluate each product in Question **3**.

5 Use a calculator to evaluate each product.

a 3.4×2.8 **b** 5.62×8.3 **c** 1.18×4.6

d 2.7×6.72 **e** 1.9×3.15 **f** 5.07×0.45

g 0.016×4.7 **h** 6.02×8.12 **i** 305.8×9.2

6 If $2.6 \times 0.8 = 2.08$, then what would be 0.26×0.08?

7 If $8.4 \times 0.06 = 0.504$, then what would be 84×0.006?

8 If $28 \times 0.07 = 1.96$, then what would be 0.28×0.007?

9 Brooke uses the following materials to build a doll's house.

12 m of timber at \$4.90/m
2 packets of nails at \$5.80/packet
4 L paint at \$23.50/L
4 sheets of roofing at \$12.80/sheet
6 packs of furniture at \$15.75/pack

a How much will it cost Brooke to build the doll's house?

b How much change will she get from \$400?

10 Ray's water bill arrived and the charges are listed below.

Water service: \$6.282
Sewer service: \$158.249
Water usage: 56.00 kL @ \$1.725 per kL

What is the total amount of this bill, correct to 2 decimal places?

ISBN 9780170350990

5–06 Dividing decimals

To **divide a decimal by a whole number**, use short division, then write the answer with the decimal point in the same column as in the original decimal.

EXAMPLE 11

Evaluate each quotient.

a 67.2 ÷ 8

b 785.5 ÷ 4

SOLUTION

a $8\overline{)67.{}^{3}2}$ = 8.4

Decimal points are underneath each other.

67.2 ÷ 8 = 8.4

Check by estimating:

67.2 ÷ 8 ≈ 70 ÷ 10 = 7

(8.4 is close to 7)

b $4\overline{)7{}^{3}8{}^{2}5.{}^{1}5{}^{3}0{}^{2}0}$ = 196.375

Decimal points are underneath each other.

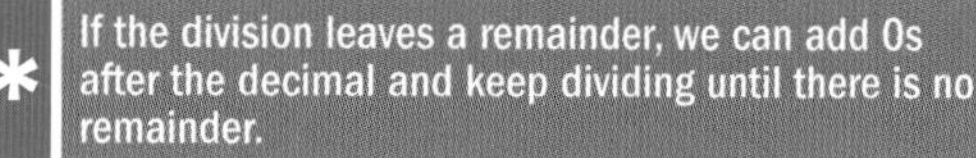

785.5 ÷ 4 = 196.375

785.5 ÷ 4 ≈ 800 ÷ 4 = 200

(196.375 is close to 200)

To **divide a decimal by another decimal**, move the points in both decimals the same number of places to the right so that you are dividing by a **whole number**.

EXAMPLE 12

How many places would you need to move the decimal point to the right in each decimal to evaluate 5.64 ÷ 0.3?

SOLUTION

We need to make 0.3 into the whole number 3.

To do this, we will need to multiply by 10.

To keep the question the same, we must multiply both numbers by 10, which means moving the decimal point one place to the right.

5.64 ÷ 0.3 = 56.4 ÷ 3

EXAMPLE 13

Evaluate each quotient.

a 5.64 ÷ 0.3

b 4.86 ÷ 0.04

SOLUTION

a 5.64 ÷ 0.3 = 56.4 ÷ 3 ← Move both decimal points 1 right, so 0.3 is the whole number 3.

$3\overline{)56.4}$ = 18.8

5.64 ÷ 0.3 = 18.8

Check by estimating: 56.4 ÷ 3 ≈ 60 ÷ 3 = 20 ← 18.8 is close to 20.

ISBN 9780170350990

b $4.86 \div 0.04 = 486 \div 4$ ← Move both decimal points 2 right, so 0.04 is the whole number 4.

$$4\overline{)486.^{2}0}$$ = 121.5

$4.86 \div 0.04 = 121.5$

Check by estimating: $486 \div 4 \approx 480 \div 4 = 120$ ← 121.5 is close to 120.

EXERCISE 5-06

1 Evaluate $28.45 \div 5$. Select the correct answer **A**, **B**, **C** or **D**.

A 5.9 **B** 5.61 **C** 5.09 **D** 5.69

2 How many places would you need to move the decimal point to the right to divide 78.655 by 0.05? Select **A**, **B**, **C** or **D**.

A 1 **B** 2 **C** 3 **D** 4

3 Which expression is the same as $6.82 \div 0.2$? Select **A**, **B**, **C** or **D**.

A $0.682 \div 2$ **B** $6.82 \div 2$ **C** $68.2 \div 2$ **D** $682 \div 2$

4 Evaluate each quotient.

a $28.6 \div 2$ **b** $106.5 \div 5$ **c** $38.4 \div 3$
d $8.68 \div 4$ **e** $387.9 \div 9$ **f** $68.24 \div 8$
g $218.4 \div 7$ **h** $48.4 \div 11$ **i** $278.4 \div 6$
j $45.2 \div 8$ **k** $108.543 \div 6$ **l** $228.954 \div 5$

5 Find the mistake in each equation, correct the mistake and evaluate the quotient.

a $48.68 \div 0.4 = 486.8 \div 40$ **b** $375.584 \div 0.5 = 37\,558.4 \div 5$
c $34.653 \div 0.03 = 346.53 \div 3$ **d** $569.64 \div 0.4 = 56\,964 \div 4$

6 Evaluate each quotient.

a $58.84 \div 0.4$ **b** $275.58 \div 0.5$ **c** $346.53 \div 0.03$
d $69.93 \div 0.6$ **e** $228.6 \div 0.09$ **f** $1089 \div 0.009$

7 Josh had 689.36 m of copper pipe that had to be cut into pieces 0.4 m long.

a How many pieces could he make?

b Was there any copper pipe left over?

c How much money did Josh's company make from selling each copper piece for $28.50?

5-07 Best buys

WORDBANK

best buy A brand or size of product that is the best value for money.

unit cost The cost of one item or metric unit; for example, 1 kg.

Unit cost = cost ÷ number of items or units.

EXAMPLE 14

What is the best buy for different sizes of punnets of strawberries?

A 250 g for \$3.50 **B** 120 g for \$2.45 **C** \$14.25 per kg

SOLUTION

Compare the price of 1 g of strawberries.

* This is called the **unit cost.**

A 1 g costs \$3.50 ÷ 250 = \$0.014

B 1 g costs \$2.45 ÷ 120 = \$0.0204

C 1 g costs \$14.25 ÷ 1000 = \$0.01425 ← 1 kg = 1000 g

The best buy is **A** at 250 g for \$3.50. ← \$0.014/g is the cheapest

EXAMPLE 15

Select the best buy among the following brands of ice-cream.

A 2 L for \$3.29 **B** 5 L for \$5.80 **C** 500 mL for \$1.25

Shutterstock.com/Kitch Bain

SOLUTION

Compare the price of 1 L of ice-cream for each brand.

A 1 L costs \$3.29 ÷ 2 = \$1.645

B 1 L costs \$5.80 ÷ 5 = \$1.16

C 1 L costs \$1.25 × 2 = \$2.50 ← 1 L = 1000 mL = 2 × 500 mL

The best buy is **B** at 5 L for \$5.80. ← \$1.16/L is the cheapest

ISBN 9780170350990

EXERCISE 5-07

1 Which is the best buy: 5 apples for $1.20, 2 apples for 60c, 4 apples for $1 or 35c each? Select the correct answer **A**, **B**, **C** or **D**.

A 5 apples **B** 2 apples **C** 4 apples **D** 1 apple

2 If I could buy 12 apples for $2.50, which would be the best buy now? Select **A**, **B**, **C** or **D**.

A 5 apples **B** 2 apples **C** 4 apples **D** 12 apples

3 Find the best buy for the following different sizes of milk.

500 mL for $1.50 $2.20 per litre $5.30 for 4 L

4 Mala went to the local market and found the following prices for apples. $3.85 per kg, $1.40 for 250 g or $12.50 for a 5 kg tray. If they were all the same quality, which was the best buy?

5 Find the best buy for each group of items.

a 2 kg tomatoes for $3.75, 4 kg for $6.20 or 5.5 kg for $8.50

b 240 g leg ham for $4.50, 400 g for $8.60 or 1 kg for $18.20

c 2 L of cream for $3.80, 1.8 L for $3.50 or 5 L for $7.90

d 1.2 kg lamb for $14.50, 800 g for $12.60 or 2.5 kg for $22.90

e 8 bread rolls for $3.60, 5 bread rolls for $2.60 or 12 bread rolls for $5.50

6 Hugo owns a delicatessen and needs to order the best-value meat.
Hugo placed the order shown below, choosing what he thought was the best buy.
Which products did Hugo order incorrectly?

Hugo's Order	Choice 1	Choice 2
Sam's salami	Sam: $18.50 per kg	Bob: $64 per 5 kg box
Bill's bacon	Greg: $85 per 6 kg box	Bill: $8.50 per 500 g
Pam's pastrami	Pam: $52 per 3 kg	Sue: $1.60 per 100 g
Mark's mortadella	Frank: $28.60 per 2.5 kg	Mark: $1.02 per 100 g

5–08 Terminating and recurring decimals

WORDBANK

recurring decimal A decimal with digits that repeat or recur, such as $0.\dot{3} = 0.3333\ldots$ and $10.\dot{2}\dot{6} = 10.262626\ldots$

terminating decimal A decimal that ends or terminates, such as 0.512 or 2.96.

Some fractions convert to decimals that have digits that repeat endlessly. For example:

$\frac{1}{3} = 0.33333\ldots$ $\frac{1}{6} = 0.16666\ldots$ $\frac{1}{11} = 0.090909\ldots$

These are called **recurring decimals**. Recur means 'to repeat'.

Recurring decimals are written with **dots** placed over the repeating digits. If a **series** of digits are repeated, then a dot is placed over the **first** and **last** digits in the series.

EXAMPLE 16

Write each recurring decimal using dot notation.

a 0.272727... **b** 2.34444... **c** 18.236236...

SOLUTION

a $0.272727\ldots = 0.\dot{2}\dot{7}$ **b** $2.34444\ldots = 2.3\dot{4}$ **c** $18.236236\ldots = 18.\dot{2}3\dot{6}$

EXAMPLE 17

Evaluate each quotient and state whether the decimal is terminating or recurring.

a $6.824 \div 3$ **b** $98.26 \div 0.8$

SOLUTION

a

$$\begin{array}{r} 2.2\,7\,4\,6\,6\,6\,6\ldots \\ 3\overline{)6.8^22^14^20^20^20^20\ldots} \end{array}$$

✱ It is necessary to add 4 or more zeros to see a repeating pattern.

$6.824 \div 3 = 2.274\dot{6}$, a recurring decimal.

b $98.26 \div 0.8 = 982.6 \div 8$

$$\begin{array}{r} 1\,2\,2.\,8\,2\,5 \\ 8\overline{)9^18^22.^66^20^40} \end{array}$$

$98.26 \div 0.8 = 122.825$, a terminating decimal.

EXAMPLE 18

Convert each fraction to a decimal.

a $\frac{1}{8}$ **b** $\frac{2}{3}$

SOLUTION

a $\frac{1}{8}$ means $1 \div 8$

$$\begin{array}{r} 0.1\,2\,5 \\ 8\overline{)1.0^20^40} \end{array}$$

$\frac{1}{8} = 0.125$

b $\frac{2}{3}$ means $2 \div 3$

$$\begin{array}{r} 0.6\,6\,6\,6\ldots \\ 3\overline{)2.0^20^20^20\ldots} \end{array}$$

$\frac{2}{3} = 0.\dot{6}$

ISBN 9780170350990

EXERCISE 5-08

1 Write 0.454545... as a recurring decimal. Select the correct answer **A**, **B**, **C** or **D**.

A 0.45 **B** $0.4\dot{5}$ **C** $0.\dot{4}5$ **D** $0.\dot{4}\dot{5}$

2 Write 0.625625625... as a recurring decimal. Select **A**, **B**, **C** or **D**.

A 0.625 **B** $0.\dot{6}2\dot{5}$ **C** $0.62\dot{5}$ **D** $0.\dot{6}\dot{2}\dot{5}$

3 Write each decimal as a recurring decimal.

a 0.5555… **b** 0.343434… **c** 2.68888…

d 6.43333… **e** 28.22222… **f** 6.252525…

g 12.213213… **h** 1.0452452… **i** 72.74566666…

4 Evaluate each quotient and state whether the decimal is terminating or recurring.

a $2.45 \div 3$ **b** $45.5 \div 5$ **c** $128.4 \div 4$

d $214.6 \div 6$ **e** $32.86 \div 0.4$ **f** $186.3 \div 0.09$

5 Use a calculator to evaluate each quotient.

a $568.4 \div 1.4$ **b** $23.8 \div 2.2$ **c** $186.45 \div 0.5$

d $128.6 \div 1.2$ **e** $10.92 \div 0.8$ **f** $2137.8 \div 0.5$

6 Is each equation true or false?

a $0.444 = 0.\dot{4}$ **b** $6.232323\ldots = 6.\dot{2}\dot{3}$

c $12.135135\ldots = 12.\dot{1}\dot{3}5$ **d** $7.248888\ldots = 7.24\dot{8}$

7 **a** How do you change $\frac{3}{4}$ into a decimal without the use of a calculator?

b Convert $\frac{3}{4}$ to a decimal and check your answer using a calculator.

c What type of decimal is it?

8 Convert each fraction below to a decimal, then check your answer using a calculator.

a $\frac{4}{5}$ **b** $\frac{1}{3}$ **c** $\frac{3}{8}$ **d** $\frac{5}{6}$

e $\frac{7}{8}$ **f** $\frac{1}{7}$ **g** $\frac{5}{9}$ **h** $\frac{4}{8}$

9 What number am I?

I am a decimal
I am less than 1
My recurring digit is even
My digits after the decimal point are all the same
I am greater than 0.7

DECODE PUZZLE

For the clues below, list the letters, in order, to spell out a two-word phrase relating to this topic.

The first 0.2 of DESCENDING

The first 0.2 of CELLOPHANE

The first 0.1 of PRODUCTION

The first 0.1 of TRANSLATES

The middle 0.2 of CAPTIVATED

The last 0.1 of DELIBERATE

The first 0.2 of DEVASTATED

The 4th letter of FASCINATED

The 8th letter of INSPECTION

The first 0.2 of MANAGEMENT

The 8th letter of SPECTACLES

The last 0.1 of DIRECTIONS

ISBN 9780170350990

PRACTICE TEST 5

Part A General topics

Calculators are not allowed.

1 Find 20% of $450.

2 Complete 22 L = _____ mL.

3 List the first four multiples of 6.

4 Evaluate 18×5.

5 What is 54.756 rounded to 2 decimal places?

6 Evaluate $\frac{3}{4} - \frac{2}{3}$.

7 Evaluate $3m - 2n$ if $m = 4$ and $n = -2$.

8 Find the area of the rectangle.

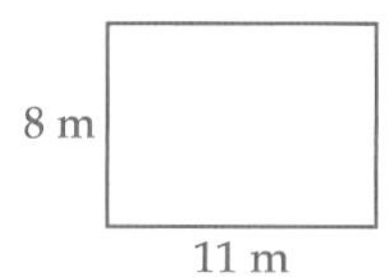

9 What is the probability of rolling a 3 or a 4 on a die?

10 Copy and complete: $\frac{5}{7} = \frac{}{42}$.

Part B Decimals

Calculators are allowed.

5-01 Decimals

11 Which fraction is the same as 0.73? Select the correct answer **A**, **B**, **C** or **D**.

A $\frac{73}{10}$ **B** $\frac{73}{100}$ **C** $\frac{73}{1000}$ **D** $\frac{73}{10\,000}$

12 Convert 0.004 to a simplified fraction. Select **A**, **B**, **C** or **D**.

A $\frac{4}{10}$ **B** $\frac{2}{500}$ **C** $\frac{1}{250}$ **D** $\frac{4}{1000}$

5-02 Ordering decimals

13 Which of these decimals lies between 0.81 and 0.86? Select **A**, **B**, **C** or **D**.

A 0.88 **B** 0.87 **C** 0.805 **D** 0.821

14 Write these decimals in ascending order.

0.02, 0.201, 0.211, 0.002

5-03 Rounding decimals and money

15 Round 183.5628 to 2 decimal places. Select **A**, **B**, **C** or **D**.

A 183.57 **B** 183.56 **C** 183.563 **D** 183.66

5-04 Adding and subtracting decimals

16 Evaluate each expression.

a $24.68 + 520.9$

b $124.7 - 58.95$

5-05 Multiplying decimals

17 Evaluate 4.7×0.23.

5-06 Dividing decimals

18 Which expression gives the same value as $48.92 \div 0.8$? Select **A**, **B**, **C** or **D**.

A $489.2 \div 0.8$ **B** $48.92 \div 8$ **C** $4.892 \div 8$ **D** $489.2 \div 8$

5-07 Best buys

19 Which is the best buy for bread rolls?

Shop 1: $4.20/ dozen
Shop 2: 50c each
Shop 3: $3 for 8
Shop 4: $2.50 for half a dozen

5-08 Terminating and recurring decimals

20 Write 195.466666... as a recurring decimal using dot notation.

21 Convert each fraction to a decimal.

a $\frac{3}{8}$

b $\frac{2}{9}$

 ISBN 9780170350990

ALGEBRA

6

WHAT'S IN CHAPTER 6?

IN THIS CHAPTER YOU WILL:

- use variables to write general rules involving numbers
- use algebraic abbreviations to simplify expressions
- convert worded descriptions into algebraic expressions
- substitute values into algebraic expressions
- add and subtract algebraic terms
- multiply and divide algebraic terms

* Shutterstock.com/solarseven

6-01 Variables

A **variable** or **pronumeral** is a symbol or letter of the alphabet used to represent a number. The value of the variable can change or **vary**, which is where the name comes from.
In algebra, we use variables to write general rules for numbers once a pattern has been discovered.

EXAMPLE 1

For each number pattern below write a general rule using a variable.

a $2 + 0 = 2$
$5 + 0 = 5$
$9 + 0 = 9$

b $3 \div 1 = 3$
$6 \div 1 = 6$
$8 \div 1 = 8$

SOLUTION

a $n + 0 = n$

b $r \div 1 = r$

* For the variable, we can use any letter of the alphabet.

In algebra, we can use abbreviations when writing variables

- $1 \times b = b$ ← '1' not needed
- $2 \times a = 2a$ ← '×' not needed
- $3 \times a \times b \times 4 = 12ab$ ← numbers multiplied together and written first
- $2 \times a \times a = 2a^2$ ← $a \times a = a^2$

EXAMPLE 2

Simplify each expression.

a $a + a + a + a$

b $5 \times m \times 4 \times n$

c $x + x + x + y + y$

d $3 \times d \times d \times 2$

SOLUTION

a $a + a + a + a = 4a$

b $5 \times m \times 4 \times n = 5 \times 4 \times m \times n$
$= 20mn$

c $x + x + x + y + y = 3x + 2y$

d $3 \times d \times d \times 2 = 3 \times 2 \times d \times d$
$= 6d^2$

ISBN 9780170350990

EXERCISE 6-01

1 Simplify $m + m + n + n + n$. Select the correct answer **A**, **B**, **C** or **D**.

A $2m + n + n$ **B** $3m + 2n$ **C** $2m + 3n$ **D** $m + m + 3n$

2 Simplify $u \times 4 \times v \times 5 \times v$. Select **A**, **B**, **C** or **D**.

A $20vu$ **B** $40uvv$ **C** $20uv$ **D** $20uv^2$

3 Write which of the following could be used as a variable.

5, a, −7, −, *, e, +, %, k, 0.9, ÷, $, u, b

4 Investigate the number patterns below and write a general rule for each one using a variable.

a $3 - 3 = 0$
$7 - 7 = 0$
$9 - 9 = 0$

b $3 \times 1 = 3$
$6 \times 1 = 6$
$8 \times 1 = 8$

c $5 + 5 = 2 \times 5$
$3 + 3 = 2 \times 3$
$7 + 7 = 2 \times 7$

5 Simplify each expression.

a $b + b + b$ **b** $5 \times m$ **c** $w + w + w + w + w$

d $1 \times a + 1 \times b$ **e** $3 \times n \times 4$ **f** $a + b + a - b - a + b$

g $5 \times c \times 2 \times c$ **h** $3 \times s + 4 \times t$ **i** $m + m + m - n - m + 2 \times n$

j $2 \times b \times 4 \times b \times 3$ **k** $4a + b - 2a$ **l** $5 \times m - 3 \times n + 4 \times p$

6 Is each equation true or false?

a $a \times a = 2a$ **b** $1 \times b \times b = b^2$ **c** $2 \times a + 3 \times b = 6ab$

d $3 \times m - 2 \times n = mn$ **e** $4 \times g \times 2 = 8g$ **f** $v + v + 2v - v = 2v$

g $8 - 2 \times a = 6a$ **h** $2 + 5 \times c = 7c$ **i** $6 \times s + 4 \times t = 6s + 4t$

j $a + a + a + a - a = 3a$ **k** $3 \times a \times 5 \times a = 15a$ **l** $20 - 4 \times b \times b = 16b^2$

m $5 \times m - 4 \times n = 5m - 4n$

7 Use the order of operations to simplify each expression.

a $5 + 2 \times m$ **b** $3 \times a \times b - 2$ **c** $20 - 3 \times n$

d $4 \times v - 3 \times w$ **e** $8 + 5 \times a$ **f** $2 \times d - d + 6$

g $20 - a \times a$ **h** $14 + 1 \times n + n$ **i** $8 \times m - 6 \times m + 5$

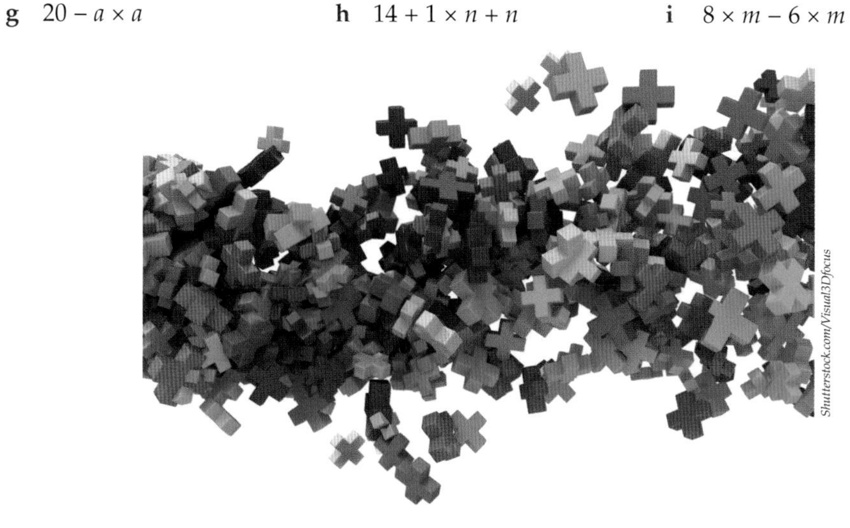

Shutterstock.com/Visual3Dfocus

ISBN 9780170350990

6-02 From words to algebraic expressions

To solve problems using algebra we need to be able to convert worded descriptions into **algebraic expressions** involving variables and numbers.

EXAMPLE 3

Write each statement as an algebraic expression. Use the variable b to stand for the number.

a The sum of a number and 4

b The product of 5 and a number

c The difference between a number and 6

d The quotient of b and 8

SOLUTION

a $b + 4$

b $5 \times b = 5b$

c $b - 6$

d $\frac{b}{8}$

EXAMPLE 4

Translate each worded description into an algebraic expression.

a Twice n minus 8

b Increase m by 7

c Triple the sum of w and 6

d Decrease the product of u and v by 9

SOLUTION

a $2n - 8$ ← 'Twice' means double, to multiply by 2.

b $m + 7$ ← 'Increase' means to add, make bigger.

c $3(w + 6)$ ← 'Triple' means to multiply by 3.

* Brackets are necessary as the sum of w and 6 must be done first.

d $uv - 9$ ← 'Decrease' means to subtract, make smaller.

ISBN 9780170350990

EXERCISE 6–02

1 Find an algebraic expression for twice the difference between m and n. Select the correct answer **A**, **B**, **C** or **D**.

A $2m - n$ **B** $m - 2n$ **C** $2(m - n)$ **D** $2 - m - n$

2 Find an algebraic expression for the sum of $5x$ and $3y$. Select **A**, **B**, **C** or **D**.

A $5x + 3y$ **B** $5x - 3y$ **C** $5x \times 3y$ **D** $5x \div 3y$

3 Match each word to a mathematical symbol +, −, ×, ÷.

a difference **b** increase **c** product

d sum **e** quotient **f** twice

g square **h** decrease **i** triple

4 Write each statement as an algebraic expression. Use n for the number.

a the product of 3 and the number

b the difference between the number and 8

c increase the number by 7

d the quotient of the number and 6

e the sum of the number and 2

f decrease the number by 12

g twice the number less 5

h the square of the number

i the number cubed

j triple the number plus 9

5 Write in words the meaning of each algebraic expression.

a $m + n$ **b** $6a$ **c** $2b + 4$

d $\frac{3n}{4}$ **e** $8 - b$ **f** $3v - 2$

g $4m - n$ **h** $\frac{2m}{n}$ **i** $3(a - b)$

6 Write the following as algebraic expressions.

a triple w minus 5

b decrease the sum of a and b by 6

c twice the quotient of d and 9

d increase 8 by the product of m and n

e square the sum of r and s

f cube the difference of c and 4

g the product of a, b and c

h triple the sum of w, v and u

6-03 Substitution

WORDBANK

substitution Replacing a variable with a number in an algebraic expression to find the value of the expression.

EXAMPLE 5

If $m = 6$ and $n = -2$, evaluate each expression.

a $4m - n$ **b** $5n + 3m$ **c** $8mn$

SOLUTION

a $4m - n = 4 \times 6 - (-2)$
$= 24 + 2$
$= 26$

b $5n + 3m = 5 \times (-2) + 3 \times 6$
$= -10 + 18$
$= 8$

c $8mn = 8 \times 6 \times (-2)$
$= -96$

It is also important to remember the **order of operations** when substituting:

- brackets ( ) first
- then powers (x^y) and square roots ($\sqrt{\ }$)
- then multiplication (×) and division (÷) from left to right
- then addition (+) and subtraction (−) from left to right.

EXAMPLE 6

Substitute $u = -4$ and $v = 0.8$ to evaluate each expression.

a $20 - uv$ **b** $5v + 2u^2$ **c** $v + \frac{7v}{u}$

SOLUTION

a $20 - uv = 20 - (-4) \times 0.8$
$= 20 - (-3.2)$ ← × first
$= 23.2$

b $5v + 2u^2 = 5 \times 0.8 + 2 \times (-4)^2$
$= 4 + 2 \times 16$ ← $(-4)^2$ first
$= 36$ ← × next

c $v + \frac{7v}{u} = 0.8 + \frac{7 \times 0.8}{-4}$
$= 0.8 + \frac{5.6}{-4}$ ← × first
$= 0.8 + (-1.4)$ ← ÷ next
$= -0.6$

ISBN 9780170350990

EXERCISE 6-03

1 If $a = -4$ and $b = 5$, evaluate $3a + b$. Select the correct answer **A**, **B**, **C** or **D**.

A -17 B -7 C 11 D 17

2 Evaluate $4ab - 2a$ if $a = 6$ and $b = -2$. Select **A**, **B**, **C** or **D**.

A -52 B 44 C -60 D -44

3 Evaluate each expression if $m = 9$, $n = -3$ and $p = 6$.

a $3mn$ b $2m - n$ c $4p + 5$ d $12 - 4n$
e $2mn + p$ f $5np - m$ g $16 - 3mn$ h $2mp + n$
i $18 - 2np$ j $20np + 6$ k $5mn + 2p$ l $4mnp$

4 State which operation you would do first if substituting $w = -6$ and $v = 8$ into:

a $5 + wv$ b $3w - v$ c $8v + w$ d $\frac{3v}{w} + 2$
e $12 - vw$ f $6v + 4w$ g $\frac{8w}{v} - 4$ h $24 - 2wv$

5 Evaluate each expression in Question **4**.

6 Use the table to evaluate each algebraic expression below.

a	b	c	d	e	f
5	−4	0.3	6	−8	0

a $3ab$ b $12 - cd$
c $8e - 2a$ d $4bc + d$
e $20abc$ f $\frac{2ab}{e}$
g $bcd - 12$ h $8cd + 3f$
i $6af - 3d$ j $2de + 4ab$
k $\frac{3de}{2b}$ l $20 - cef$

7 If $a = 8$ and $b = 0.6$, state whether each equation is true or false.

a $4ab = 19.2$ b $18 - 2b = 6$ c $5b - a = -5$ d $2ba + 8 = 104$

6–04 Adding and subtracting terms

WORDBANK

term The parts of an algebraic expression. For example, $2x + 3y - 5$ has three terms: $2x$, $3y$ and 5.

like terms Terms with exactly the same variables; for example, $2a$ and $3a$, $7y$ and $-3y$, $4ab$ and $-2ba$.

We know that $a + a + a + a = 4a$ if a stands for the same number.
We also know that $x + x + x + y + y = 3x + 2y$ if x and y stand for different numbers.
We can add the x's and we can add the y's separately but we cannot add x and y together because they are **unlike** terms that stand for different numbers.

Examples of **like terms** are: $5n, -3n, \frac{n}{4}$

Examples of **unlike terms** are: $5n, 2m, 5a, 12bc$

- Only like terms can be added and subtracted.
- The sign in front of a term belongs to it.
- x means $1x$, the '1' does not need to be written.

EXAMPLE 7

Simplify each expression.

a $12a + 3a$ **b** $8x - 6x$ **c** $4b + 2b - 5b$

d $10w - w + 4z$ **e** $7m + 2n - 6m + 3n$ **f** $12p^2 + 4p + p - 5p^2$

SOLUTION

a $12a + 3a = 15a$ ← 12a and 3a are like terms so we can add.

b $8x - 6x = 2x$ ← 8x and 6x are like terms so we can subtract.

c $4b + 2b - 5b = 1b$ ← 1b is the same as b on its own.
$= b$

d $10w - w + 4z = 9w + 4z$ ← 10w and w are like terms but 4z is not.

e $\underline{7m} + 2n \underline{- 6m} + 3n = 7m - 6m + 2n + 3n$ ← collecting like terms together
$= m + 5n$

 Group together like terms, including the sign in front of it: $+2n$, $-6m$, $+3n$.

f $\underline{12p^2} + 4p + p \underline{- 5p^2} = 12p^2 - 5p^2 + 4p + p$ ← p^2 and p are not like terms.
$= 7p^2 + 5p$

ISBN 9780170350990

EXERCISE 6–04

1 Simplify $7m - 3m + 5m$. Select the correct answer **A**, **B**, **C** or **D**.

A $17m$ **B** $14m$ **C** $9m$ **D** cannot be simplified

2 Simplify $8a - 2a - 4$. Select **A**, **B**, **C** or **D**.

A $2a$ **B** $6a - 4$ **C** $6a$ **D** cannot be simplified

3 Write the like terms in this list:

$2ab, 5a, 6b, 4m, -3a, 7n, 2w, 12a, 3mn, \frac{a}{3}, 8ac, \frac{2n}{5}$

4 Simplify each expression.

a $6b + 3b$ **b** $4a - 3a$ **c** $8m + 3m$

d $9a - 4a$ **e** $3x + 2x$ **f** $4x - 2x$

g $5b - 2b$ **h** $2w + 3w$ **i** $7b - 7b$

j $5m + (-3m)$ **k** $6w - (-2w)$ **l** $2ab - 5ab$

m $2y - 3y$ **n** $4m - 9m$ **o** $3b + 4b - 2b$

p $5n - 4n + n$ **q** $8ab - 2ab$ **r** $5r^2 - 3r^2 + r^2$

s $12mn - 3mn + 2mn$ **t** $3a - 3 + 5a$ **u** $6v - 3v + 2w$

v $7s + 8st - 4st - 2s$ **w** $6p + 8p^2 - p + 2p^2$ **x** $4a - 5 + (-8a) + 3$

5 Is each equation true or false?

a $3uv = 3vu$ **b** $5a - 6a = a$

c $12ab + ab - 3 = 13ab - 3$ **d** $-4b - 3b = 7b$

e $-3a + 5a = -8a$ **f** $4mn - mn + 3m = 3mn + 3m$

g $5ac - 6ca = -ac$ **h** $20 - y^2 - 5y^2 = 14a$

i $2ab - ba + 4ab = 6ab$

6 Write an algebraic expression for the perimeter of this rectangle.

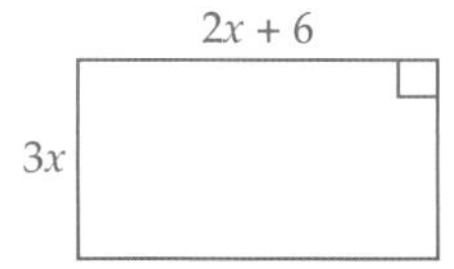

7 Simplify each algebraic expression.

a $12ab - 7ba$ **b** $8mn + mn - 8$

c $24 - 12uv - 12$ **d** $7rs - 2sr - 7$

e $-5n - 6n + n$ **f** $22a - 11b + 6a - b$

g $28 - 15y + 6y$ **h** $3ab - ba + 8ab$

i $36 - 6w + 12w - 12$

6-05 Multiplying terms

To multiply algebraic terms:

- multiply the numbers first
- then multiply the variables
- write the variables in alphabetical order.

✱ They do not have to be like terms!

EXAMPLE 8

Simplify each expression.

a $5 \times x \times 2 \times y$
b $b \times (-3) \times (-8) \times c$
c $2a \times (-4b)$
d $4m \times 5n \times 3$
e $12a \times (-9bc)$
f $3v \times (-2v) \times 4w$

SOLUTION

a $5 \times x \times 2 \times y = 5 \times 2 \times x \times y$ ← Multiply the numbers and the variables separately.
$= 10xy$

b $b \times (-3) \times (-8) \times c = -3 \times (-8) \times b \times c$
$= 24bc$

c $2a \times (-4b) = 2 \times (-4) \times a \times b$
$= -8ab$

d $4m \times 5n \times 3 = 4 \times 5 \times 3 \times m \times n$
$= 60mn$

e $12a \times (-9bc) = 12 \times (-9) \times a \times bc$
$= -108abc$

f $3v \times (-2v) \times 4w = 3 \times (-2) \times 4 \times v \times v \times w$
$= -24v^2w$ ← $v \times v = v^2$

iStockphoto/HadelProductions

ISBN 9780170350990

EXERCISE 6-05

1 Simplify $3a \times (-2b)$. Select the correct answer **A, B, C** or **D**.

A $-6ab$ B $-32ab$ C ab D $5ab$

2 Simplify $-5m \times 4np$. Select **A, B, C** or **D**.

A $20mnp$ B $-54mnp$ C $-20mnp$ D $-mnp$

3 Simplify each expression.

a $2 \times a$ **b** $5 \times m$ **c** $-4 \times n$

d $7 \times u \times v$ **e** $8 \times r \times (-2) \times s$ **f** $12 \times a \times b \times (-4)$

g $-3 \times c \times (-5) \times b$ **h** $11 \times d \times e$ **i** $5 \times y \times (-4) \times z \times (-1)$

j $-6 \times a \times 5$ **k** $2 \times w \times (-3)$ **l** $-12a \times 2c$

m $3ab \times -4c$ **n** $-4n \times (-3n)$ **o** $6m \times (-3m) \times (-2)$

p $5w \times 3u \times (-4)$ **q** $2ba \times 5ab$ **r** $-3a \times 2a \times (-4a)$

4 Find an algebraic expression for the area of each shape.

a

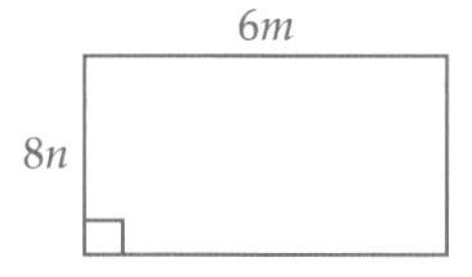

b

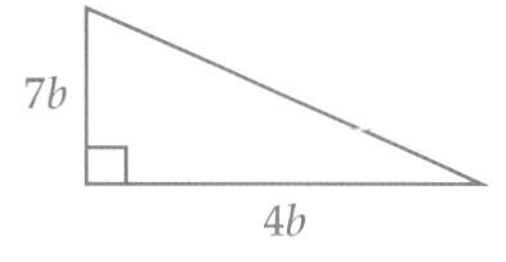

5 Simplify each algebraic expression.

a $5m \times (-3n) \times 4$ **b** $-6a \times 3b \times (-c)$ **c** $5m \times (-2n) \times 3q$

d $4ab \times (-ba) \times 3$ **e** $-2bc \times 5d \times (-3)$ **f** $12a \times (-2a) \times a$

g $6t \times 3t \times (-2t)$ **h** $5e \times (-3e) \times e$ **i** $4c \times 8cd \times (-3de)$

6 Renee earns $20 per hour and works 6 hours each day. How much would she earn if she worked for:

a 5 days

b m days

c 4 days per week for 5 weeks

d m days per week for w weeks?

ISBN 9780170350990

6-06 Dividing terms

When dividing algebraic terms, the answer is written in fraction form; for example, $2a \div 5b = \frac{2a}{5b}$.

To divide algebraic terms:

- write each quotient as a fraction and simplify
- divide the numbers first
- then divide the variables
- write the answer in fraction form.

They do not have to be like terms!

Expressions may be simplified by dividing both numerator and denominator by any common factors, preferably the highest common factor (HCF).

EXAMPLE 9

Simplify each expression.

a $30m \div 3n$ **b** $6a \div 2b$ **c** $-4mn \div 16$

d $20ab \div 4b$ **e** $\frac{-12m^2n}{8mn}$ **f** $\frac{15w}{-3uvw}$

SOLUTION

a $30m \div 3n = \frac{30m}{3n}$

$= \frac{10m}{n}$ ← dividing numerator and denominator by the HCF 3

b $6a \div 2b = \frac{6a}{2b}$

$= \frac{3a}{b}$ ← dividing numerator and denominator by the HCF 2

c $-4mn \div 16 = \frac{-4mn}{16}$

$= \frac{-mn}{4}$ ← dividing by 4

d $20ab \div 4b = \frac{20ab}{4b}$ ← dividing by $4b$

$= 5a$ $\frac{20}{4} = 5$, $\frac{ab}{b} = a$

e $\frac{-12m^2n}{8mn} = \frac{-3m}{2}$ ← dividing by $4mn$

$\frac{-12}{8} = \frac{-3}{2}$, $\frac{m^2n}{mn} = m$

f $\frac{15w}{-3uvw} = -\frac{5}{uv}$ ← dividing by $3w$

$\frac{15}{-3} = -5$, $\frac{w}{uvw} = \frac{1}{uv}$

ISBN 9780170350990

EXERCISE 6–06

1 Simplify $16ab \div 8b$. Select the correct answer **A**, **B**, **C** or **D**.

A $2ab$ **B** $2a$ **C** $8a$ **D** $8ab$

2 Simplify $\frac{-24mn}{-3m}$. Select **A**, **B**, **C** or **D**.

A $8n$ **B** $-8m$ **C** $6n$ **D** $-6m$

3 Is each equation true or false?

a $48a \div (-6) = 8a$ **b** $-12n \div 4 = -3n$ **c** $8ab \div 4b = 2ab$

d $-9m \div 3n = -3$ **e** $27w \div (-9) = -3w$ **f** $12bc \div (-6c) = 2b$

g $18q \div (-6) = -3q$ **h** $32mn \div 8n = 4n$ **i** $0de \div (-5e) = -4d$

j $72ab \div (-9) = 8ab$ **k** $56w \div (-8w) = 7w$ **l** $36rt \div (-6t) = 6r$

m $\frac{18a}{9b} = 2ab$ **n** $\frac{-15ab}{3b} = -5b$ **o** $\frac{28ab}{-7b} = -4a$

p $\frac{25mn}{-5n} = -5m$ **q** $\frac{24m}{6n} = 4mn$ **r** $\frac{35xy}{7xz} = \frac{5y}{z}$

4 Simplify each expression.

a $4m \div 2$ **b** $15a \div (-5)$ **c** $-20mn \div 4$

d $24ab \div (-6a)$ **e** $-30mn \div (-6)$ **f** $12w \div (-6v)$

g $16bc \div 8b$ **h** $-12r \div 6s$ **i** $-18st \div (-9)$

j $\frac{24ab}{4a}$ **k** $\frac{-12mn}{6n}$ **l** $\frac{3bc}{6cd}$

m $\frac{-24rs}{-6st}$ **n** $\frac{18v}{-9w}$ **o** $\frac{-56y}{7wy}$

p $\frac{48ew}{-6w}$ **q** $\frac{4ef}{-12fg}$ **r** $\frac{42abc}{7cd}$

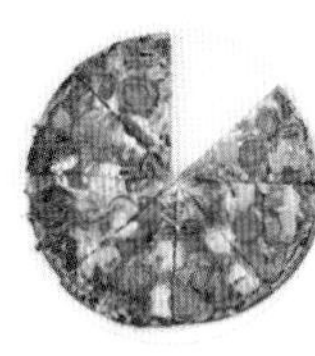
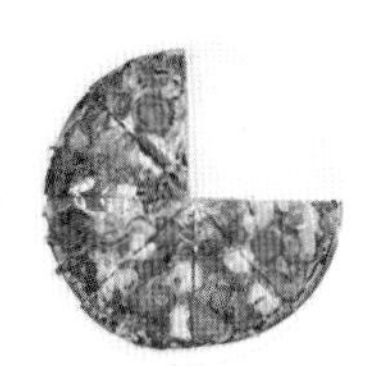

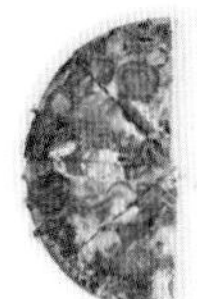
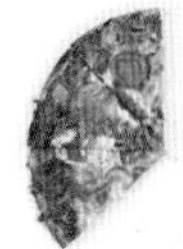

iStockphoto/Urs Siedentop

ISBN 9780170350990

MIX AND MATCH

Match each question on the left with a simplified expression on the right.

1 Twice the sum of x and 5

2 The product of $4a$ and $2b$

3 The value of $2x - y$ if $x = 2$ and $y = -1$

4 The value of $3ab$ if $a = -2$ and $b = 6$

5 $4a - 2b + 5a - 3b$

6 $5x - 7y - 6x + 3y$

7 $2m \times 3n$

8 $5a \times 2b$

9 $3x - 3 + 5x - 9$

10 $4a^2 + 4a + 5a + 2a^2$

11 $3a + 7b - 20b + a$

A $9a - 5b$

B $6mn$

C $2(x + 5)$

E $9a + 6a^2$

G 5

I $10ab$

L $-x - 4y$

N $4a - 13b$

R -36

S $8ab$

T $8x - 12$

Match the letter of the correct answer with each question number to decode this phrase:

5–7–2–9–4–5–1–9   5–6–3–10–7–4–5   5–11–9–8–1–2

iStockphoto/Oshi

ISBN 9780170350990

PRACTICE TEST 6

Part A General topics

Calculators are not allowed.

1 Evaluate $\sqrt{16}$.

2 Complete 3 days = ________ hours.

3 Evaluate $(-3)^3$.

4 Evaluate $168 \div 3$.

5 Find the range of these scores: 11, 8, 5, 6, 8, 4, 5, 7.

6 Given that $33 \times 6 = 198$, evaluate 3.3×6.

7 Find $\frac{3}{4}$ of \$820.

8 Find the area of this triangle.

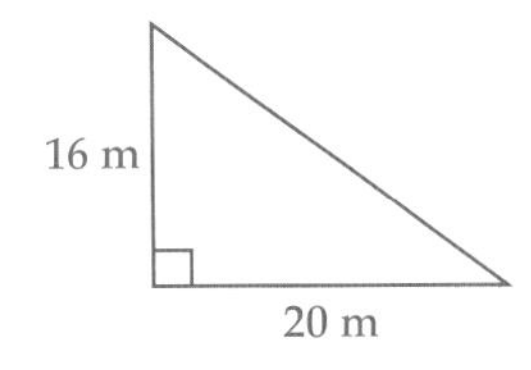

9 Evaluate $-6.2 + 8.9$.

10 What is the probability of selecting a red sock from a drawer containing 6 red and 8 blue socks?

Part B Algebra

Calculators are allowed.

6-01 Variables

11 Simplify $m + m + n - n + n + m$. Select the correct answer **A**, **B**, **C** or **D**.

A $2m + n + m$ B $3m + 2n$ C $2m + 3n$ D $3m + n$

12 Simplify $u \times (-2) \times v \times 5$. Select **A**, **B**, **C** or **D**.

A $3vu$ B $-10uv$ C $-10uv$ D $10uv$

6-02 From words to algebraic expressions

13 Which is the algebraic expression for '20 less triple b plus c'? Select **A**, **B**, **C** or **D**.

A $3b + c - 20$ B $20 - b + c$ C $3b + c + 20$ D $20 - 3b + c$

14 Write an algebraic expression for twice the sum of a and b.

6-03 Substitution

15 If $m = -5$ and $n = 7$, evaluate each expression.

a $20 - mn$ b $4n - 3m$

6-04 Adding and subtracting terms

16 Simplify each expression.

a $12w - 6w + 4w$ b $9a - 3b - 6a + b$

c $4ab - 6ba + ab$ d $-5bc + 8cb - 12$

6-05 Multiplying terms

17 Simplify each expression.

a $-4a \times (-5b)$ **b** $12mn \times 4n$

6-06 Dividing terms

18 Simplify each expression.

a $48st \div (-8tr)$ **b** $\frac{-54ab}{-9bc}$

ISBN 9780170350990

ANGLES AND SYMMETRY

WHAT'S IN CHAPTER 7?

IN THIS CHAPTER YOU WILL:

- name angles using three letters; for example, $\angle ABC$
- classify angles as acute, right, obtuse, straight, reflex and revolution
- use a protractor to measure and draw angles
- solve geometry problems involving right angles, angles on a straight line, angles at a point and vertically opposite angles
- identify parallel and perpendicular lines
- solve geometry problems involving corresponding angles, alternate angles and co-interior angles on parallel lines, including proving that two lines are parallel
- draw and count the axes of symmetry of different shapes
- identify the rotational symmetry of different shapes
- perform transformations: translate, reflect and rotate shapes
- perform transformations on the number plane

7-01 Angles

WORDBANK

arm One side or line of an angle.

vertex The corner of an angle.

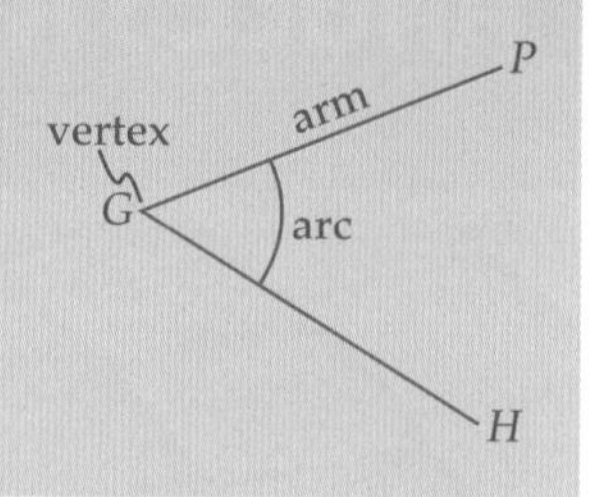

An **angle** measures how much an object turns or spins, and is measured in **degrees** (°). An angle is named using three letters, with its vertex being the middle letter. The angle above is named ∠*PGH* or ∠*HGP*.

Just think of the order of letters when you draw the angle: *P-G-H*.

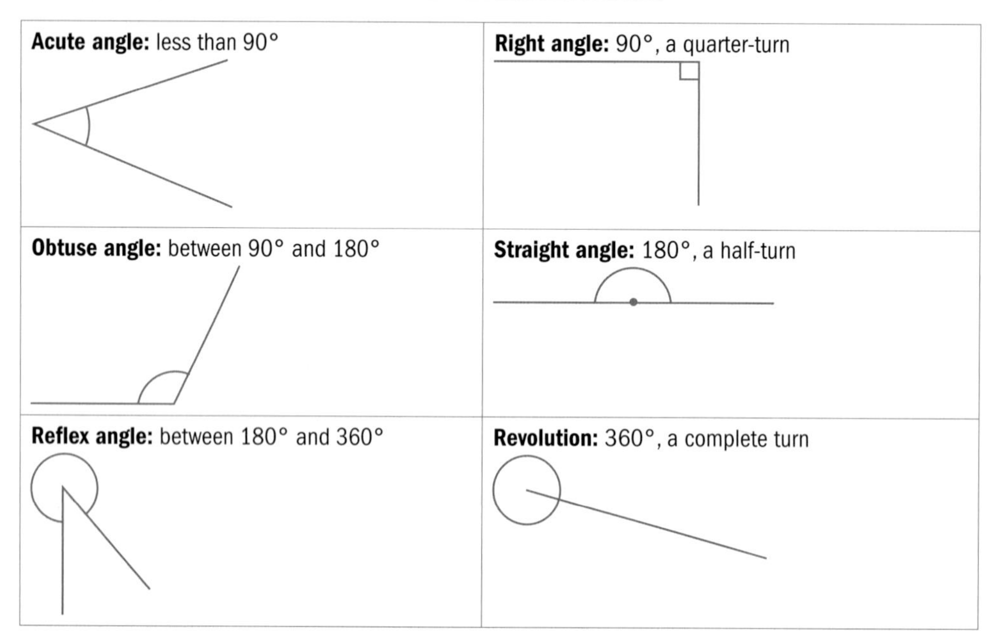

EXAMPLE 1

Name and classify each angle.

a

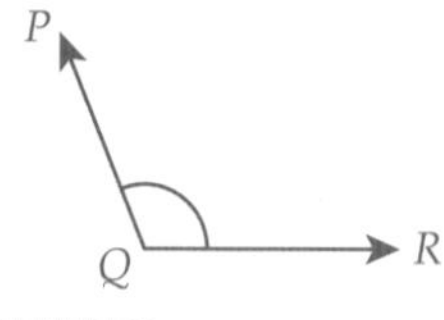

b

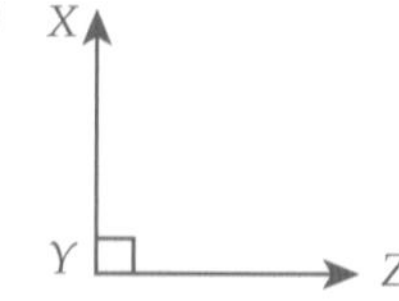

SOLUTION

a ∠*PQR* (or ∠*RQP*), obtuse angle

b ∠*XYZ* (or ∠*ZYX*), right angle

ISBN 9780170350990

EXERCISE 7-01

1 What is a vertex? Select the correct answer **A**, **B**, **C** or **D**.

A part of a line　　**B** a rotation

C the turn between two lines　　**D** a corner

2 Name this angle. Select **A**, **B**, **C** or **D**.

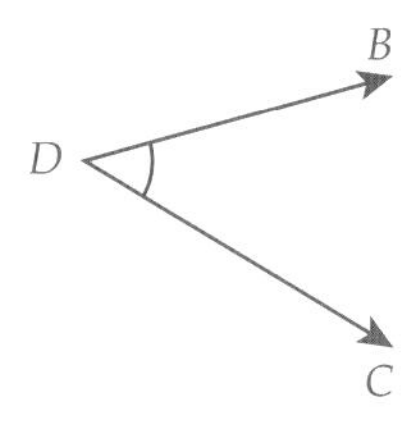

A $\angle BCD$　　**B** $\angle CDB$　　**C** $\angle DCB$　　**D** $\angle CBD$

3 Name and classify each angle.

a

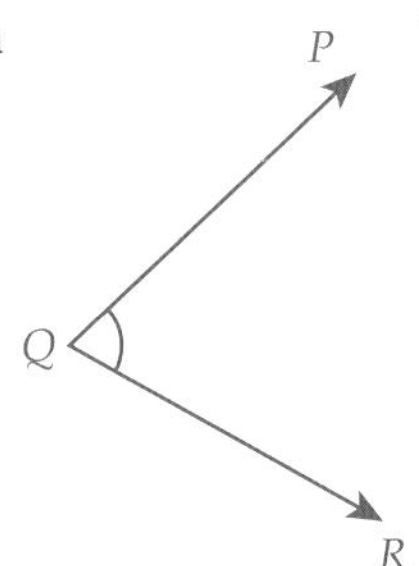

b

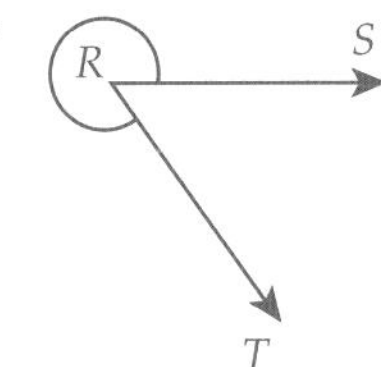

c

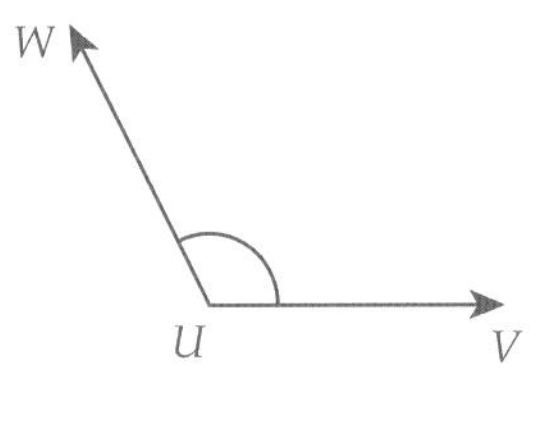

d

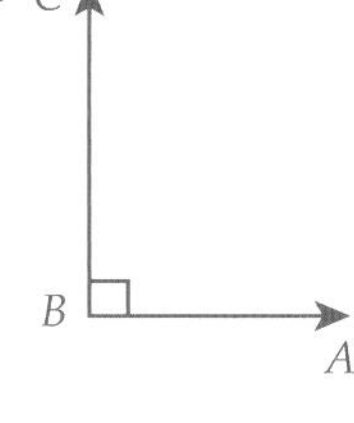

4 What type of angle is:

a between 180° and 360°　　**b** 360°　　**c** less than 90°

d between 90° and 180°　　**e** 180°　　**f** 90°?

5 Name the vertex in each angle, then classify the angle.

a

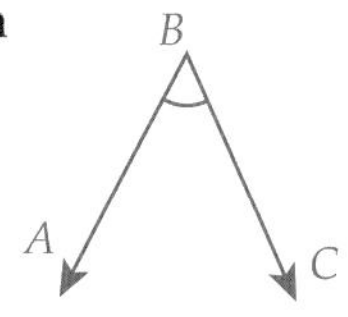

b

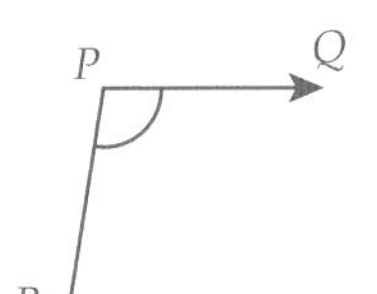

c

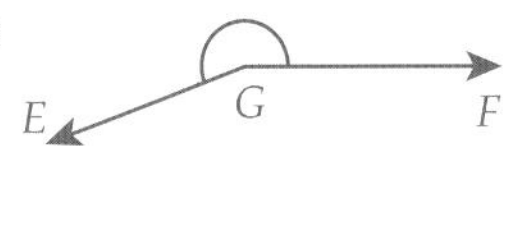

6 What type of angle is 89°? Select **A**, **B**, **C** or **D**.

A reflex　　**B** obtuse　　**C** right　　**D** acute

7 Draw a diagram showing an acute angle and a right angle.

8 Classify each angle size.

a 140°　　**b** 275°　　**c** 60°　　**d** 200°

7-02 Measuring and drawing angles

Angles are measured in degrees (°) using a **protractor**.

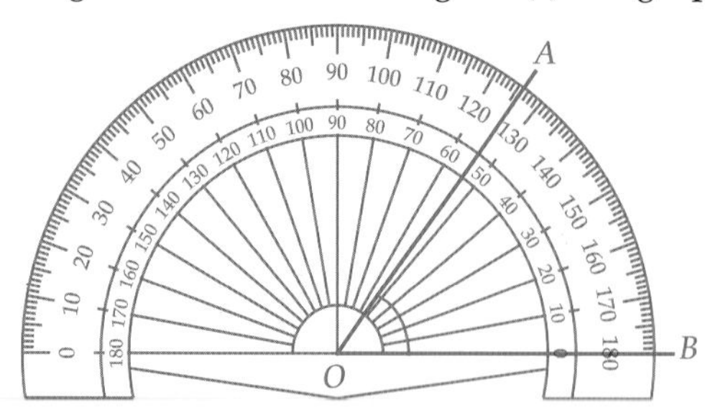

$\angle AOB$ is measured to be 54°.

$\angle PMQ$ is measured to be 155°.

> **To measure an angle with a protractor:**
> - line up the base line of the protractor with one arm of the angle
> - position the centre of the protractor on the vertex of the angle
> - use the scale that begins with 0° to read off the angle size from the other arm.

EXAMPLE 2

Construct an angle $\angle KPM$ of size 76°.

SOLUTION

Draw a base line PM, measure 76° from one end of the line with your protractor and make a mark. Join this mark to point P and label it K.

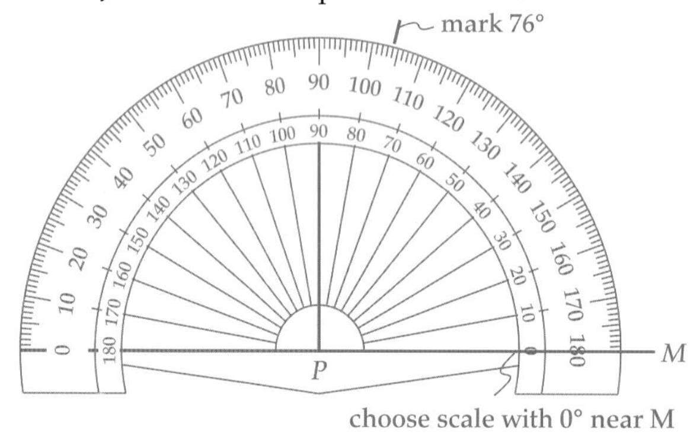

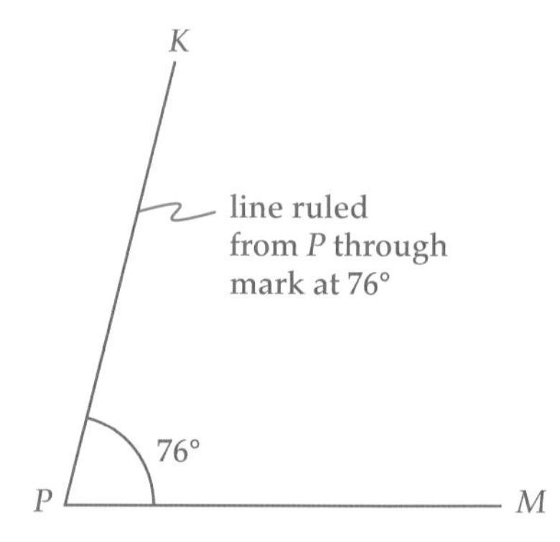

ISBN 9780170350990

EXERCISE 7-02

1 What instrument is used to measure angles? Select the correct answer **A**, **B**, **C** or **D**.

A compass **B** protractor **C** ruler **D** set square

2 Estimate the size of this angle. Select **A**, **B**, **C** or **D**.

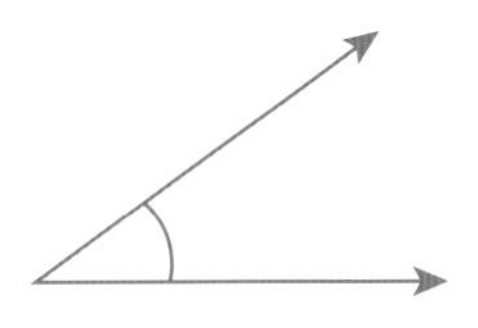

A 20° **B** 40° **C** 35° **D** 45°

3 Use a protractor to measure each angle.

a

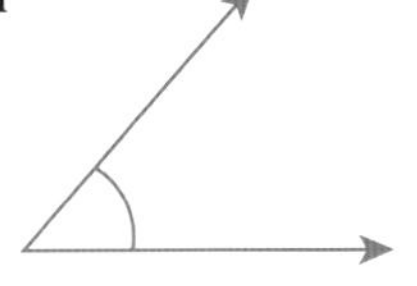

b

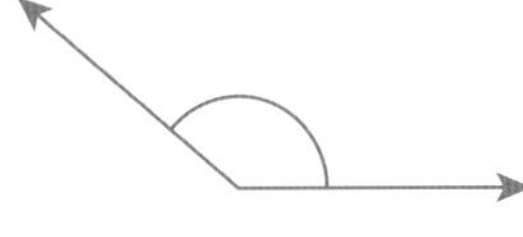

c

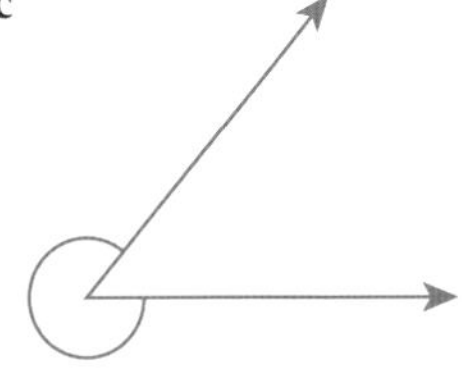

d

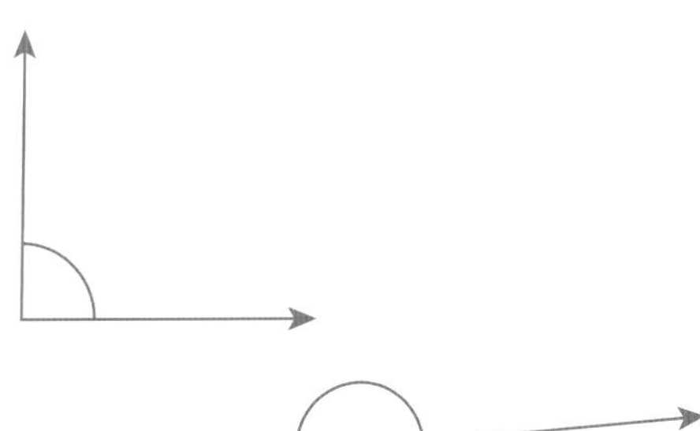

e

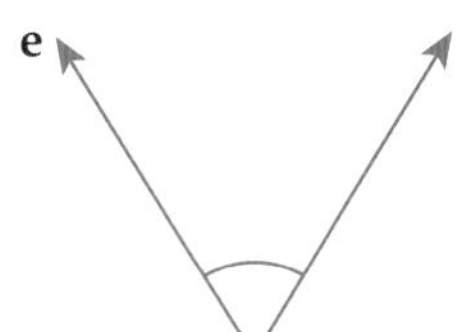

f

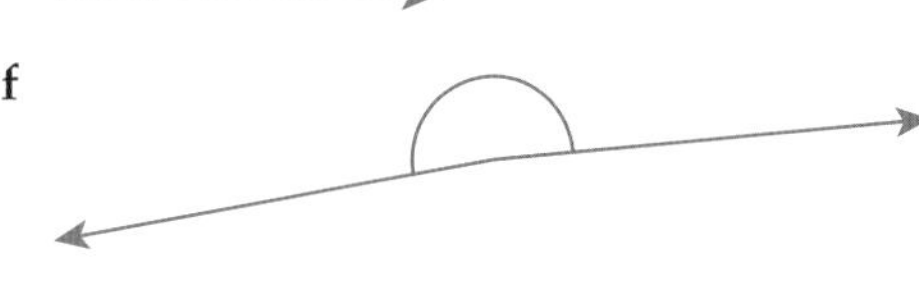

4 Construct an angle for each angle size.

a 45° **b** 82° **c** 106° **d** 152°

e 170° **f** 90° **g** 74° **h** 179°

5 To construct angles greater than 180°, it is easier to subtract the number of degrees from 360° (a revolution). For example, to construct an angle of 220°, construct 360° – 220° = 140° and mark the other side of the angle as 220°.

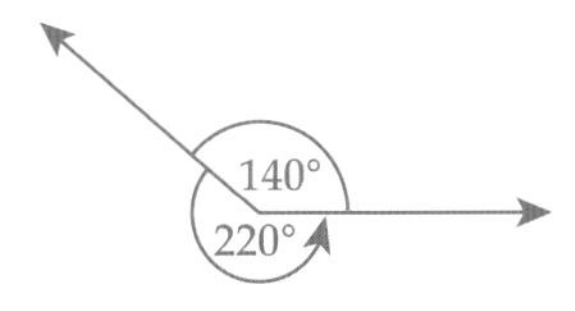

Use this method to construct an angle of size:

a 260° **b** 285° **c** 310°

7-03 Angle geometry

WORDBANK

adjacent angles Angles that are next to each other. They share a common arm.

complementary angles Two angles that add to 90°.

supplementary angles Two angles that add to 180°.

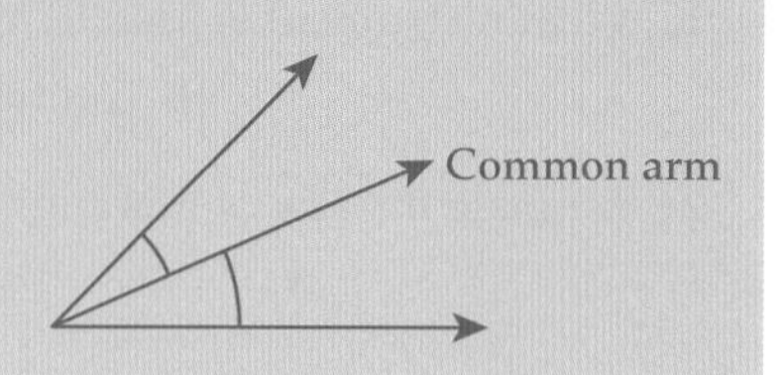

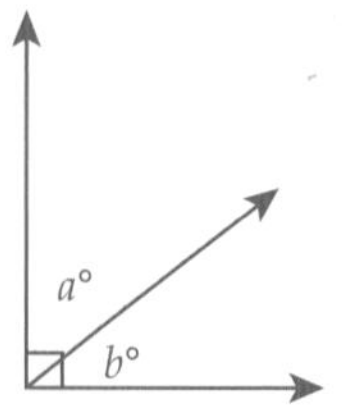

Angles in a right angle are complementary (add to 90°).

$a + b = 90$

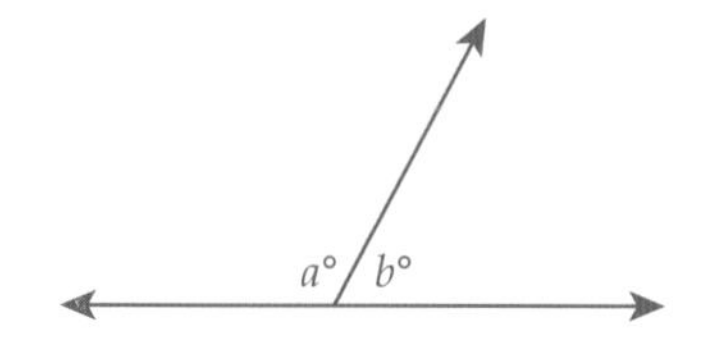

Angles on a straight line are supplementary (add to 180°).

$a + b = 180$

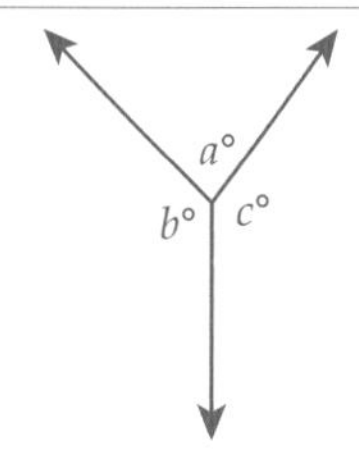

Angles at a point (in a revolution) add to 360°.

$a + b + c = 360$

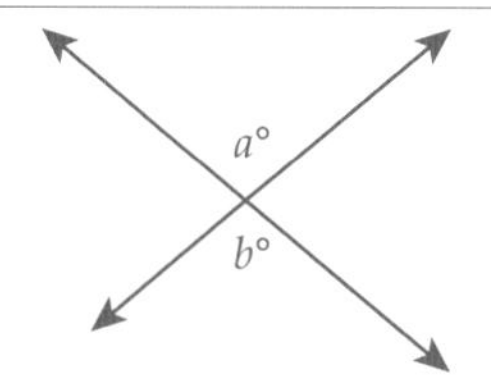

Vertically opposite angles are equal.

$a = b$

EXAMPLE 3

Find the value of each pronumeral, giving a reason.

a

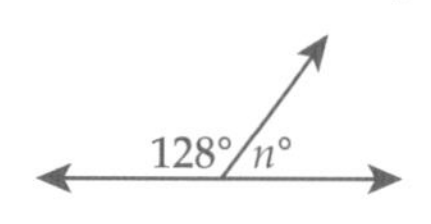

b

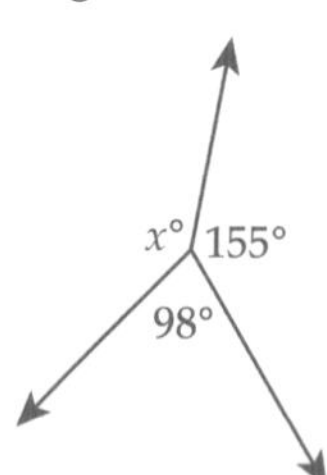

c

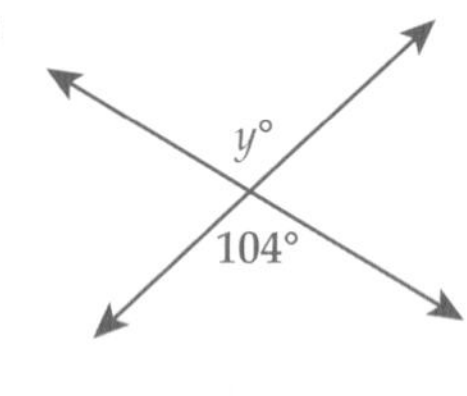

SOLUTION

a $n + 128 = 180$ (angles on a straight line)
$n = 180 - 128$
$n = 52$

b $x + 155 + 98 = 360$ (angles at a point)
$x = 360 - 155 - 98$
$x = 107$

c $y = 104$ (vertically opposite angles)

ISBN 9780170350990

EXERCISE 7–03

1 What do supplementary angles add to? Select the correct answer **A**, **B**, **C** or **D**.

A 360° **B** 90° **C** 180° **D** 270°

2 Which of the following is a true statement about vertically opposite angles? Select **A**, **B**, **C** or **D**.

A They are complementary. **B** They are equal.

C They are supplementary. **D** They are adjacent.

3 **a** Describe the rule about angles in a right angle and draw an example of one.

b Describe the rule about angles at a point and draw an example of one.

4 Is each statement is true or false?

a Vertically opposite angles are complementary.

b Angles at a point add to 360°.

c Two adjacent right angles make a straight angle.

d The lines must be parallel for vertically opposite angles to be formed.

5 Find the value of each pronumeral, giving a reason.

a

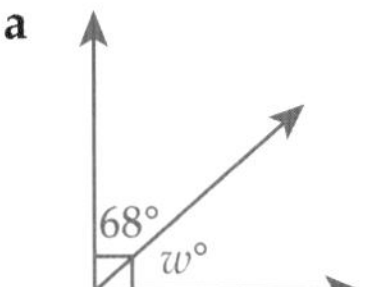

b

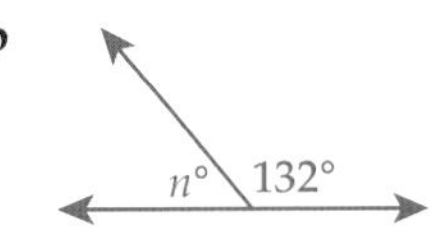

c

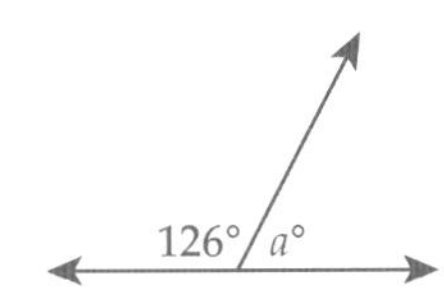

d

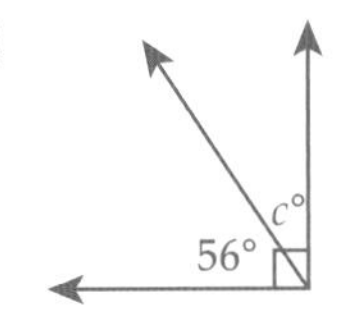

e

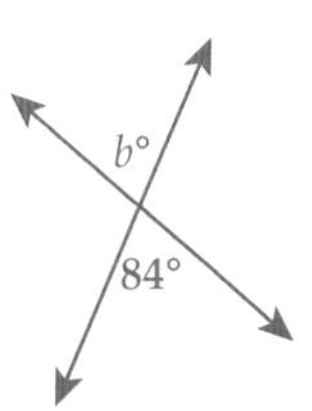

f

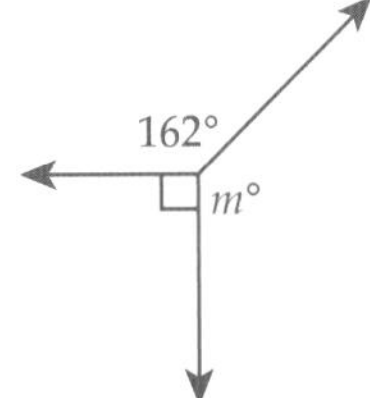

g

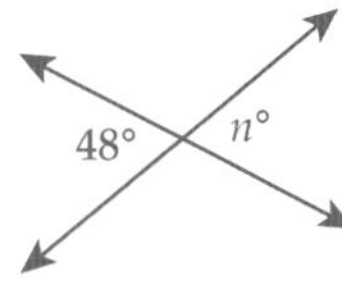

h

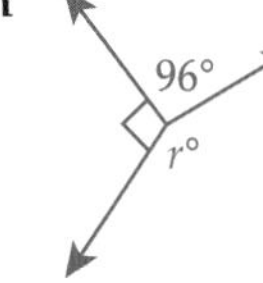

i

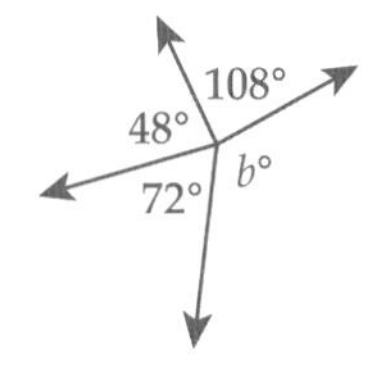

7-04 Parallel and perpendicular lines

WORDBANK

parallel lines Lines that point in the same direction and which never meet.

perpendicular lines Lines that cross at right angles (90°).

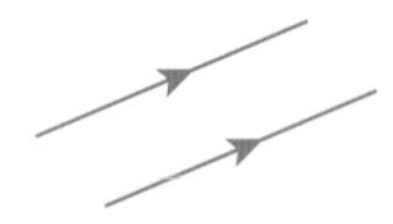

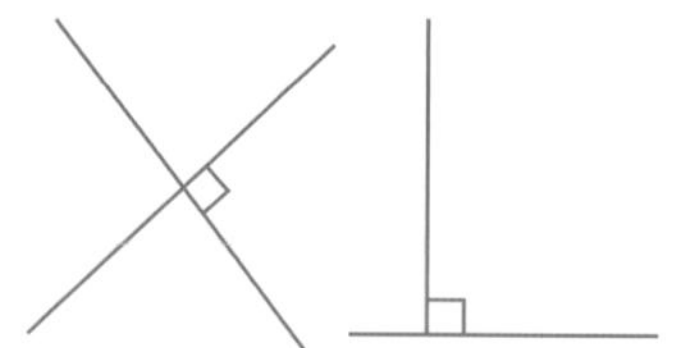

Parallel lines are marked with arrows to show that the lines run together and are always the same distance apart. The tracks on a railway line are parallel.
The symbol for 'is parallel to' is ||.

Perpendicular lines are marked with the right angle 'box' symbol.
The symbol for 'is perpendicular to' is ⊥.

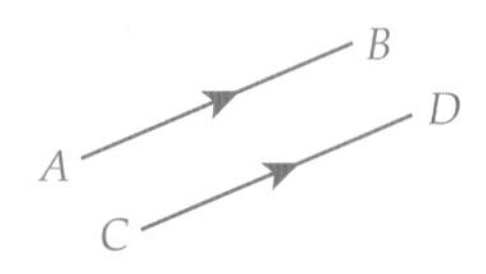

Here, $AB \parallel CD$, meaning 'AB is parallel to CD'.

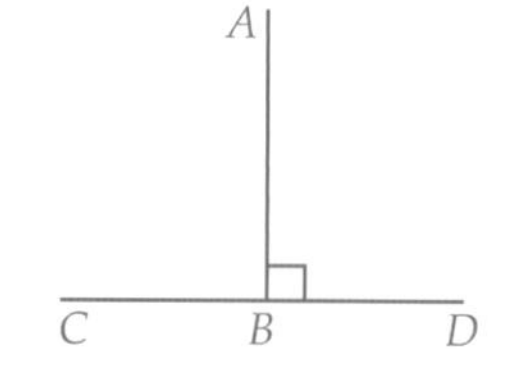

Here, $AB \perp CD$, meaning 'AB is perpendicular to CD'.

EXAMPLE 4

Name pairs of parallel and perpendicular sides in this figure.

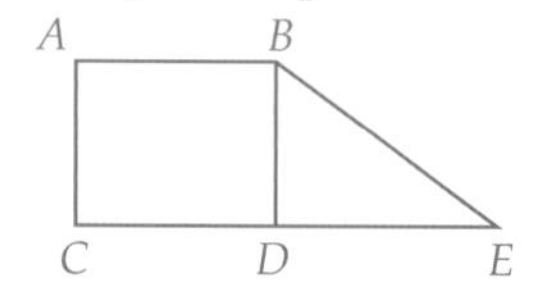

SOLUTION

$AB \parallel DE$

$AC \perp CE$

iStockphoto/lopurice

ISBN 9780170350990

EXERCISE 7-04

1 Which statement about parallel lines is true? Select the correct answer **A**, **B**, **C** or **D**.

A They are equal. **B** They cross at 90°.

C They never cross. **D** They are reflex.

2 Which statement about perpendicular lines is true? Select **A**, **B**, **C** or **D**.

A They are equal. **B** They cross at 90°.

C They never cross. **D** They are reflex.

3 Draw a pair of:

a parallel lines **b** perpendicular lines.

4 Copy each diagram, mark two parallel sides in red and two perpendicular sides in black.

a

b

c

d

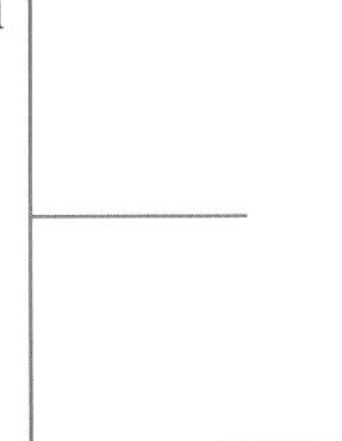

e

f

g

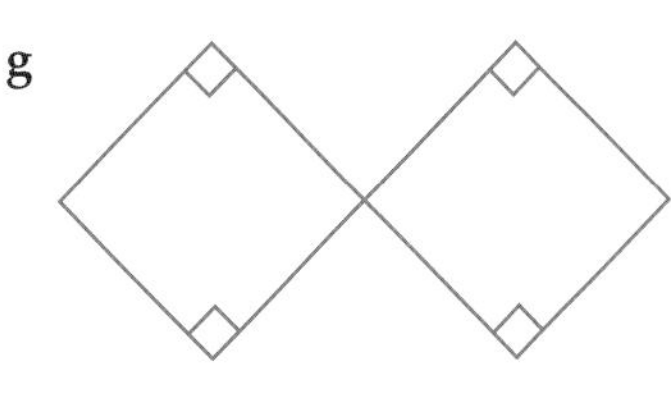

h

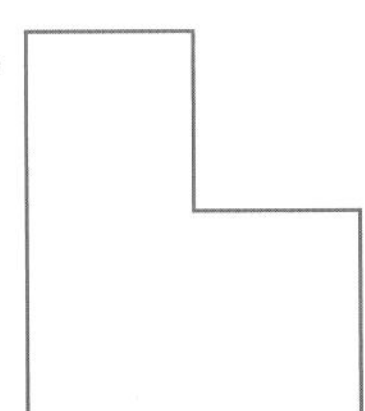

i

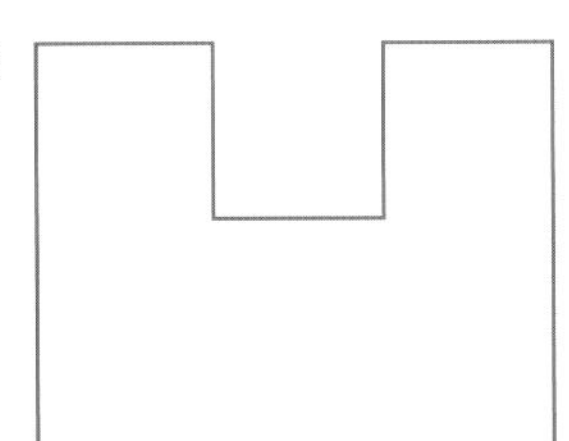

j

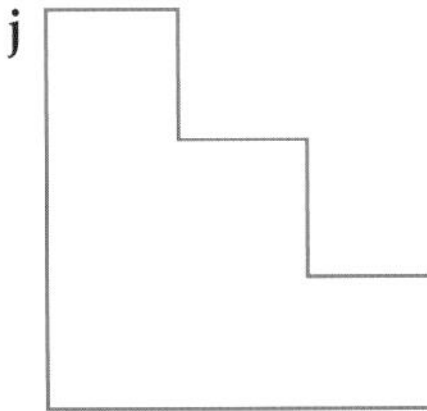

5 Copy and complete each statement.

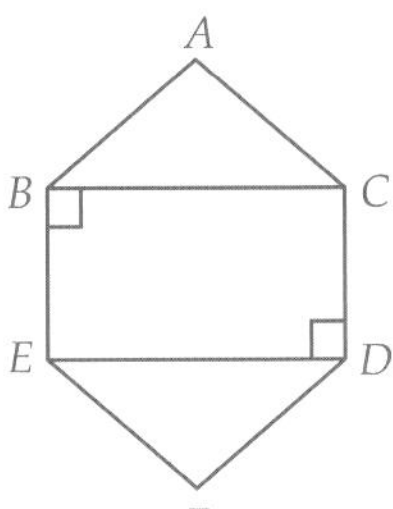

a BC is parallel to _______.

b CD is perpendicular to _______.

c $BE \parallel$ _______.

d $BE \perp$ _______.

7-05 Angles on parallel lines

WORDBANK

transversal A line that cuts across two or more lines.

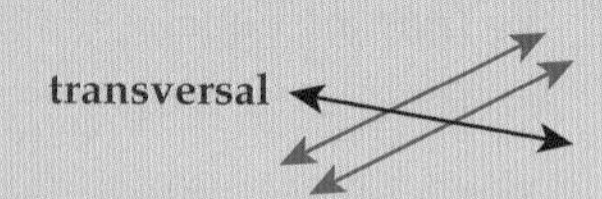

corresponding angles Angles on the same side of the transversal and in the same position on the parallel lines. Corresponding angles form the letter F.

alternate angles Angles on opposite sides of the transversal and in between the parallel lines. Alternate angles form the letter Z.

co-interior angles Angles on the same side of the transversal and in between the parallel lines. They form the letter C.

Corresponding angles on parallel lines are equal.	Alternate angles on parallel lines are equal.	Co-interior angles on parallel lines are supplementary (add to 180°).
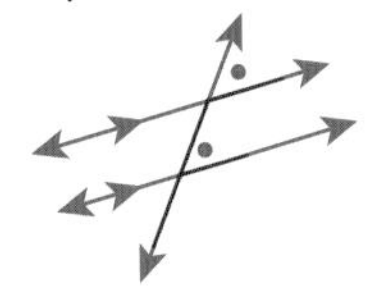	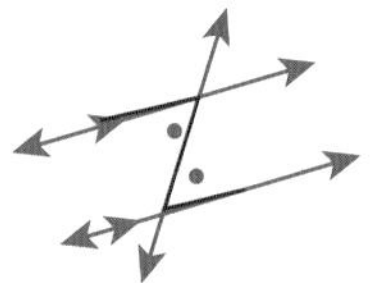	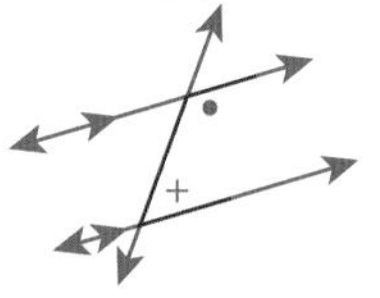

EXAMPLE 5

Find the value of each pronumeral, giving reasons.

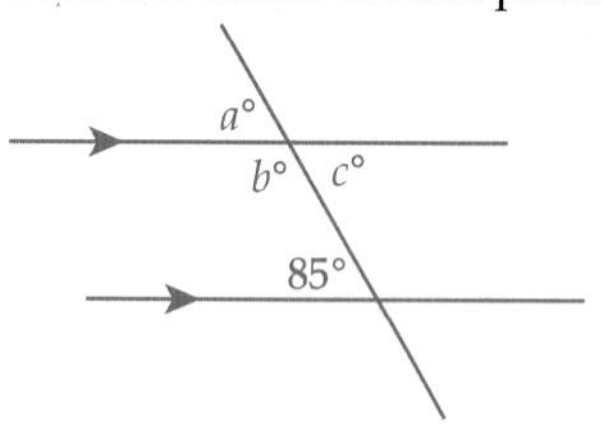

SOLUTION

$a = 85$ (corresponding angles on parallel lines)

$b = 180 - 85$ (co-interior angles on parallel lines)
$= 95$

$c = 85$ (alternate angles on parallel lines)

 ISBN 9780170350990

EXERCISE 7–05

1 Complete: Co-interior angles on parallel lines are __________. Select the correct answer **A**, **B**, **C** or **D**.

A complementary **B** equal **C** supplementary **D** opposite

2 Complete: Alternate angles on parallel lines are __________. Select **A**, **B**, **C** or **D**.

A complementary **B** equal **C** supplementary **D** opposite

3 Copy this diagram and mark both pairs of alternate angles. Mark each pair with a different symbol.

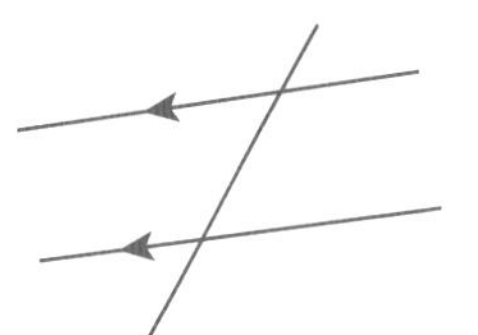

4 Find the value of each pronumeral, giving reasons.

a

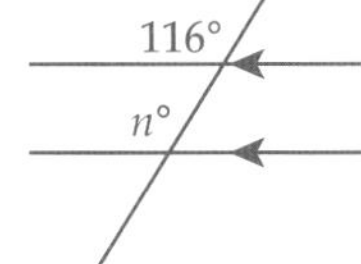

b

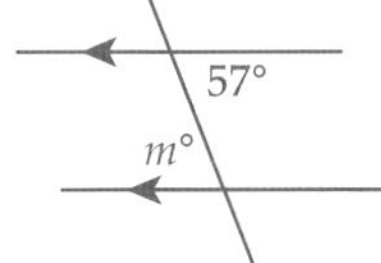

c

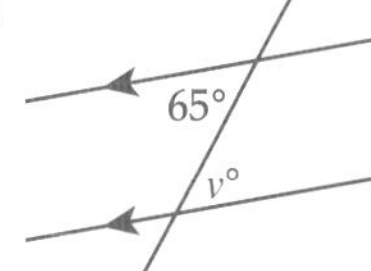

d

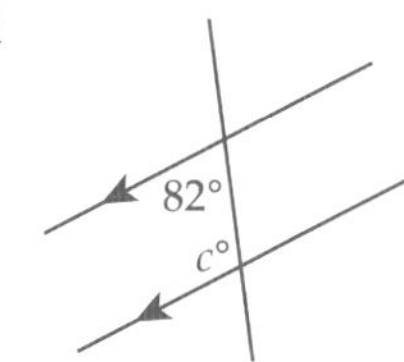

e

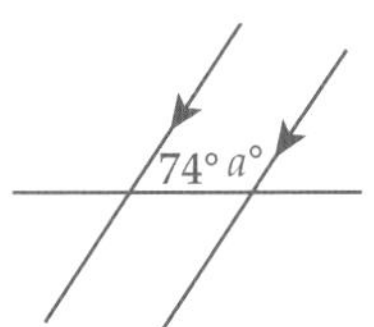

f

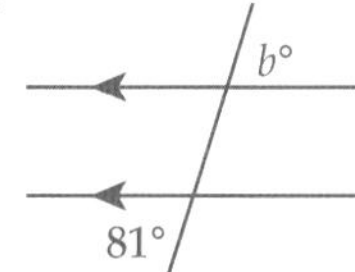

g

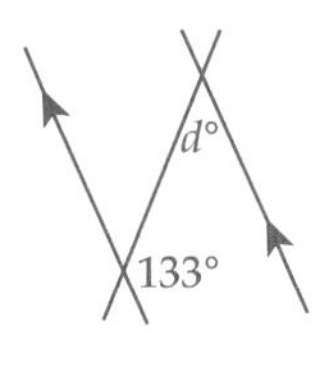

h

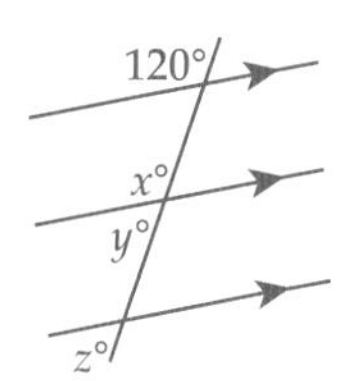

i

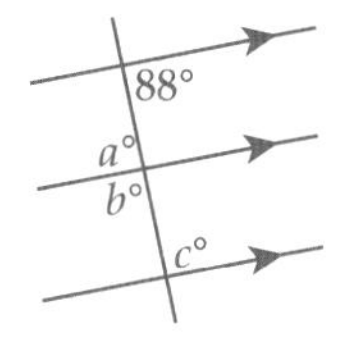

5 Draw two parallel lines crossed by a transversal. Mark one of the acute angles 48° and then find the sizes of the other seven angles.

6 Is each statement true or false?

a A transversal is a line that crosses two or more other lines.

b Parallel lines never meet.

c Alternate angles are in matching positions.

d Alternate angles are equal if the lines are parallel.

e Corresponding angles are supplementary if the lines are parallel.

f Co-interior angles are on the same side of the transversal.

7 Find the value of each pronumeral.

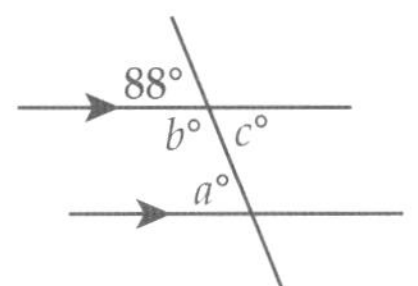

7-06 Proving parallel lines

We can use what we know about angles and parallel lines to prove that two lines are parallel.

Two lines are parallel if:
- corresponding angles are equal, or alternate angles are equal
- co-interior angles are supplementary (add to 180°).

So, for example, if alternate angles between two lines are equal, then the lines are parallel. However, if alternate angles between two lines are not equal, then the lines are not parallel.

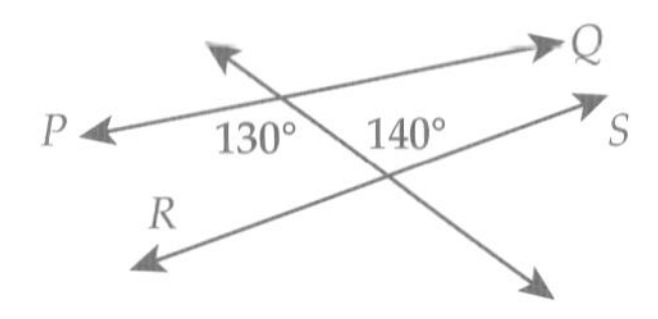

In this diagram, alternate angles 130° and 140° are not equal so PQ and RS are not parallel.

EXAMPLE 6

Prove whether each pair of lines are parallel.

a

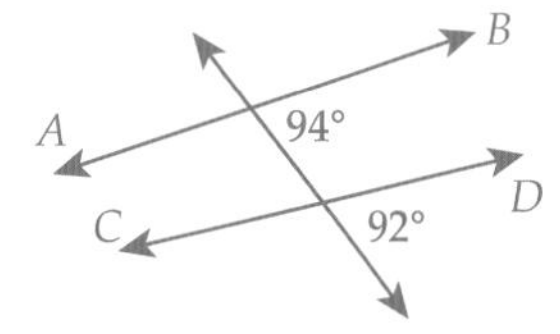

b

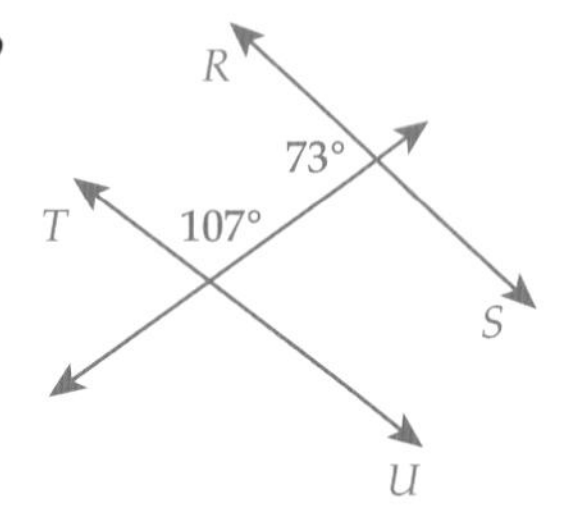

SOLUTION

a 94° and 92° are corresponding angles that are not equal, so AB and CD are not parallel.

b 107° and 73° are co-interior angles that are supplementary (107° + 73° = 180°), so $RS \parallel TU$.

ISBN 9780170350990

EXERCISE 7-06

1 If a pair of corresponding angles on two lines are equal, then what is true about the lines? Select the correct answer **A**, **B**, **C** or **D**.

A They are perpendicular.
B They are not parallel.
C They are not perpendicular.
D They are parallel.

2 If a pair of co-interior angles between two lines are equal, then what is true about the lines? Select **A**, **B**, **C** or **D**.

A They are perpendicular.
B They are not parallel.
C They are not perpendicular.
D They are parallel.

3 Prove whether each pair of lines are parallel.

a

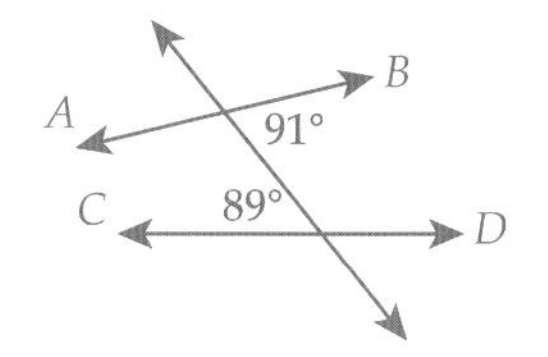

b

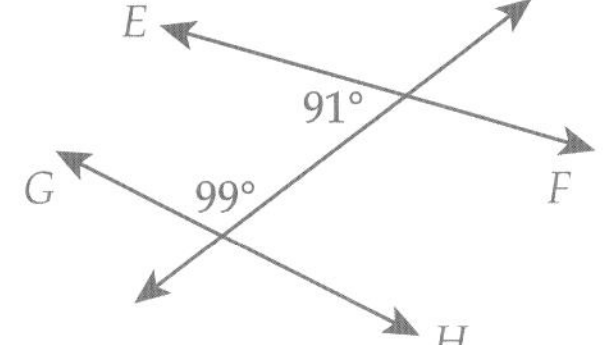

c

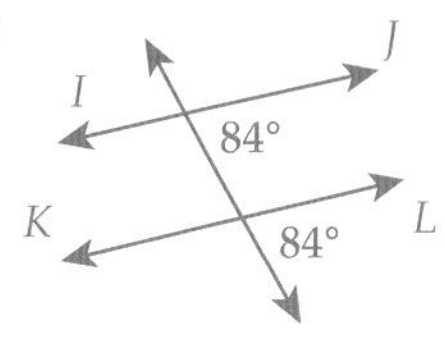

d

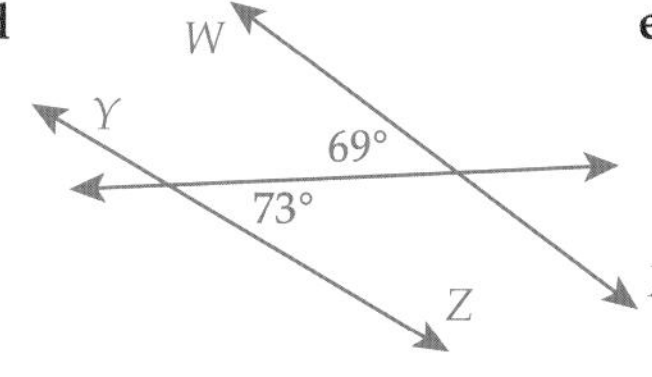

e

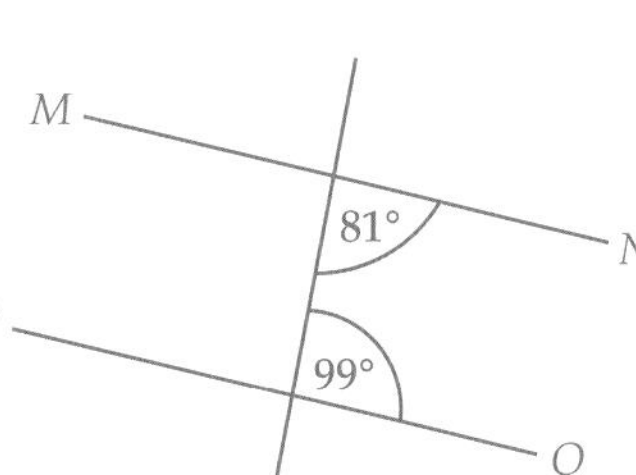

f

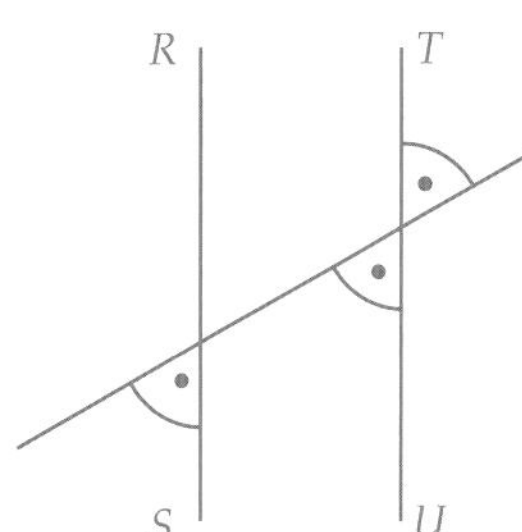

4 Is each statement true or false?

a Corresponding angles are always equal.
b Alternate angles are sometimes equal.
c If co-interior angles are supplementary, then the lines are parallel.
d If alternate angles are different, then the lines are perpendicular.

5 What reason can be used to prove that $GC \parallel HE$? Select **A**, **B**, **C** or **D**.

A $\angle ABC = \angle HDF$ (alternate angles)
B $\angle CBD = \angle BDH$ (alternate angles)
C $\angle ADE = 91°$ (corresponding angles)
D $\angle BDE = \angle FDH$ (vertically opposite angles)

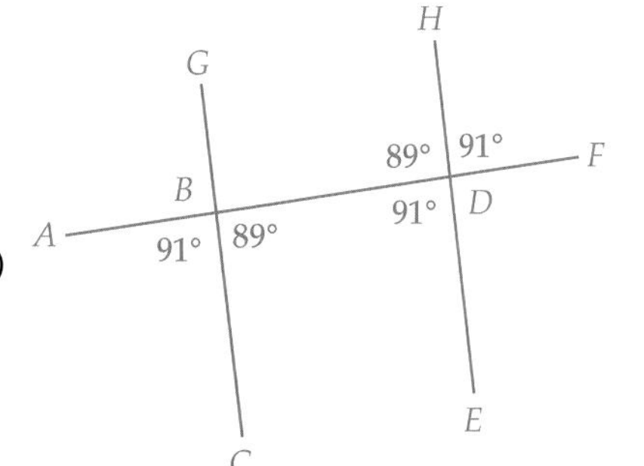

6 **a** Draw two non-parallel lines crossed by a transversal and label the sizes of a pair of corresponding angles.

b Draw two parallel lines crossed by a transversal and label the sizes of a pair of co-interior angles.

7-07 Line and rotational symmetry

WORDBANK

line symmetry A shape has line symmetry if one half exactly folds onto the other half. One half is the mirror image of the other half.

axis of symmetry The line that divides a symmetrical shape in half. 'Axis' means line (plural is 'axes').

rotational symmetry A shape has rotational symmetry if it can be spun around its centre so that it fits onto itself before a complete revolution.

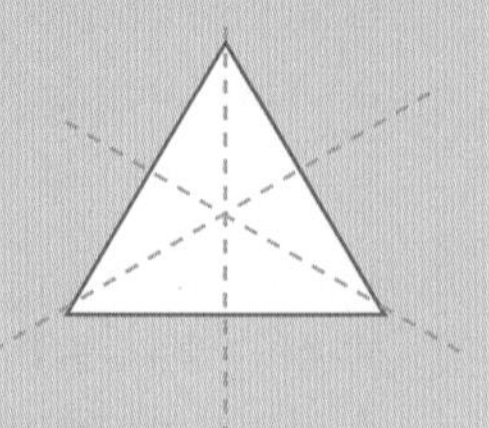

This regular hexagon has **rotational symmetry** because it fits onto itself 6 times when spun around its centre during a full rotation of 360°. It has **rotational symmetry of order 6**.

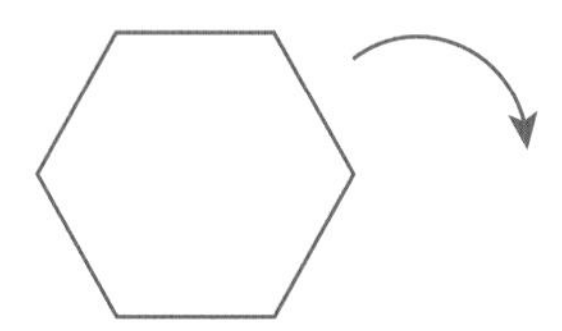

EXAMPLE 7

For each symmetrical shape, draw the axes of symmetry.

a 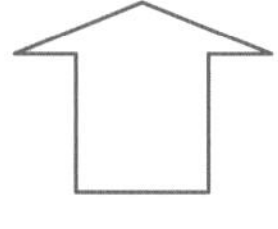

b 

c 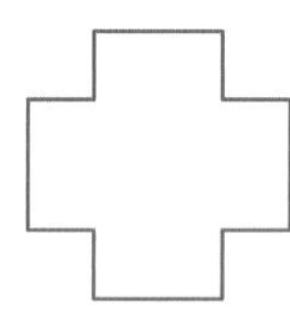

SOLUTION

a 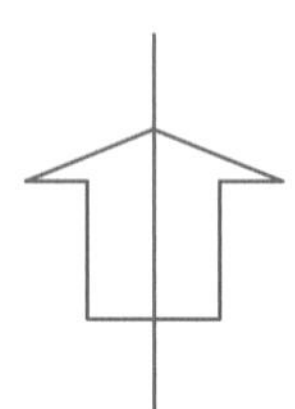

b 

c 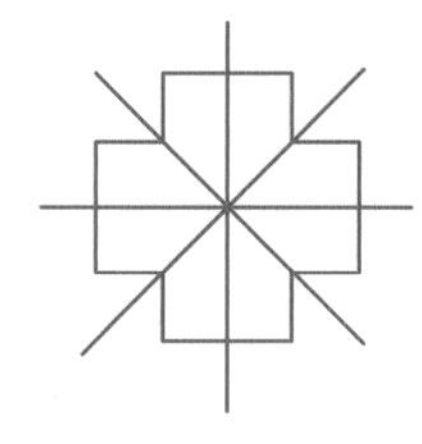

EXAMPLE 8

Does each figure have rotational symmetry? If so, state the order of rotational symmetry.

a 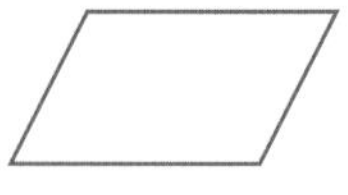

b 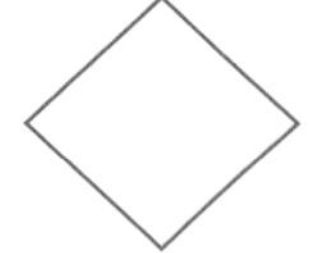

c 

SOLUTION

a Yes, if turned through 180°. Order is 2.

b Yes, if rotated through 90°. Order is 4.

c No

ISBN 9780170350990

EXERCISE 7–07

1 How many axes of symmetry does a rectangle have? Select the correct answer **A**, **B**, **C** or **D**.

A 0 **B** 1 **C** 2 **D** 4

2 What order of rotational symmetry does a rectangle have? Select **A**, **B**, **C** or **D**.

A 0 **B** 1 **C** 2 **D** 4

3 Copy each shape and draw its axis (or axes) of symmetry.

a 

b 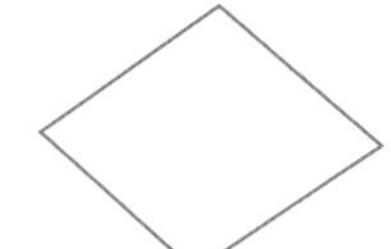

c 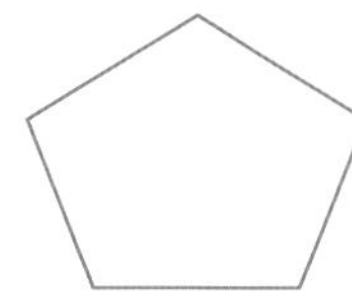

d 

e 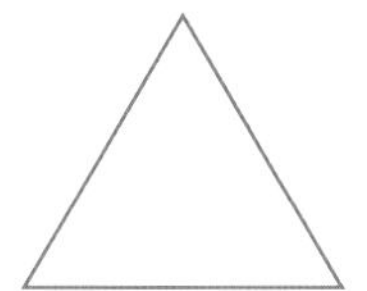

f 

4 Does each shape have rotational symmetry? If yes, state the order of rotational symmetry.

a 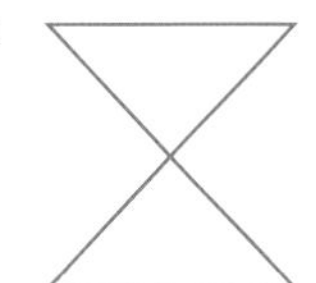

b 

c 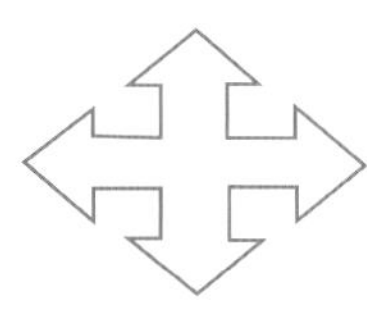

d 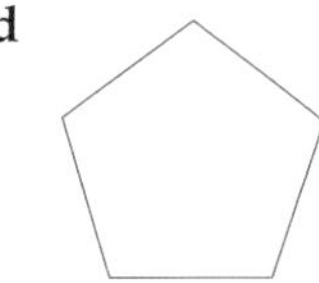

e 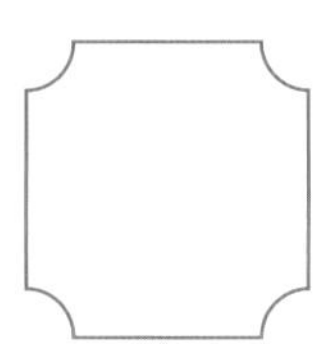

f 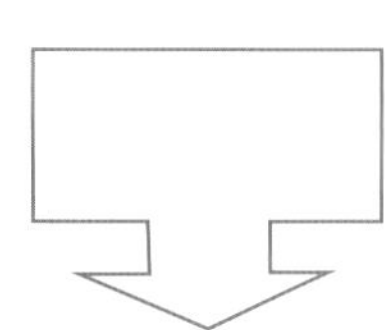

g 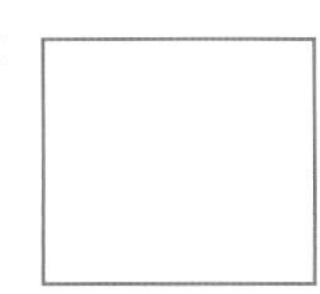

h 

i 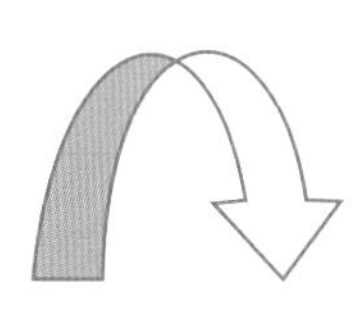

5 For each shape with rotational symmetry in Question **4**:

a write how many degrees it must be turned first to fit onto itself

b its order of rotational symmetry.

6 Is each statement true or false?

a A square has 2 axes of symmetry.

b A circle does not have rotational symmetry.

c An isosceles triangle has 1 axis of symmetry.

d A regular octagon has rotational symmetry of order 8.

7-08 Transformations

WORDBANK

transformation The process of moving or changing a shape by translation, reflection or rotation.

translation The process of 'sliding' a shape: moving it up, down, left or right.

reflection The process of 'flipping' a shape: making it back-to-front as in a mirror.

rotation The process of 'spinning' a shape around a point: tilting it sideways or upside-down.

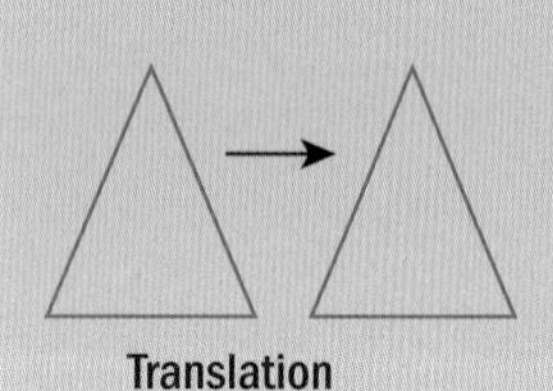

Translation

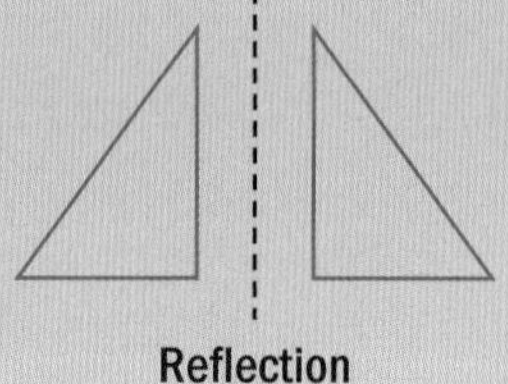

Reflection

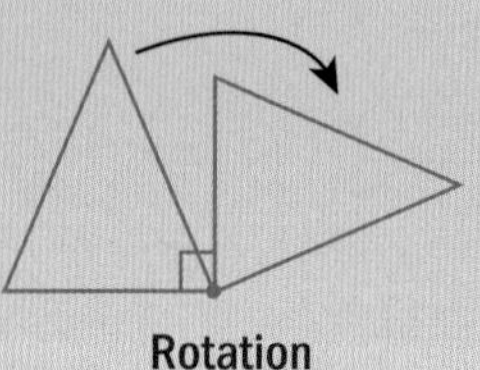

Rotation

A **composite transformation** is a combination of two or more transformations on the one shape, such as a reflection followed by a rotation.
The dark parallelogram below has been translated 5 units left and then reflected in the line AB.

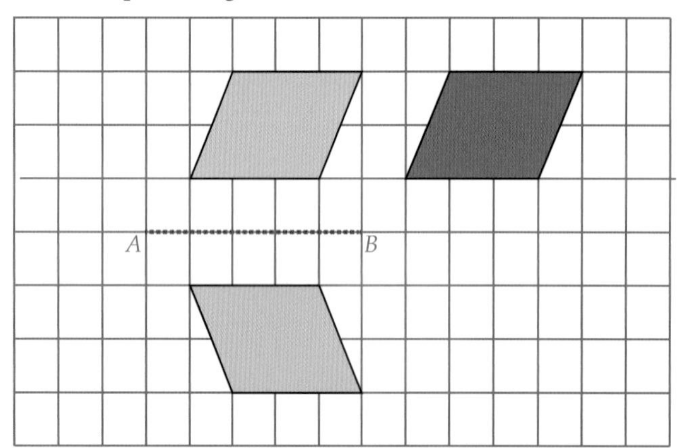

The dark T-shape below has been rotated 90° clockwise about point O and then translated 7 units left.

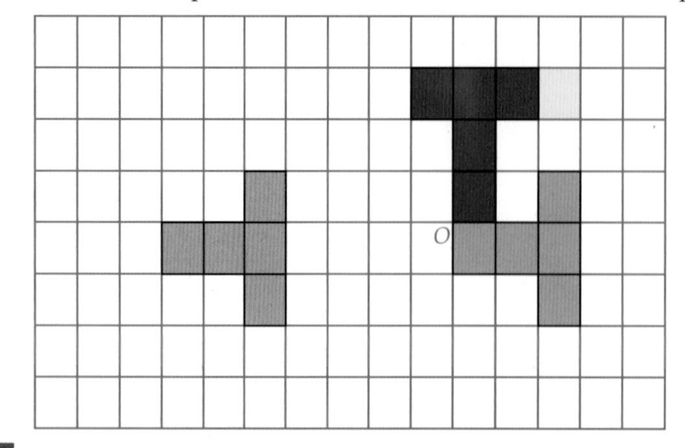

ISBN 9780170350990

EXERCISE 7-08

1 Copy this shape onto grid paper, reflect it in the line AB and then reflect that image in the line CD.

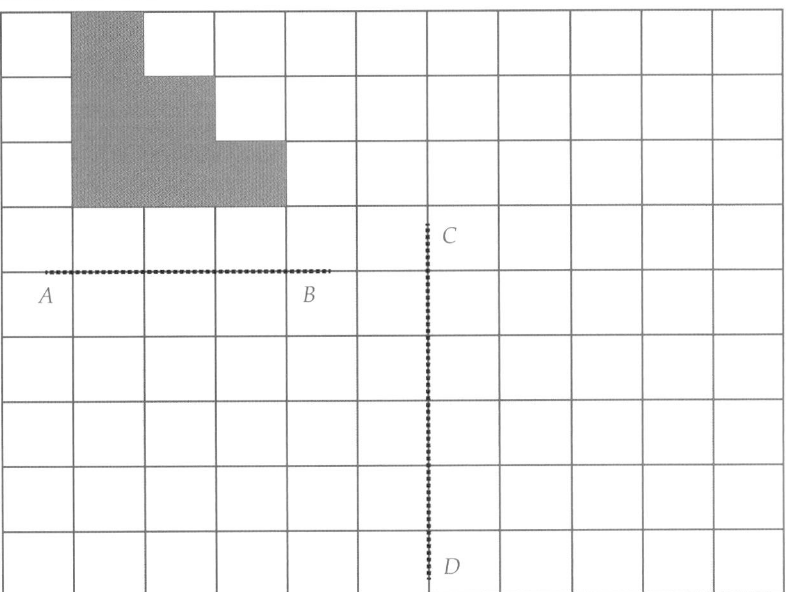

2 Copy this shape onto grid paper, rotate it 90° anticlockwise about the point O and then rotate that image 180° about the point P.

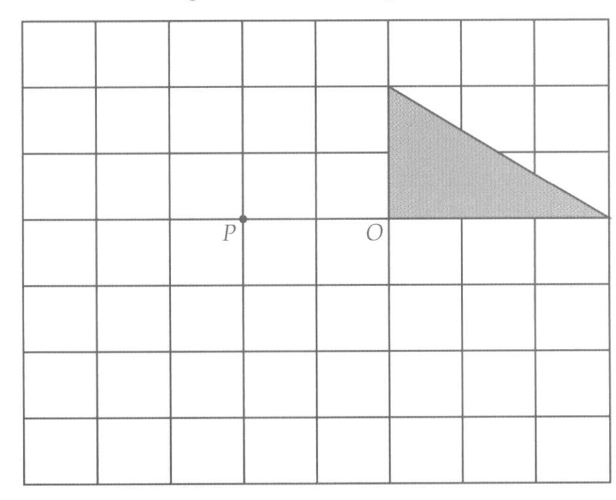

ISBN 9780170350990

EXERCISE 7–08

3 Describe the composite transformation on each dark shape.

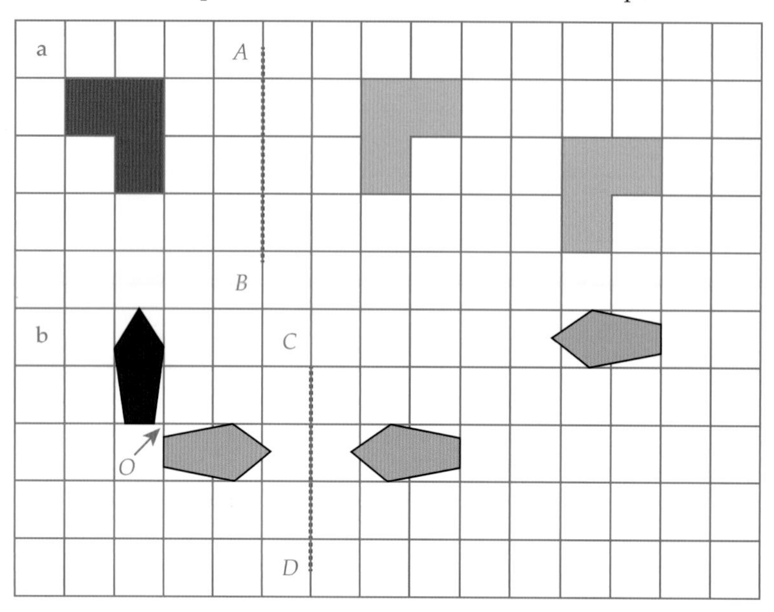

4 Copy each diagram and perform the composite transformation stated.

a Translate 2 units right and then reflect in the line AB.

b Rotate 180° anticlockwise about O and then reflect in the line CD.

c Translate 3 units right and 2 units down and then rotate 90° clockwise about P.

d Reflect in the line EF and then rotate 180° anticlockwise about Q.

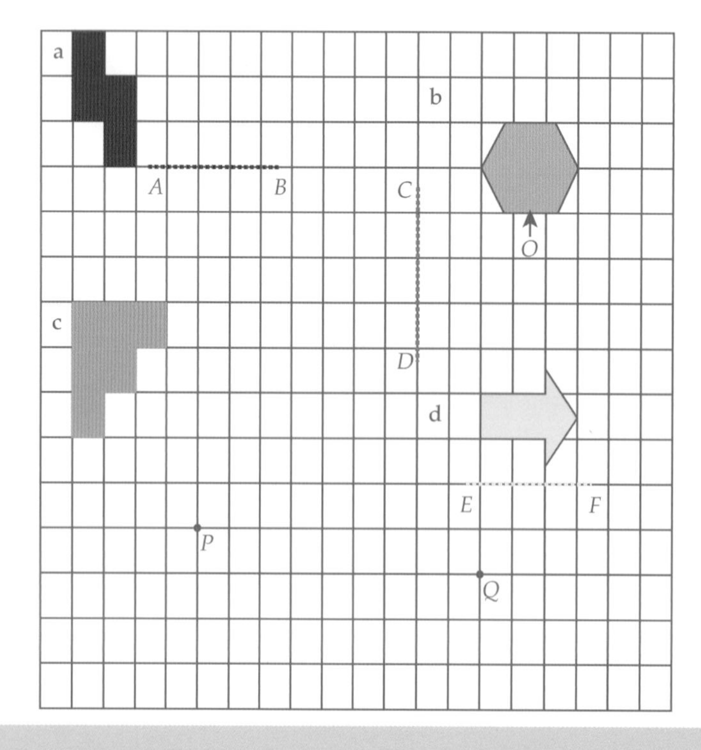

ISBN 9780170350990

7-09 Transformations on the number plane

When a shape is translated, reflected or rotated, the transformed shape is called the **image.** When a point or vertex P of an original shape is transformed, the corresponding point or vertex on the image is labelled P', pronounced 'P-dash' or 'P-prime'.

EXAMPLE 9

This diagram shows an arrowhead shape $PQRS$ being translated 4 units right and 1 unit up to create the image $P'Q'R'S'$. Compare the coordinates of P, Q, R, S to the coordinates of P', Q', R', S'.

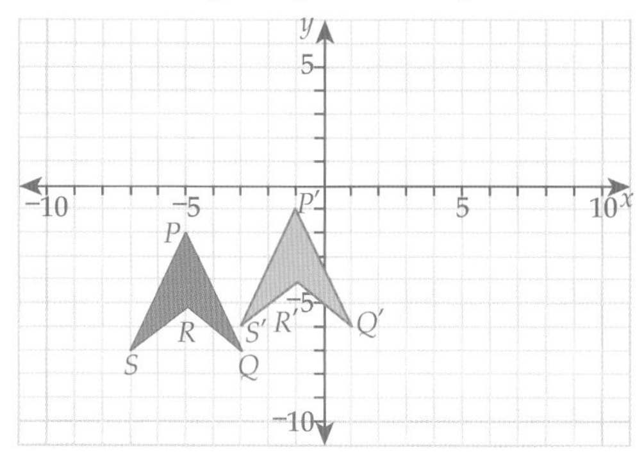

SOLUTION

$P(-5, -2) \rightarrow P'(-1, -1)$ $\quad$ $Q(-3, -7) \rightarrow Q'(1, -6)$

$R(-5, -5) \rightarrow R'(-1, -4)$ $\quad$ $S(-7, -7) \rightarrow S'(-3, -6)$

When translated 4 units right and 1 unit up, the x-coordinate of each vertex increases by 4 whereas the y-coordinate increases by 1.

EXAMPLE 10

The orange L-shape below named $ABCDEF$ has been reflected across the y-axis to create the image $A'B'C'D'E'F'$. Compare the coordinates of D and F to those of D' and F', respectively.

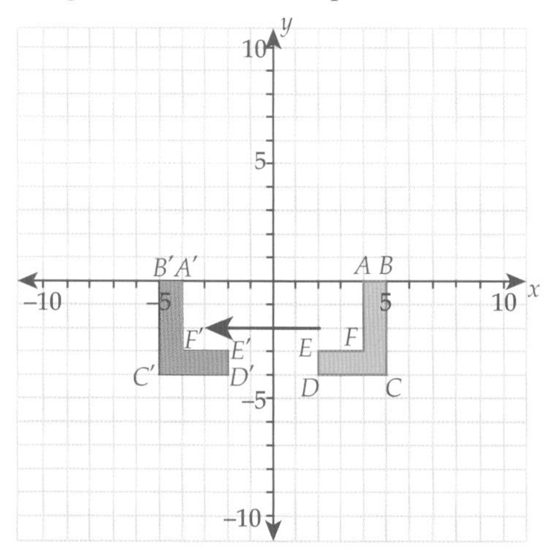

SOLUTION

$D(2, -4) \rightarrow D'(-2, -4)$ $\quad$ $F(4, -3) \rightarrow F'(-4, -3)$

When reflected across the y-axis, the x-coordinate of each vertex changes sign (positive to negative) whereas the y-coordinate stays the same.

ISBN 9780170350990

EXERCISE 7–09

1 **a** Copy this trapezium $ABCD$ onto grid paper and translate it 10 units right to create the image $A'B'C'D'$.

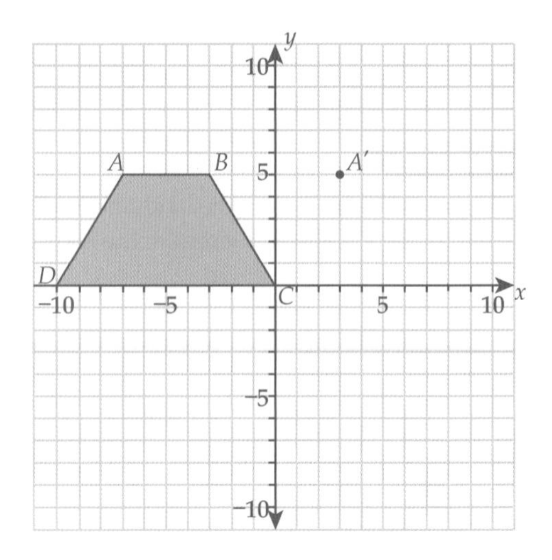

b Compare the coordinates of A, B, C, D to those of A', B', C', D'.

2 This flag shape $PQRS$ has been rotated 90° anticlockwise about the origin T to create the image $P'Q'R'S'$. Compare the coordinates of P, Q, R, S to those of P', Q', R', S'.

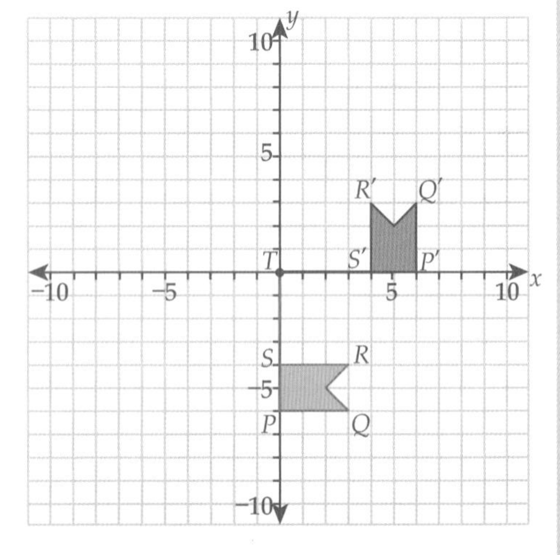

3 **a** Copy the L-shape $STUVWX$ onto grid paper and reflect it across the x-axis to create the image $S'T'U'V'W'X'$.

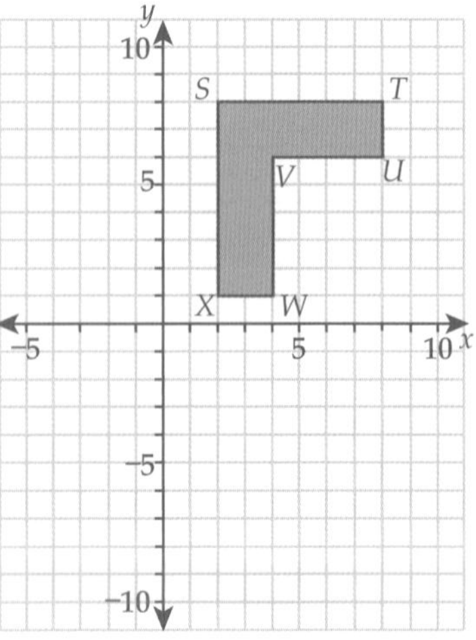

b Compare the coordinates of S, T, U, V to those of S', T', U', V'.

ISBN 9780170350990

EXERCISE 7–09

4 a Describe the translation that has been performed on triangle WXY to create the image $W'X'Y'$.

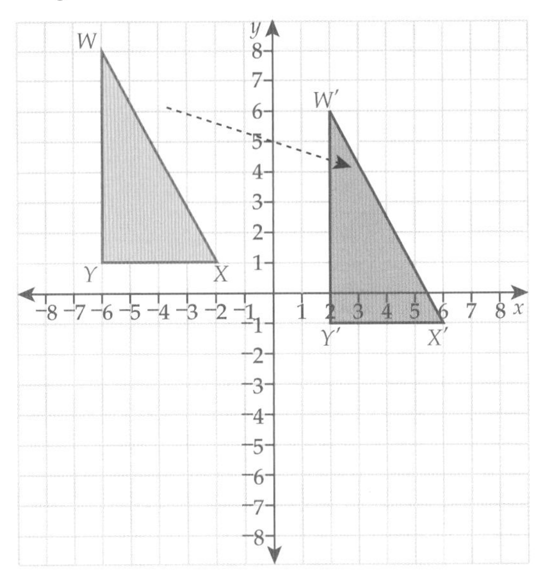

b Compare the coordinates of the original triangle to those of the image.

5 a Copy $OPQRST$ onto grid paper and rotate it 180° about O.

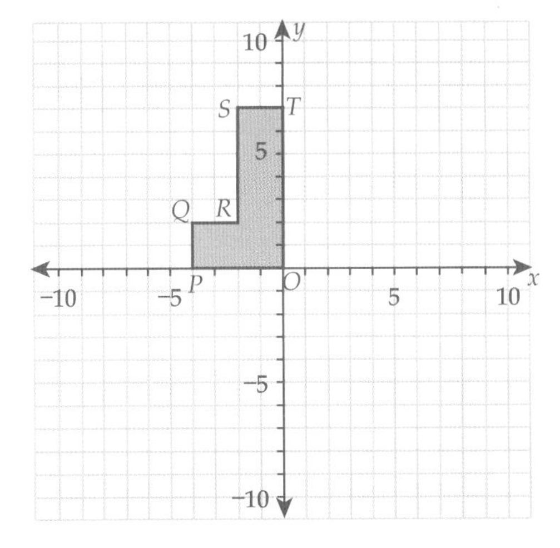

b Find the coordinates of the new positions of P, Q and S and compare them to their original coordinates.

ISBN 9780170350990

FIND-A-WORD PUZZLE

Make a copy of this puzzle, then find the words below in this grid of letters.

R	N	O	I	T	C	E	L	F	E	R	T	J	Y	M
P	A	O	L	Y	S	F	L	E	U	R	R	R	O	Q
A	H	L	C	A	W	Y	T	I	A	B	A	S	Y	H
R	C	L	U	F	N	A	M	N	N	T	N	E	R	G
A	H	O	G	C	N	I	S	M	N	E	S	L	A	L
L	Q	L	I	R	I	L	G	E	E	R	V	G	T	A
L	K	Z	E	N	A	D	M	I	O	T	E	N	N	N
E	P	T	R	T	T	E	N	T	R	G	R	A	E	O
L	L	R	I	H	L	E	A	E	A	O	S	Y	M	I
A	J	O	J	P	C	T	R	M	P	D	A	V	E	T
A	N	W	P	O	I	N	I	I	S	R	L	E	L	A
Z	N	U	C	O	U	U	H	L	O	I	E	C	P	T
B	S	D	N	E	Q	U	A	L	J	R	T	P	M	O
C	O	R	R	E	S	P	O	N	D	I	N	G	O	R
C	O	M	P	O	S	I	T	E	Z	C	I	M	C	H

ALTERNATE
ANGLES
CO-INTERIOR
COMPLEMENTARY
COMPOSITE
CORRESPONDING
EQUAL
IMAGE
LINE
ORIGINAL
PARALLEL
PERPENDICULAR
REFLECTION
ROTATION
ROTATIONAL
SUPPLEMENTARY
SYMMETRY
TRANSLATION
TRANSVERSAL

ISBN 9780170350990

PRACTICE TEST 7

Part A General topics

Calculators are not allowed.

1 Evaluate 18 × 20.

2 How many degrees in a right angle?

3 Simplify $\frac{32ab}{-4a}$.

4 Find the volume of the prism.

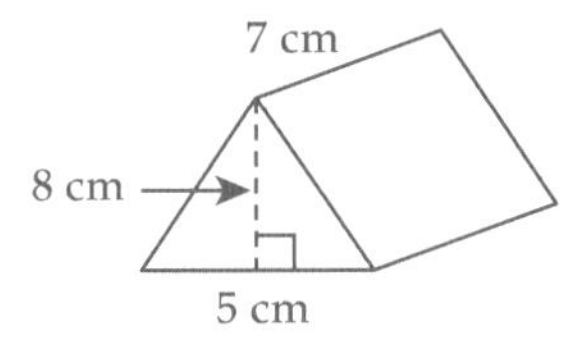

5 Find $\frac{1}{8}$ of \$120.

6 Evaluate \$14.25 + \$28.90.

7 Find the mode of 6, 3, 2, 6, 5, 4, 6.

8 How many sides has a quadrilateral?

9 Arrange these decimals in ascending order: 8.95, 8.909, 8.19, 8.9.

10 Complete this number pattern:
1, 2, 4, ___, 16, ___.

Part B Angles and symmetry

Calculators are allowed.

7-01 Angles

11 Use three letters to name this angle.

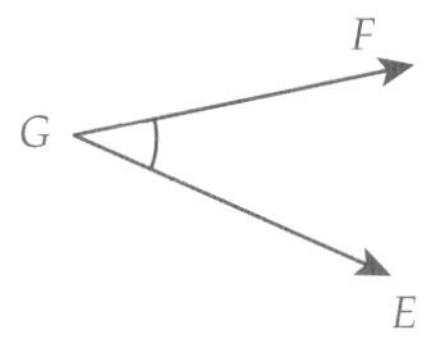

7-02 Measuring and drawing angles

12 Measure the size of this angle. Select the correct answer **A**, **B**, **C** or **D**.

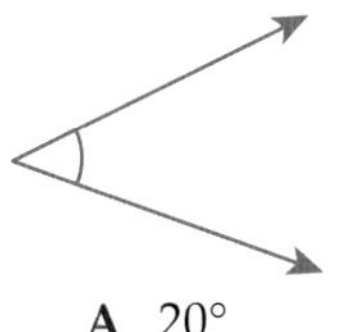

A 20° **B** 40° **C** 35° **D** 45°

7-03 Angle geometry

13 Find the value of each pronumeral, giving a reason.

a

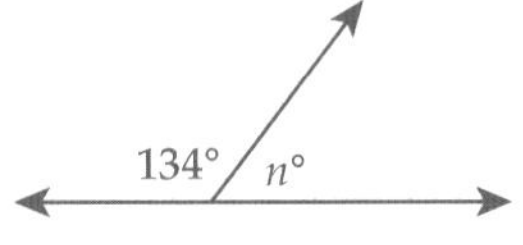

b

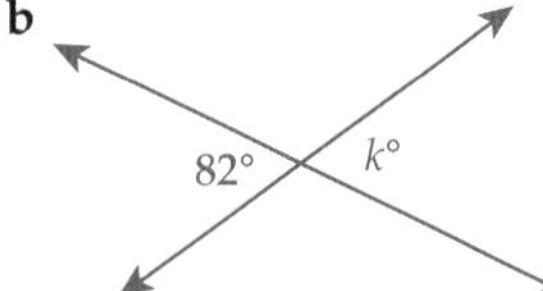

7-04 Parallel and perpendicular lines

14 For this pentagon, list one pair of parallel lines and one pair of perpendicular lines.

7-05 Angles on parallel lines

15 When a transversal crosses two parallel lines, which pair of angles are supplementary?

16 Find c.

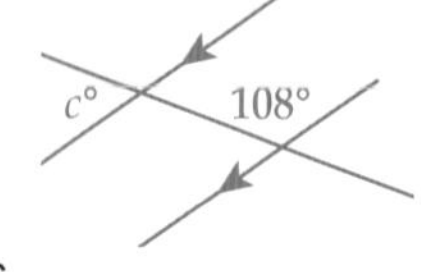

7-06 Proving parallel lines

17 Is $AB \parallel CD$? Give a reason for your answer.

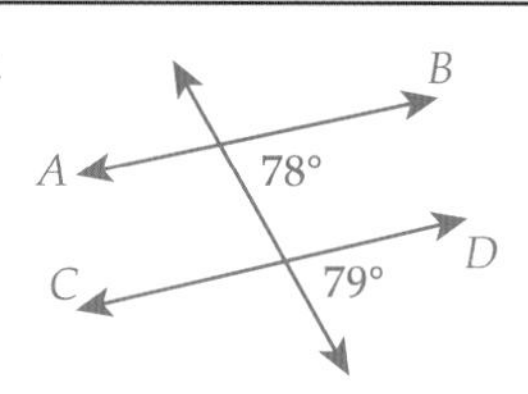

7-07 Line and rotational symmetry

18 How many axes of symmetry has this shape?

7-08 Transformations

19 What type of transformation is a turn through 180°? Select **A**, **B**, **C** or **D**.

A translation　**B** reflection　**C** rotation　**D** composite

7-09 Transformations on the number plane

20 a Describe the transformation that has been performed on rectangle $MNPQ$ to create the image $M'N'PQ'$.

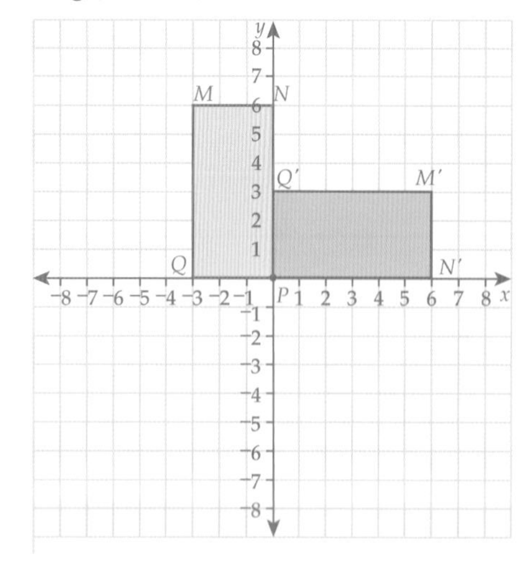

b Compare the coordinates of M, N, Q to those of M', N', Q'.

ISBN 9780170350990

TRIANGLES AND QUADRILATERALS

8

WHAT'S IN CHAPTER 8?

IN THIS CHAPTER YOU WILL:

- name and classify triangles and their properties
- find the angle sum of a triangle
- find the exterior angle of a triangle
- name and classify quadrilaterals
- find the angle sum of a quadrilateral
- find properties of the sides, angles and diagonals of quadrilaterals
- solve geometry problems involving the properties of triangles and quadrilaterals

8-01 Types of triangles

WORDBANK

triangle A shape with three straight sides.

There are three types of triangles according to the lengths of their sides:

- **scalene**: all sides different (no equal sides)
- **isosceles**: two equal sides
- **equilateral**: three equal sides.

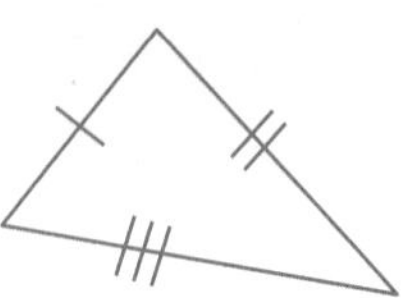

Scalene triangle

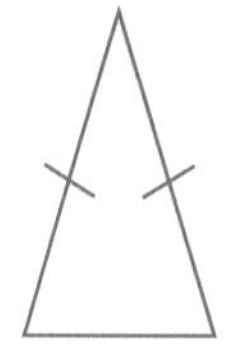

Isosceles triangle

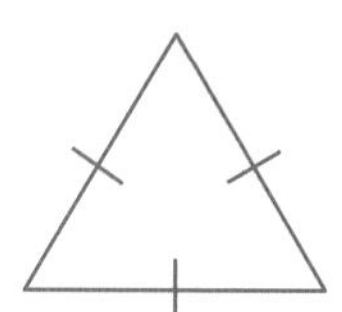

Equilateral triangle

The dashes are drawn on each side to show which sides are equal.

There are also three types of triangles according to the sizes of their angles:

- **acute-angled**: all angles are acute (less than 90°)
- **obtuse-angled**: one of the angles is obtuse (more than 90° and less than 180°)
- **right-angled**: one of the angles is a right angle (90°).

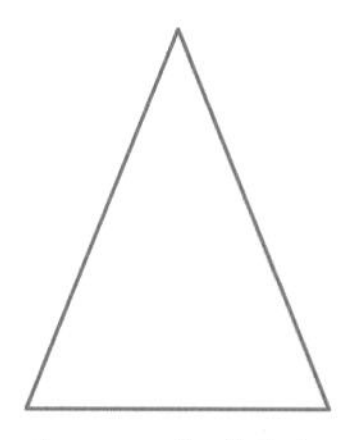

Acute-angled triangle

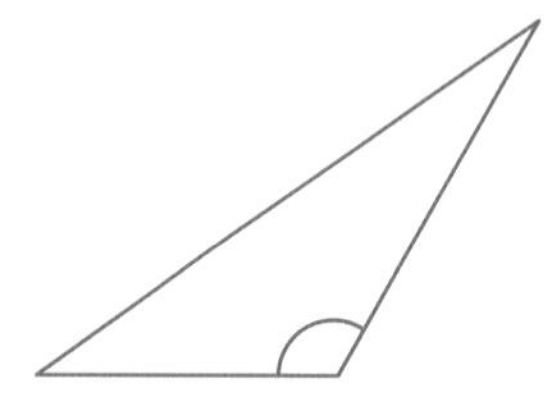

Obtuse-angled triangle

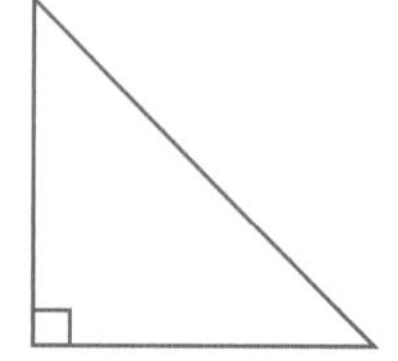

Right-angled triangle

EXAMPLE 1

Classify each triangle by side and by angle.

a

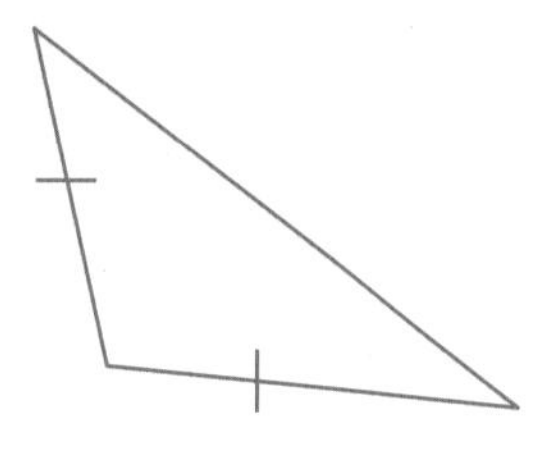

b

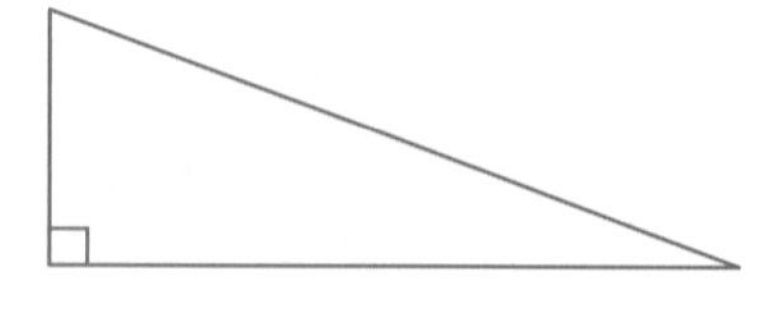

SOLUTION

a Triangle is isosceles and obtuse-angled. ← two equal sides

b Triangle is scalene and right-angled. ← no equal sides

ISBN 9780170350990

EXERCISE 8–01

1 What type of triangle has two sides equal? Select the correct answer **A**, **B**, **C** or **D**.

A isosceles **B** acute-angled **C** obtuse-angled **D** equilateral

2 What type of triangle has one angle greater than 90°? Select **A**, **B**, **C** or **D**.

A isosceles **B** acute-angled **C** obtuse-angled **D** equilateral

3 Classify each triangle according to its sides.

a

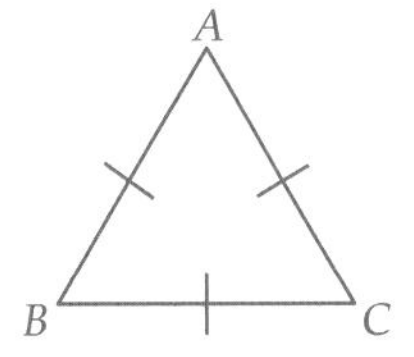

b

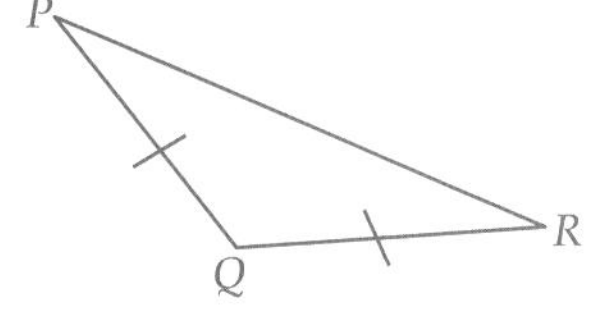

c

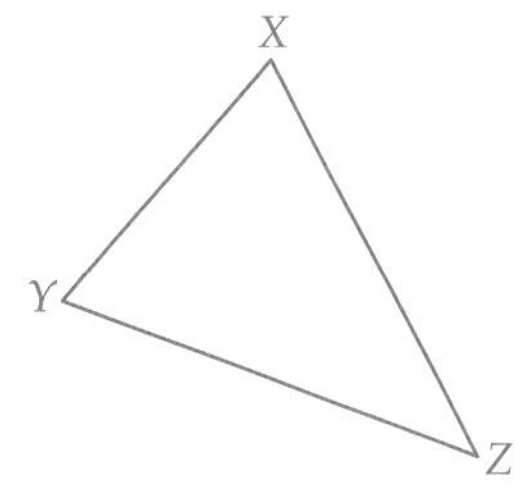

4 Classify each triangle in Question **3** according to its angles.

5 Draw each type of triangle described below.

a acute-angled and scalene **b** isosceles and obtuse-angled

c right-angled and scalene **d** equilateral and acute-angled

6 **a** Is it possible to draw an isosceles right-angled triangle?

b Why is each angle of an equilateral triangle 60°?

c Is it possible to draw an obtuse-angled equilateral triangle?

d A right-angled triangle has one 90° angle. What do the other two angles add to?

e Is it possible to draw an equilateral right-angled triangle?

7 To name a triangle we use three capital letters that are the vertices of the triangle, for example, ΔABC. Name each triangle below.

a

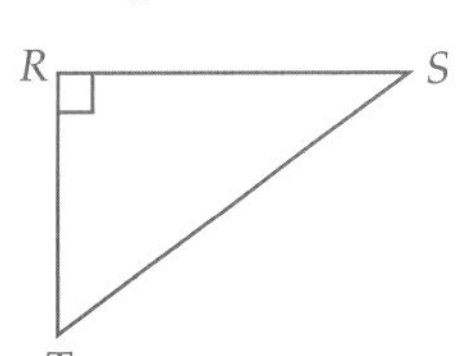

b

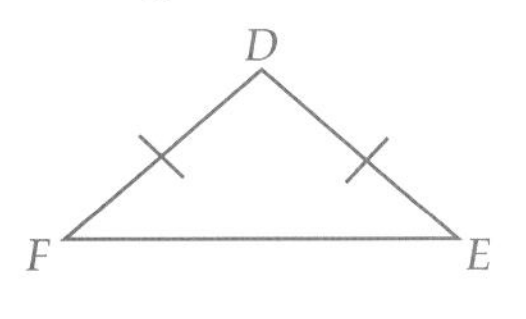

c

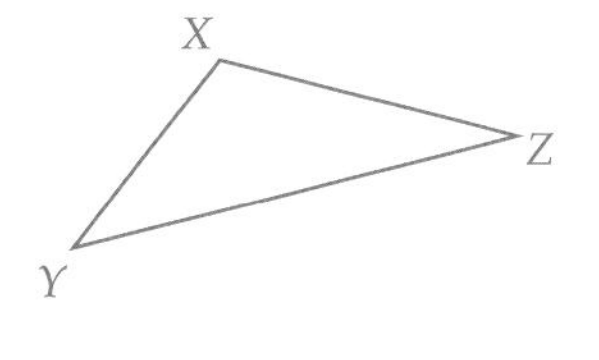

8 Classify each triangle in Question **7** by sides and by angles.

9 **a** Classify each type of triangle seen in the diagram below.

b How many triangles are there?

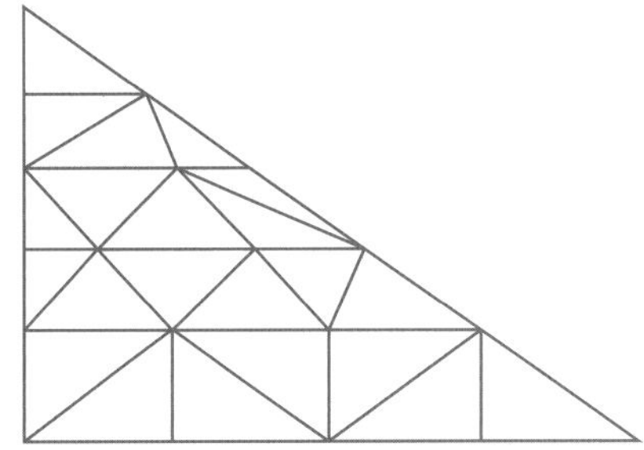

8–02 Angle sum of a triangle

The **angle sum** of a triangle is the total when its three angles are added together.

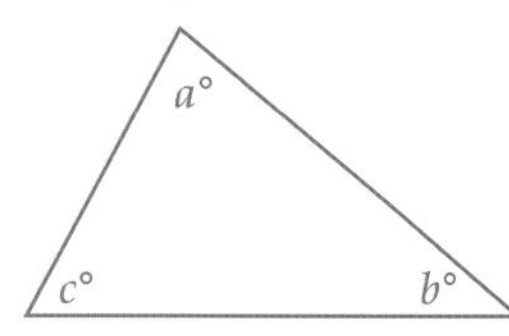

Angle sum $= a° + b° + c°$

The angle sum can easily be found by drawing any triangle on paper, cutting it out and then tearing off each angle and placing them together as shown:

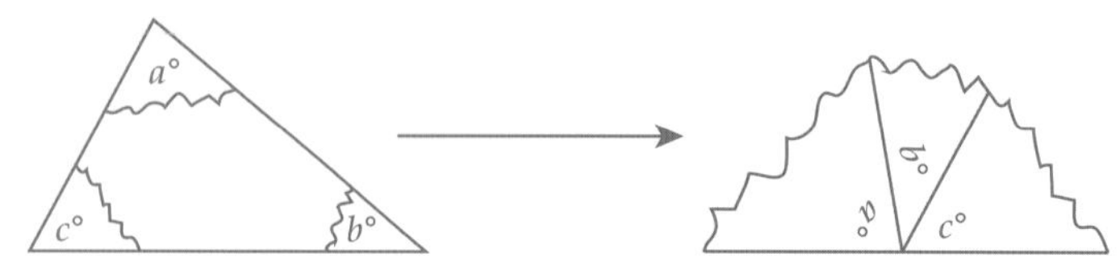

When placed together, the angles form a straight line, which is 180°.

The angle sum of any triangle is 180°.

EXAMPLE 2

Find the value of each pronumeral.

a

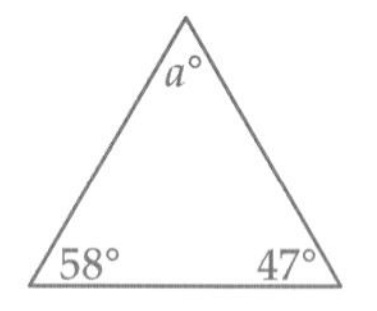

b

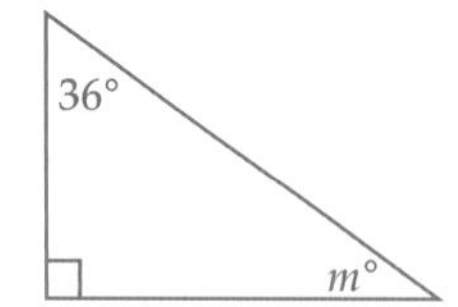

SOLUTION

a $a + 58 + 47 = 180$ ← Angle sum of a triangle is 180°.

$a + 105 = 180$

$a = 180 - 105$ ← Solve as an equation.

$a = 75$

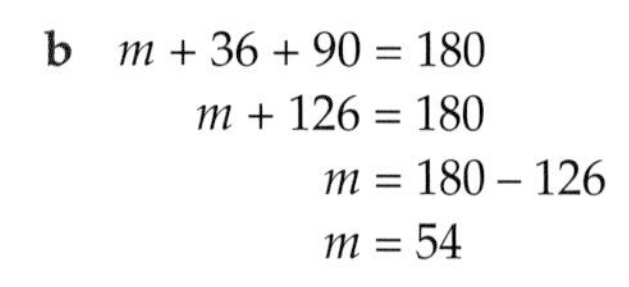

b $m + 36 + 90 = 180$

$m + 126 = 180$

$m = 180 - 126$

$m = 54$

Shutterstock.com/oover

ISBN 9780170350990

EXERCISE 8–02

1 Find n.

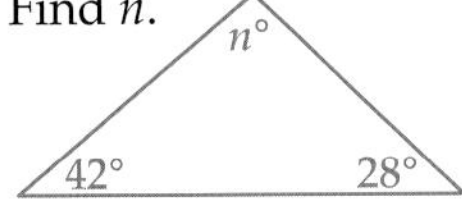

Select the correct answer **A**, **B**, **C** or **D**.

A 130 **B** 140 **C** 120 **D** 110

2 What is the angle sum of an equilateral triangle? Select **A**, **B**, **C** or **D**.

A 90° **B** 150° **C** 360° **D** 180°

3 Find the value of each pronumeral.

a

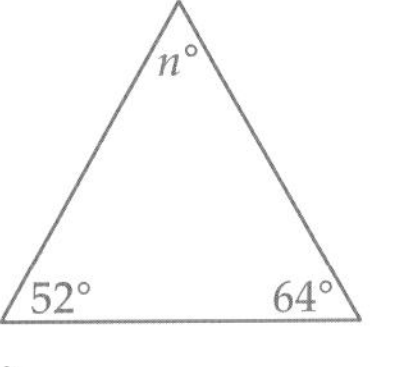

b

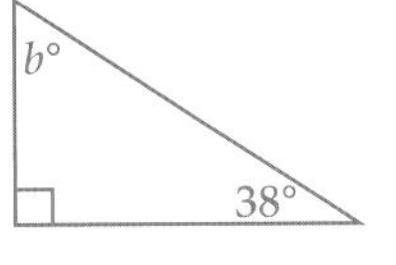

c

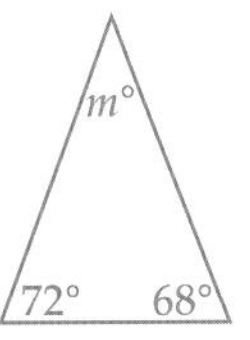

d

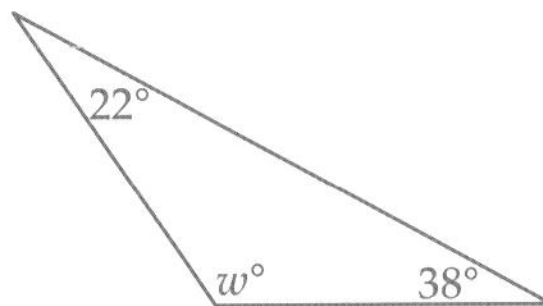

e

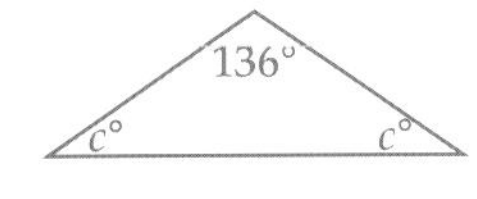

f

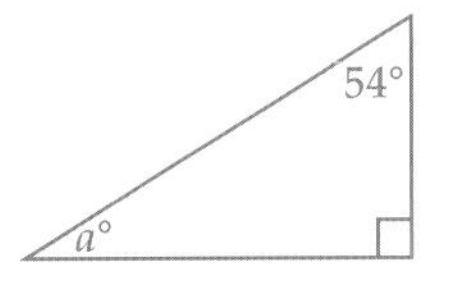

4 **a** If two angles of a triangle are 75° and 50°, what is the size of the third angle?

b If the three angles of a triangle are $p°$, $q°$ and $r°$, what rule can we write about p, q and r?

5 Describe each triangle in words.

a an equilateral triangle **b** a right triangle

c an isosceles triangle **d** a scalene triangle

6 What do you know about the angles in:

a an equilateral triangle? **b** an isosceles triangle?

7 Find the value of each pronumeral.

a

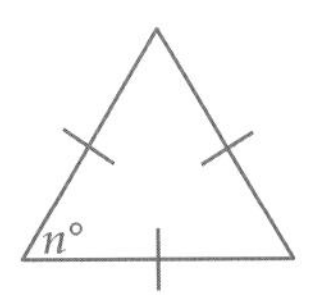

b

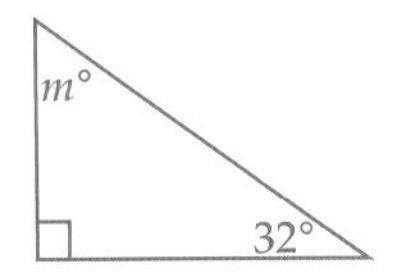

c

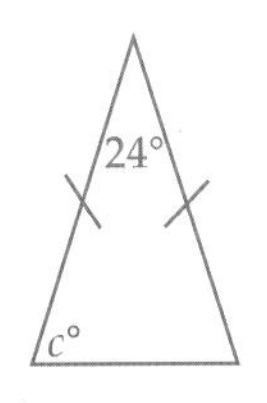

d

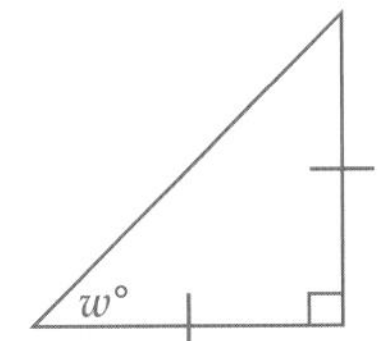

e

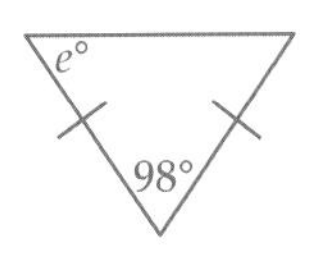

f

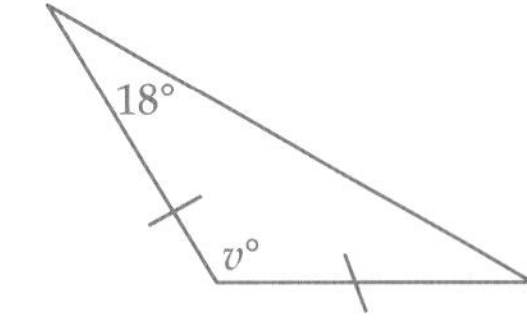

8-03 Exterior angle of a triangle

WORDBANK

exterior angle An angle outside a shape created by extending one side of the shape. The diagram below shows the exterior angle of a triangle.

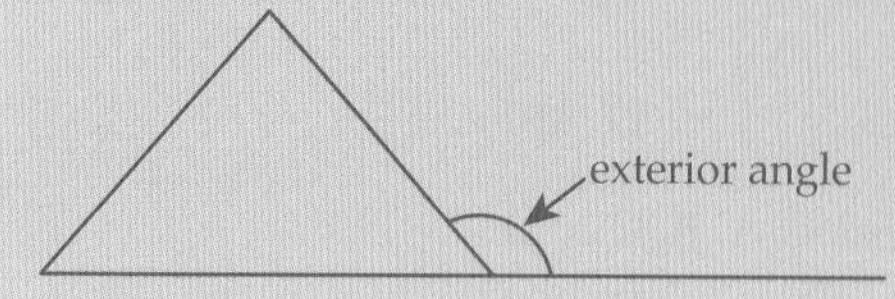

In the diagram below, the exterior angle is $d°$.

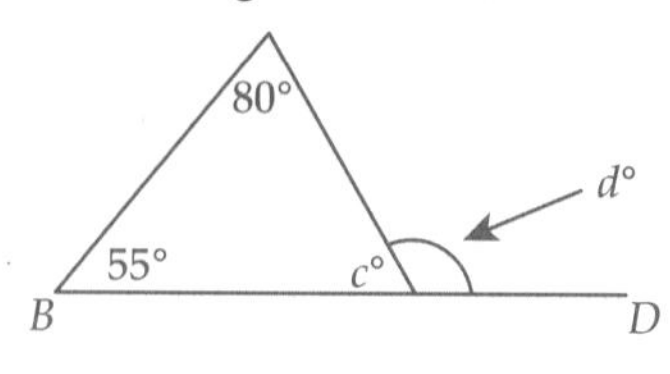

$55 + 80 + c = 180$ ⟵ angle sum of a triangle

But also $d + c = 180$ ⟵ angles on a straight line

So $d = 55 + 80$

So $d = 135$

The **exterior angle of a triangle** is equal to the sum of the two interior opposite angles (55° and 80° in the diagram above).

EXAMPLE 3

Find the value of each pronumeral.

a

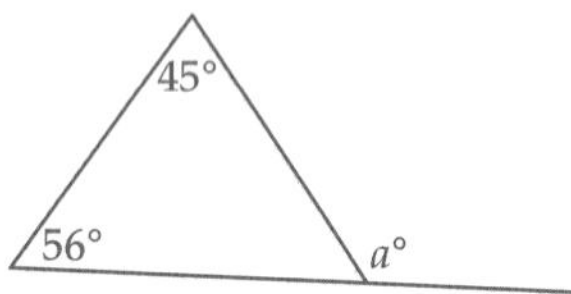

b

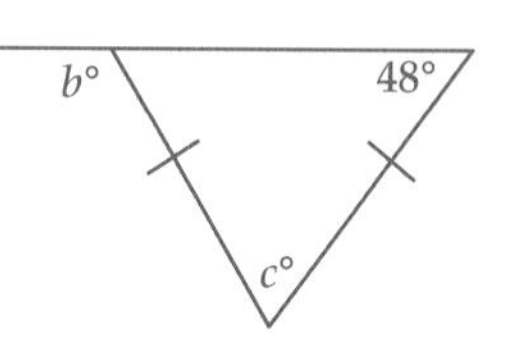

SOLUTION

a $a = 45 + 56$ ⟵ exterior angle of a triangle

$a = 101$

b $c = 180 - 2 \times 48$ ⟵ angle sum of an isosceles triangle

$c = 84$

$b = 48 + 84$ ⟵ exterior angle of a triangle

$b = 132$

iStockphoto/maticomp

ISBN 9780170350990

EXERCISE 8–03

1 The exterior angle of a triangle equals ________. Select the correct answer **A**, **B**, **C** or **D**.

A the opposite angle

B two of the interior angles

C the sum of all the angles

D the sum of the two interior opposite angles

2 What is the size of an exterior angle of an equilateral triangle? Select **A**, **B**, **C** or **D**.

A 60° **B** 120° **C** 180° **D** 90°

3 For the diagram below:

a use 3 letters to name each interior angle

b name the exterior angle

c write an equation to find the size of the exterior angle.

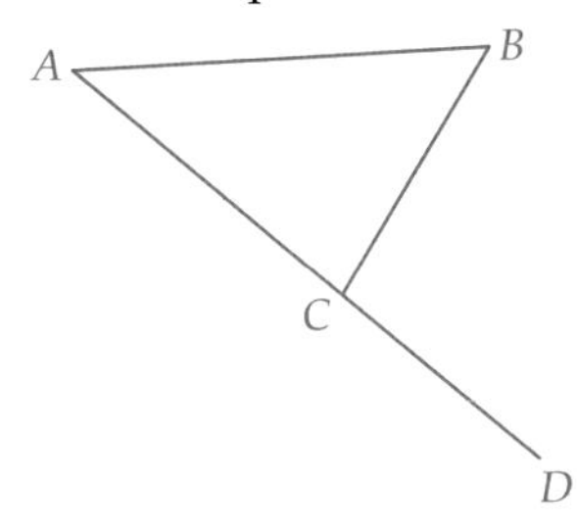

4 Find the value of each pronumeral.

a

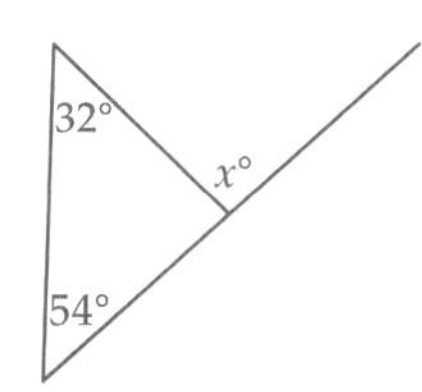

b

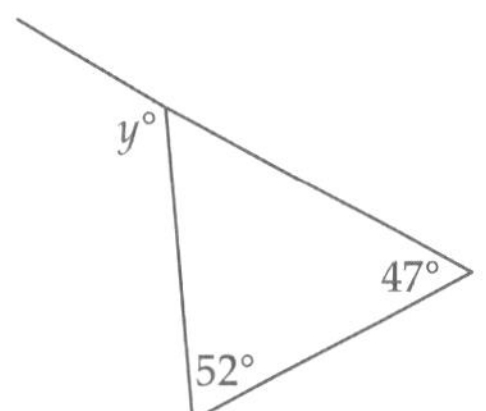

c

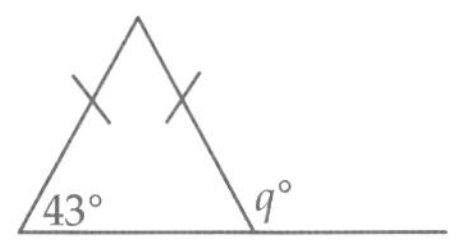

d

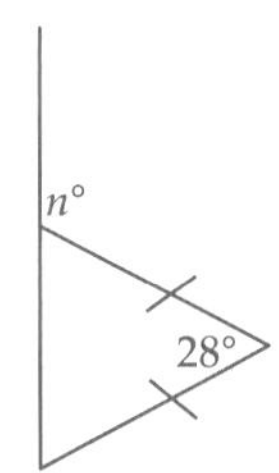

5

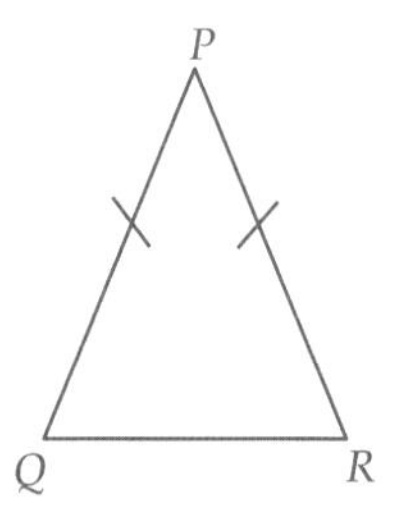

a Copy this diagram and extend PR to form an exterior angle.

b Name the two interior opposite angles.

c Calculate the size of the exterior angle if $\angle PQR = 64°$.

ISBN 9780170350990

8-04 Types of quadrilaterals

WORDBANK

quadrilateral A shape with four straight sides.

There are many types of quadrilaterals. They are displayed in the table below.

Name of quadrilateral	Diagram	Features
Parallelogram		Opposite sides parallel. Opposite sides equal. Opposite angles equal.
Rhombus		All sides equal in length. A special type of parallelogram.
Trapezium		One pair of sides parallel.
Kite		Adjacent sides equal. One pair of opposite angles equal.
Rectangle		All angles are 90°. A special type of parallelogram.
Square		All sides are equal. All angles are 90°. A special type of parallelogram and rhombus.
Convex quadrilateral		Any quadrilateral where all vertices (corners) point outwards. All diagonals lie inside the shape.
Non-convex quadrilateral		Any quadrilateral where one vertex points inwards. One diagonal lies outside the shape. One angle is more than 180°.

ISBN 9780170350990

EXERCISE 8–04

1 Which quadrilateral has one pair of opposite sides parallel? Select the correct answer **A**, **B**, **C** or **D**.

A square **B** trapezium **C** parallelogram **D** rhombus

2 Which quadrilateral has all four sides equal? Select **A**, **B**, **C** or **D**.

A rectangle **B** trapezium **C** parallelogram **D** rhombus

3 Name each type of quadrilateral.

a 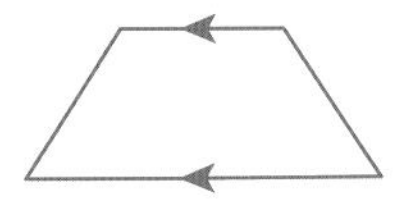

b 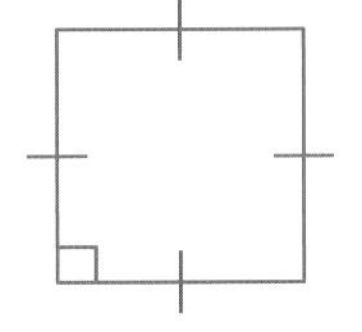

c 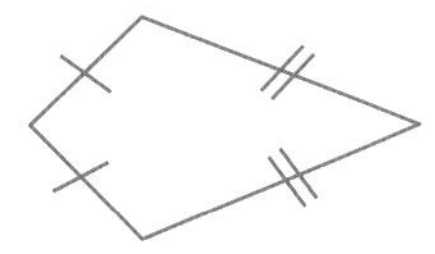

d 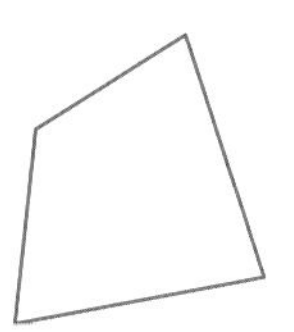

e 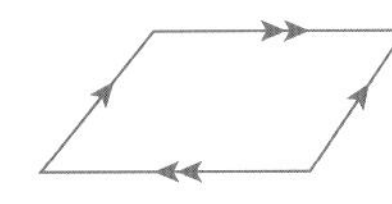

f 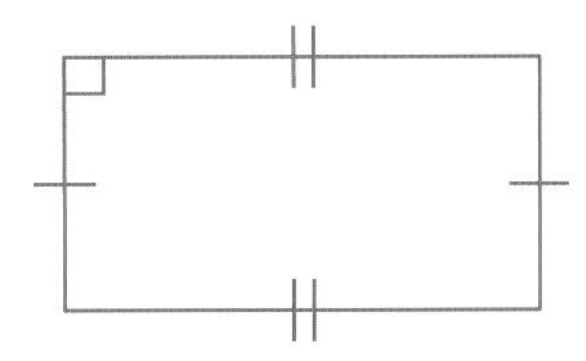

4 Draw:

a a rectangle

b a non-convex quadrilateral

c a trapezium

d an irregular quadrilateral with one obtuse angle

e a square with side lengths 5 cm

f a parallelogram with one pair of sides 8 cm and the other pair 6 cm

g a rhombus with one pair of angles equal to 65° and all sides 7 cm long.

5 Is each statement true or false?

a All rectangles are squares.

b All parallelograms are rhombuses.

c All squares are rectangles.

d All rhombuses are parallelograms.

e All kites are quadrilaterals.

f All quadrilaterals are kites.

Alamy/Gaertner

ISBN 9780170350990

8–05 Angle sum of a quadrilateral

The **angle sum** of a quadrilateral is the total when its four angles are added together.

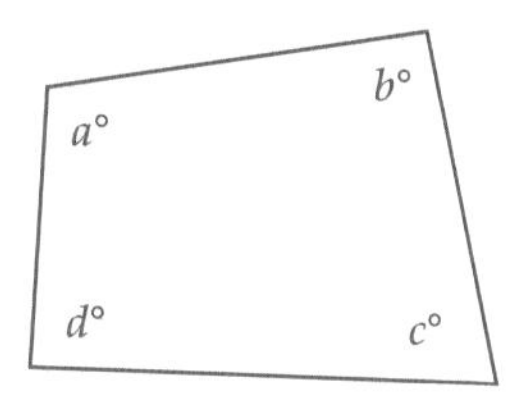

Angle sum = $a° + b° + c° + d°$

The angle sum can easily be found by drawing any quadrilateral on paper, cutting it out and then tearing off each angle and placing them together as shown:

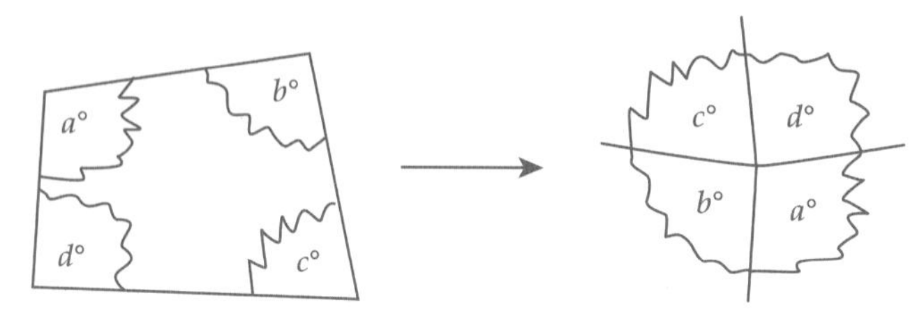

When placed together, the four angles form a revolution, which is 360°.

The angle sum of any quadrilateral is 360°.

EXAMPLE 4

Find the value of each pronumeral.

a

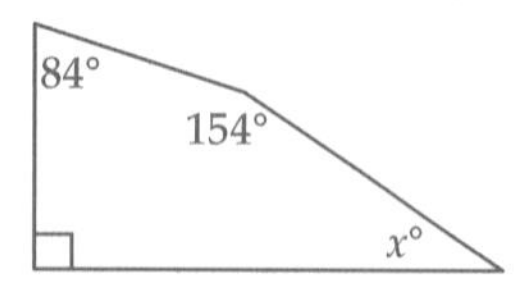

b

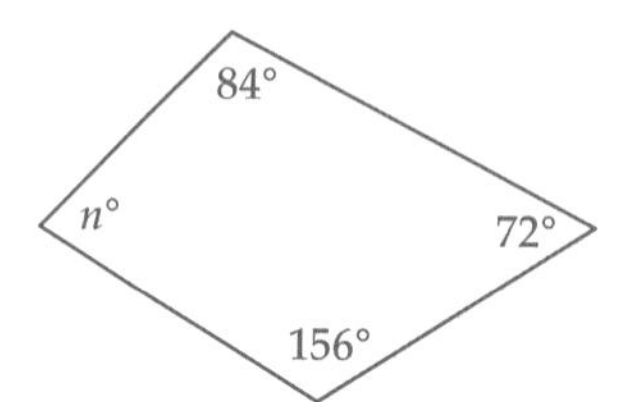

SOLUTION

a $x + 84 + 154 + 90 = 360$ ← Angle sum of a quadrilateral is 360°.

$x + 328 = 360$

$x = 360 - 328$ ← Solve as an equation.

$x = 32$

b $n + 84 + 72 + 156 = 360$

$n + 312 = 360$

$n = 360 - 312$

$n = 48$

iStockphoto/StanRohrer

ISBN 9780170350990

EXERCISE 8-05

1 What is the angle sum of a square? Select the correct answer **A**, **B**, **C** or **D**.

A 180° B 360° C 120° D 90°

2 A quadrilateral has two equal angles of 75° and a right angle. What is the size of its fourth angle? Select **A**, **B**, **C** or **D**.

A 120° B 130° C 90° D 110°

3 Is each statement true or false?

a Each angle in a parallelogram is 90°.

b All angles in a square are equal.

c Opposite angles in a rectangle are equal.

d Opposite angles in a trapezium are equal.

e Adjacent angles in a kite are equal.

4 Find the value of each pronumeral.

a 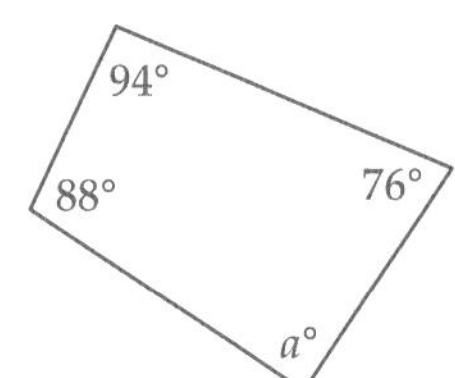

b 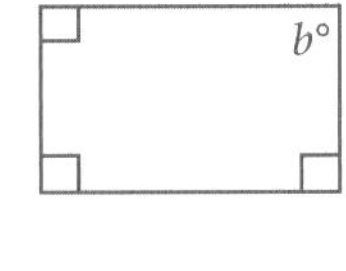

c 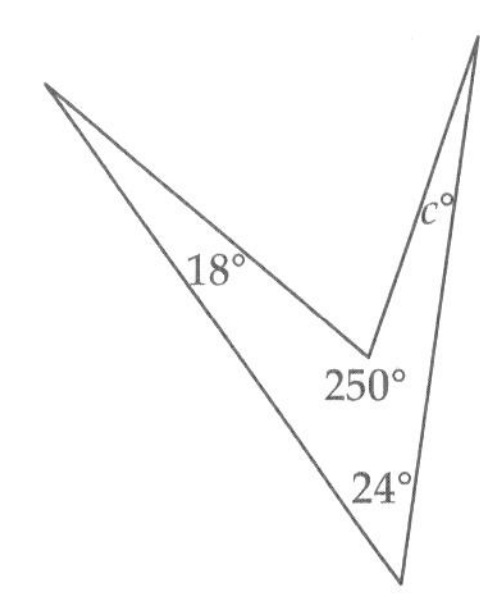

d 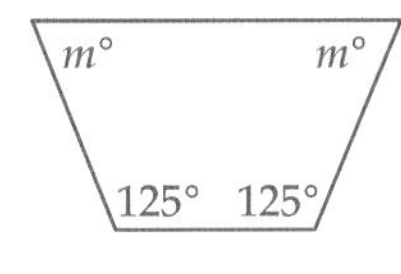

e 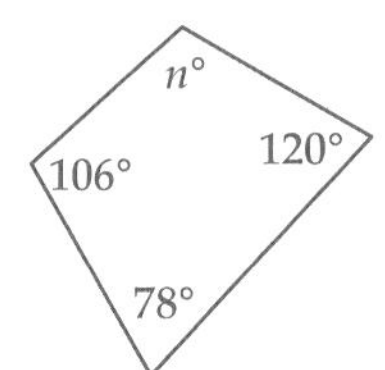

f 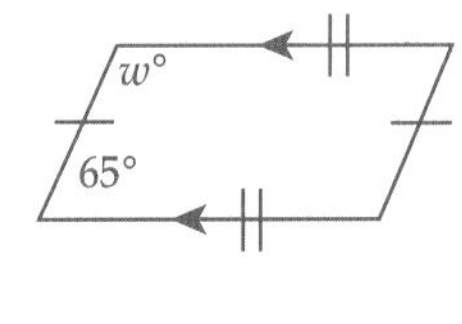

5 a Draw a parallelogram and mark the equal angles.

b What do the co-interior angles in a parallelogram add to?

c How many pairs of co-interior angles are there in a parallelogram?

d What is the angle sum of a parallelogram?

6 a What type of quadrilateral is drawn below?

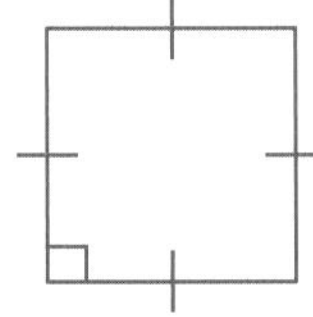

b What do you know about its four sides?

c Is this quadrilateral also a parallelogram?

d Is this quadrilateral also a rectangle?

e What is the angle sum of this quadrilateral?

f What is the size of each angle in this quadrilateral?

g What is the size of an exterior angle of this quadrilateral?

8–06 Properties of quadrilaterals

The diagram below shows how the special quadrilaterals are related to each other.

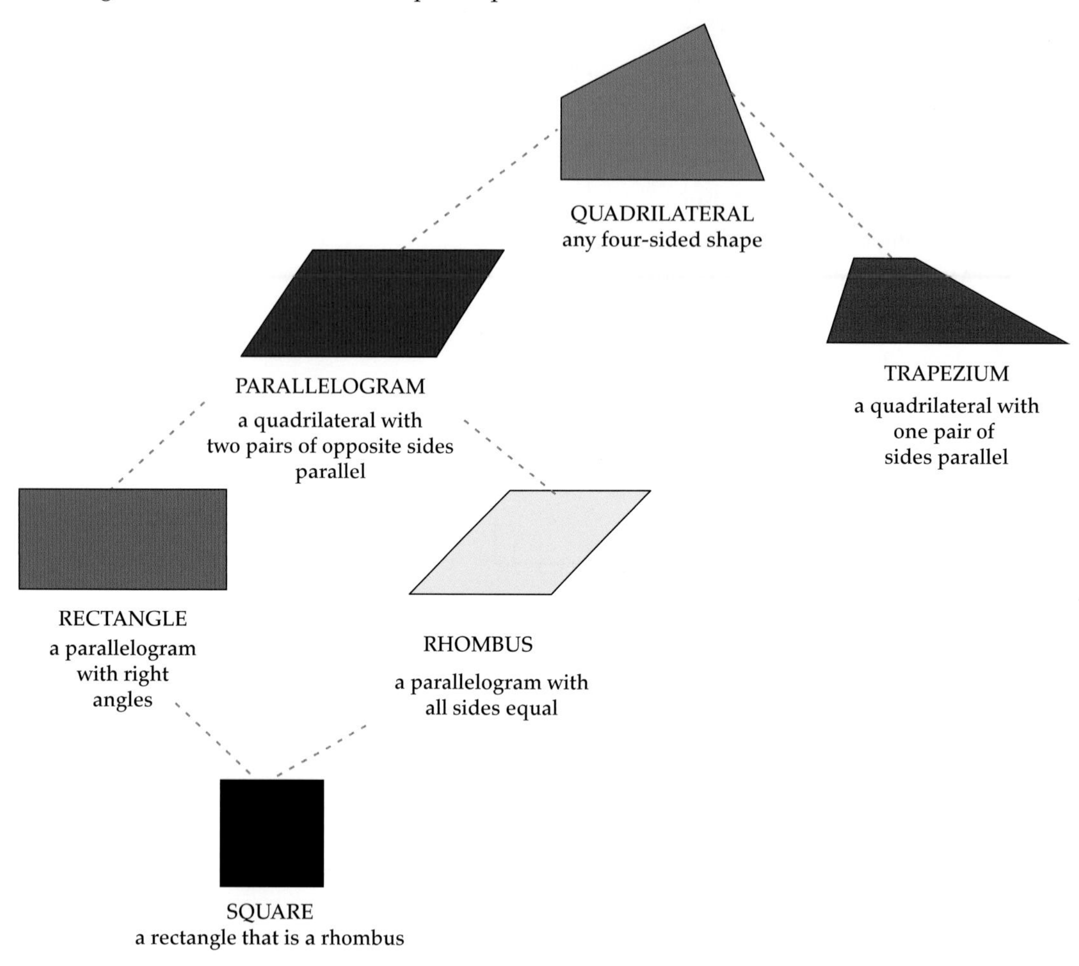

The diagram above shows us that:

- a rhombus is a special parallelogram
- a rectangle is also a special parallelogram
- a square is a special rhombus and is also a special rectangle.

EXAMPLE 5

Write the difference between:

a a rhombus and a parallelogram

b a square and a rhombus.

SOLUTION

a A rhombus has all four sides equal and a parallelogram does not.

b A square has all angles 90° and a rhombus does not.

 ISBN 9780170350990

8-06 Properties of quadrilaterals

Properties of the diagonals of the special quadrilaterals

Parallelogram	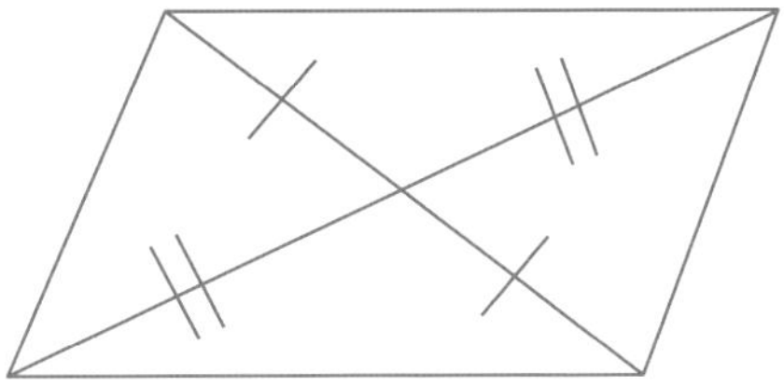	The diagonals bisect each other.
Rectangle	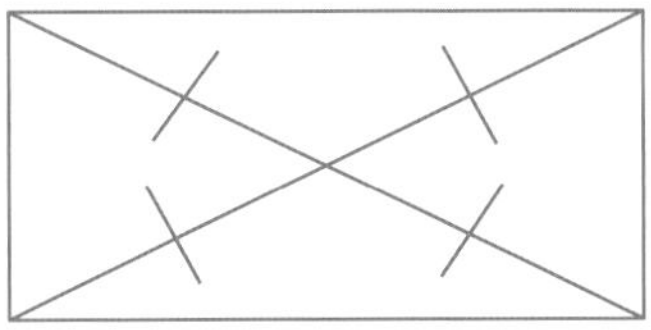	The diagonals are equal in length. The diagonals bisect each other.
Kite	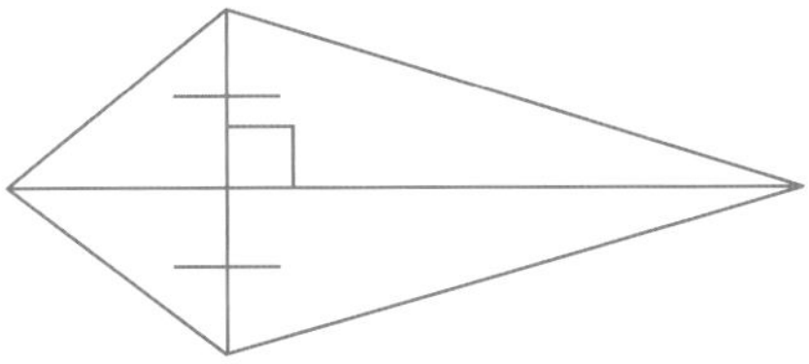	One diagonal bisects the other at right angles.
Rhombus	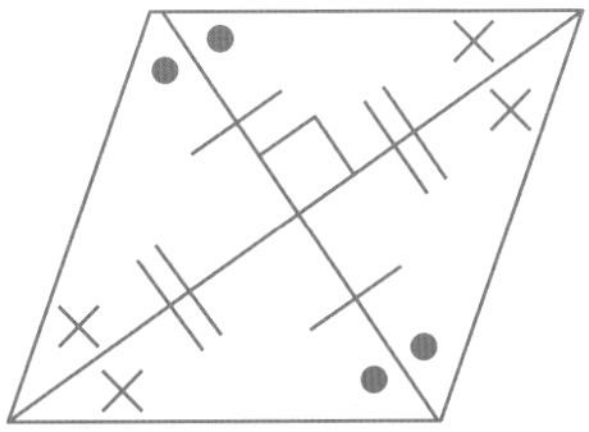	The diagonals bisect each other at right angles. The diagonals bisect the angles of the rhombus.
Square	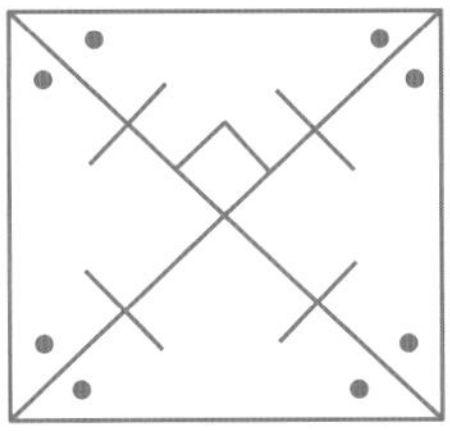	The diagonals bisect each other at right angles. The diagonals bisect the angles of the square. The diagonals are equal in length.

EXERCISE 8-06

1 Which of these quadrilaterals has diagonals that cross each other at right angles? Select the correct answer **A**, **B**, **C** or **D**.

A parallelogram **B** kite **C** rectangle **D** trapezium

2 A rhombus is also a special type of ________. Select **A**, **B**, **C** or **D**.

A parallelogram **B** square **C** pentagon **D** rectangle

3 Write the difference between:

a a square and a rectangle

b a parallelogram and a quadrilateral

c a rhombus and a parallelogram

d a parallelogram and a trapezium.

4 Is each statement true or false?

a The diagonals of a rectangle are equal.

b A rectangle is a special parallelogram.

c A square has equal diagonals that bisect each other at right angles.

d A square is a special rhombus.

e The diagonals of a parallelogram are equal and bisect each other.

f A rhombus is a special parallelogram.

iStockphoto/Viorika

ISBN 9780170350990

EXERCISE 8–06

5 Name all quadrilaterals that have:

a equal diagonals

b diagonals that bisect each other

c diagonals that cross at right angles

d diagonals that bisect the angles of the quadrilateral.

6 Draw a kite and a rectangle, showing their diagonal properties.

7 Copy and complete this table.

Quadrilateral	Angles 90°	Equal sides	Equal diagonals
Parallelogram	No		
Rhombus			No
Rectangle	Yes		
Kite		No	
Square			Yes

8 Copy and complete this table.

Quadrilateral	Sides	Angles	Diagonals
Parallelogram	Opposite sides are equal and parallel.	Opposite angles are equal.	Diagonals bisect each other.
Trapezium		All angles are different.	
Rhombus	All sides are equal.		
Rectangle		All angles are 90°.	
Square			

Shutterstock.com/Vladitto

ISBN 9780170350990

CROSSWORD PUZZLE

Make a copy of this puzzle, then complete the crossword using the clues given below.

Across

1 Quadrilateral with two pairs of adjacent sides equal.

7 Triangle with one angle 90°.

9 Triangle with all sides equal.

11 Quadrilateral with opposite sides parallel.

12 Triangle with all sides different.

13 This shape has an angle sum of 180°.

15 Quadrilateral with opposite sides equal and all angles 90°.

Down

2 Triangle with two sides equal.

3 Opposite of 'regular'.

4 General name for any shape with four sides.

5 Quadrilateral with all vertices pointing outwards.

6 Special rectangle with four equal sides.

8 Quadrilateral with one pair of sides parallel.

10 Special parallelogram with four equal sides.

14 Quadrilateral with one vertex pointing inward.

ISBN 9780170350990

PRACTICE TEST 8

Part A General topics

Calculators are not allowed.

1 Describe in words an acute angle.

2 What is the supplement of 38°?

3 List the first six multiples of 7.

4 Is 81 a prime or a composite number?

5 Find the median of the scores: 12, 8, 7, 6, 8, 7, 5, 7.

6 Is 728 divisible by 4?

7 Simplify: $\frac{7}{8}-\frac{1}{3}$.

8 Find the area of this triangle.

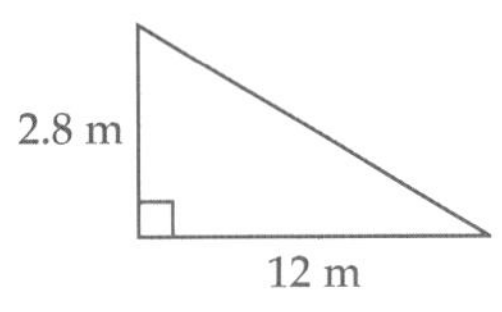

9 Simplify $4ab \times (-8bc)$.

10 Copy and complete: $\frac{4}{5}=\frac{}{35}$.

Part B Triangles and quadrilaterals

Calculators are allowed.

8-01 Types of triangles

11 What type of triangle has three sides equal? Select the correct answer **A**, **B**, **C** or **D**.

A isosceles **B** acute-angled **C** obtuse-angled **D** equilateral

12 What type of triangle has all angles less than 90°? Select **A**, **B**, **C** or **D**.

A isosceles **B** acute-angled **C** obtuse-angled **D** equilateral

8-02 Angle sum of a triangle

13 What is the size of each angle in an equilateral triangle?

14 Find m.

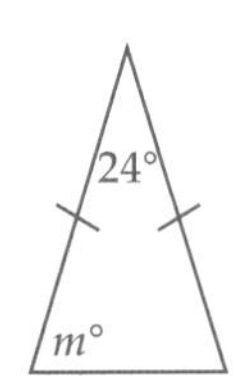

8-03 Exterior angle of a triangle

15 Find w.

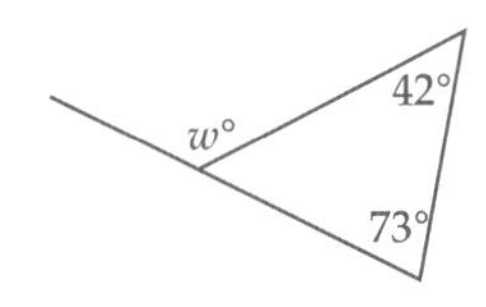

ISBN 9780170350990

16 For this diagram:

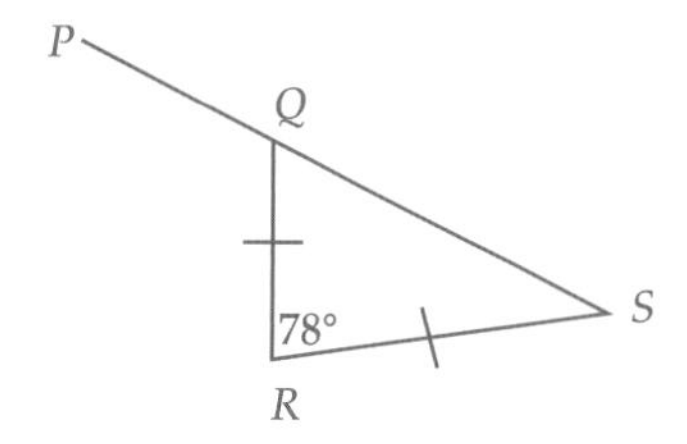

a using three letters, name the exterior angle

b find the size of the exterior angle.

8-04 Types of quadrilaterals

17 Name each quadrilateral.

a 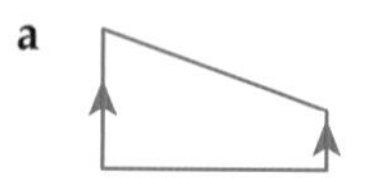

b

8-05 Angle sum of a quadrilateral

18 What is the angle sum of a rhombus?

19 Find the value of each pronumeral.

a 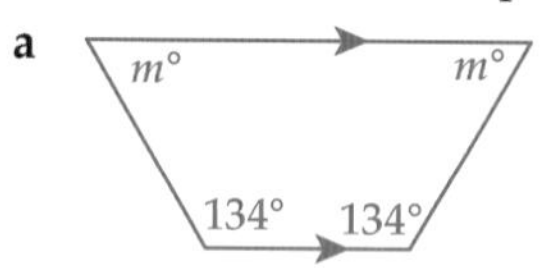

b 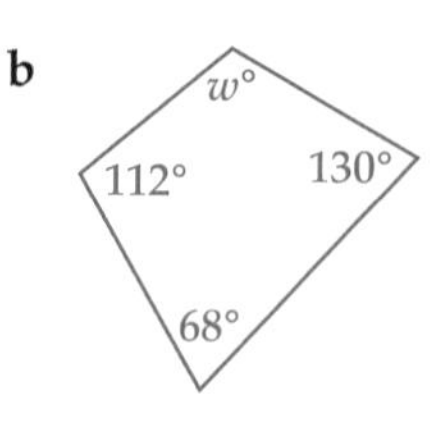

8-06 Properties of quadrilaterals

20 Write the properties of the diagonals of:

a a kite

b a rectangle.

 ISBN 9780170350990

LENGTH AND TIME

9

WHAT'S IN CHAPTER 9?

IN THIS CHAPTER YOU WILL:

- convert between metric units of length, capacity and mass
- find perimeters of shapes, including composite shapes
- name the parts of a circle
- find the circumference of a circle and circular shapes using π
- convert between metric units of time
- convert between 24-hour time and 12-hour (a.m./p.m.) time
- round times to the nearest minute or hour
- calculate time differences
- read and interpret timetables
- understand and use international time zones

* Shutterstock.com/Andrey_Kuzmin

9-01 The metric system

The metric system is based on powers of 10. Milli- means $\frac{1}{1000}$, centi- means $\frac{1}{100}$, kilo- means 1000 and mega- means 1 000 000.

Length	Size	Example
millimetre (mm)	1000 mm = 1 m	Smallest gap on your ruler
centimetre (cm)	100 cm = 1 m	Width of a pen
metre (m)	base unit	Height of a kitchen bench
kilometre (km)	1000 m = 1 km	Distance between bus stops
Capacity		
millilitre (mL)	1000 mL = 1 L	A large drop of water
litre (L)	base unit	A carton of milk
kilolitre (kL)	1000 L = 1 kL	Amount of water in a spa
megalitre (ML)	1 ML = 1 000 000 L	Water in two Olympic-sized swimming pools
Mass		
milligram (mg)	1000 mg = 1 g	A grain of salt
gram (g)	1000 g = 1 kg	A tablet
kilogram (kg)	base unit	A packet of sugar
tonne (t)	1000 kg = 1 t	A small car

To convert units, remember the following initials.

- **SOLD = Small Over to Large Divide**: to convert from small unit to large unit, divide
- **LOSM = Large Over to Small Multiply**: to convert from large unit to small unit, multiply

EXAMPLE 1

Convert:

a 27 m to cm **b** 1652 L to kL **c** 4.8 t to kg

SOLUTION

a 27 m = 27 × 100 cm
= 2700 cm ← LOSM Large Over to Small Multiply: 1 m = 100 cm.

b 1652 L = 1652 ÷ 1000 kL
= 1.652 kL ← SOLD: Small Over to Large Divide: 1000 L = 1 kL.

c 4.8 t = 4.8 × 1000 kg
= 4800 kg ← LOSM: Large Over to Small Multiply: 1 t = 1000 kg.

ISBN 9780170350990

EXERCISE 9-01

1 What is the height of a door handle closest to? Select the correct answer **A**, **B**, **C** or **D**.

A 1 mm **B** 1 cm **C** 1 m **D** 1 km

2 What is the capacity of a cup closest to? Select **A**, **B**, **C** or **D**.

A 250 mL **B** 250 L **C** 250 kL **D** 250 ML

3 Choose the most appropriate unit for each measurement.

a your height
b the length of a book
c the capacity of a can of drink
d the distance from home to school
e the time to run 100 m
f the amount of water in your swimming pool
g the amount of water in a glass
h the time to drive from Townsville to Brisbane

4 Copy and complete:

a 2000 mL = ___ L **b** 4.9 km = ___ m **c** 72 kL = ___ L
d 1440 kg = ___ t **e** 125 g = ___ mg **f** 8500 g = ___ kg
g 86.4 m = ___ km **h** 7250 mL = ___ L **i** 125 cm = ___ m
j 45 000 mm = ___ m **k** 820 g = ___ kg **l** 4.6 t = ___ kg

5 Is each statement true or false?

a 6.5 km = 650 m **b** 3600 mm = 36 m **c** 2288 mg = 2.288 g
d 7.8 kg = 780 g **e** 82 L = 8.2 kL **f** 950 mm = 95 cm

6 Convert each length to metres and then find its sum.
22.8 m, 560 cm, 3400 mm, 0.6 km, 78.25 m

7 Write these capacities in ascending order.
6.2 L, 4500 mL, 0.98 kL, 16.4 L, 2350 mL

8 Copy and complete the following table.

Millimetres	Centimetres	Metres	Kilometres
3500			
	640		
		28	
			6.5
	5200		
420 000			

9–02 Perimeter

WORDBANK

perimeter The distance around the edges of a shape.

To find the **perimeter** of a shape, add the lengths of its sides.
For a rectangle,
$P = 2 \times \text{length} + 2 \times \text{width}$
$P = 2l + 2w$

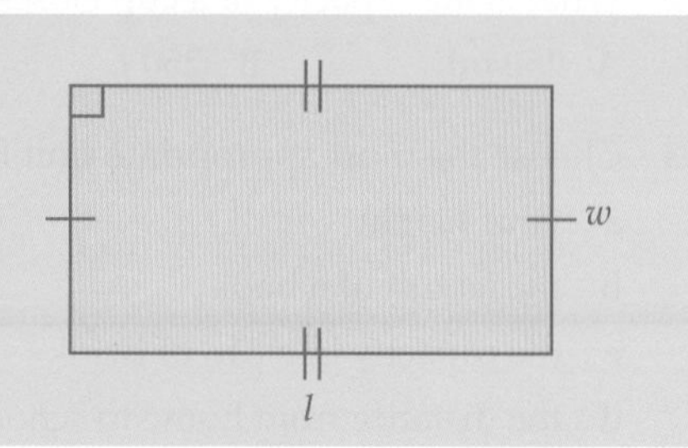

EXAMPLE 2

Find the perimeter of each shape.

a

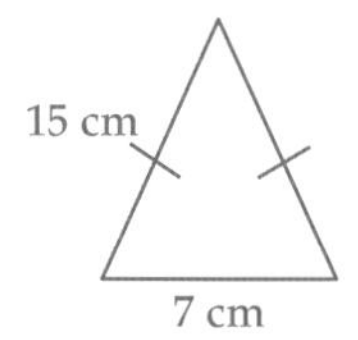

b

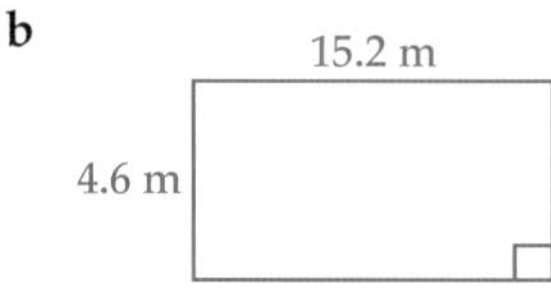

c

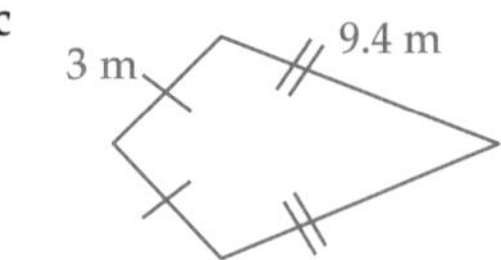

d

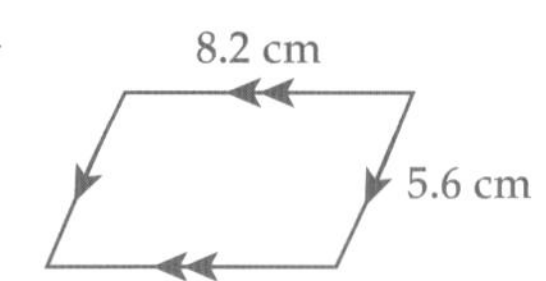

SOLUTION

a Perimeter = 15 + 15 + 7
= 37 cm

✱ Isosceles triangle, so two sides equal.

b Perimeter = 2 × 15.2 + 2 × 4.6
$P = 2l + 2w$
= 39.6 m

c Perimeter = 3 + 3 + 9.4 + 9.4
= 24.8 m

 Kite, so adjacent sides equal.

d Perimeter = 2 × 8.2 + 2 × 5.6
= 27.6 cm

 Parallelogram, so opposite sides equal.

EXAMPLE 3

Find the perimeter of this shape.

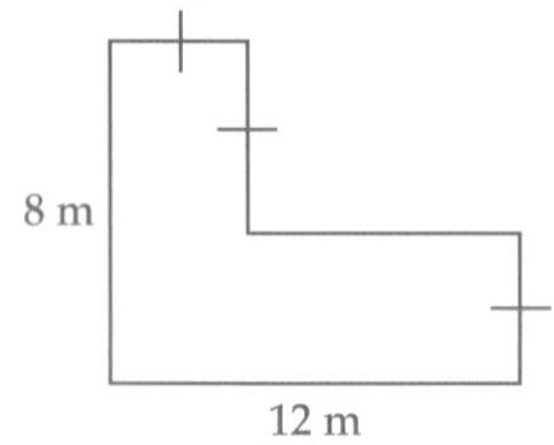

ISBN 9780170350990

SOLUTION

The two vertical sides with the dash are $\frac{1}{2} \times 8 = 4$ m.

The top horizontal side with the dash is also 4 m.

The horizontal side without a dash is $12 - 4 = 8$ m.

$$\begin{aligned}\text{Perimeter} &= 8 + 4 + 4 + 8 + 4 + 12 \\ &= 40 \text{ m}\end{aligned}$$

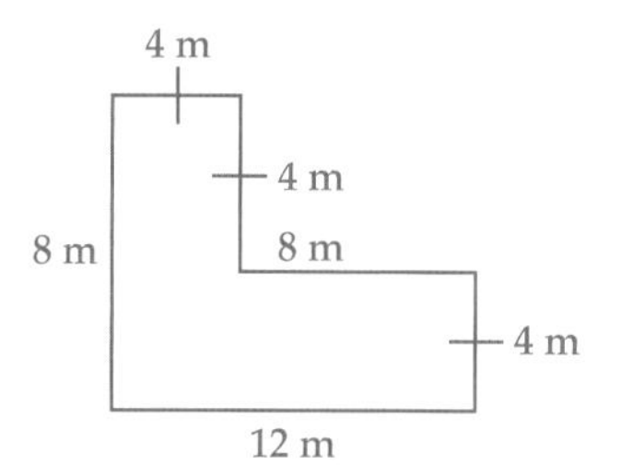

Can you see why the perimeter of this L-shape can also be found using $2 \times 8 + 2 \times 12 = 40$ m?

EXERCISE 9-02

1 Find the perimeter of a rectangle with length 8 mm and width 5 mm. Select the correct answer **A**, **B**, **C** or **D**.

A 21 mm **B** 13 mm **C** 26 mm **D** 18 mm

2 Find the perimeter of a rhombus of side length 6 cm. Select **A**, **B**, **C** or **D**.

A 12 cm **B** 30 cm **C** 18 cm **D** 24 cm

3 Find the perimeter of each shape.

a

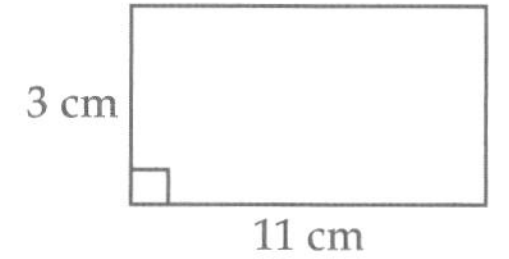

b

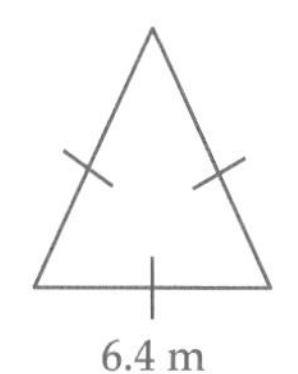

c

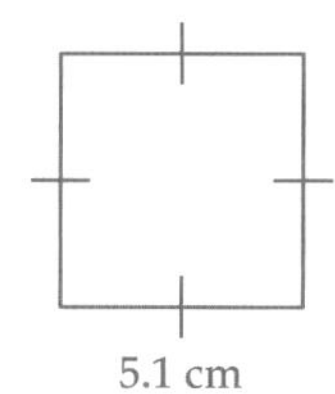

d

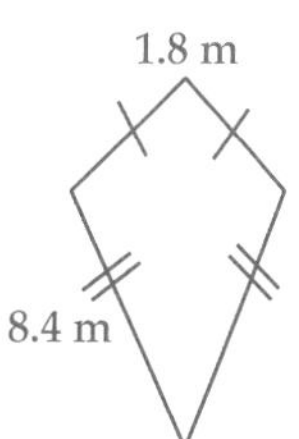

e

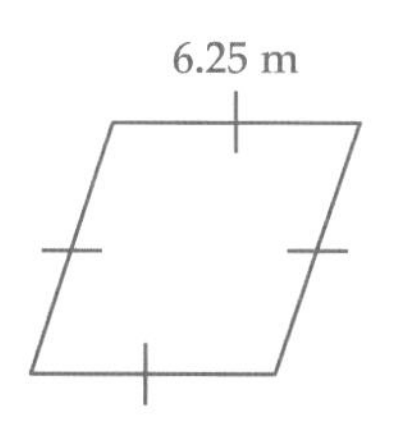

f

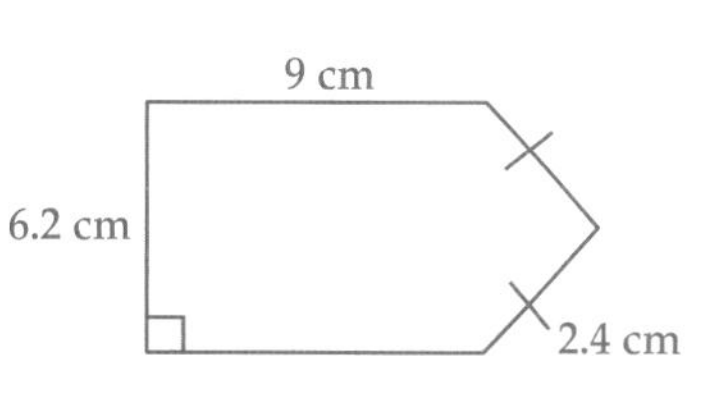

g

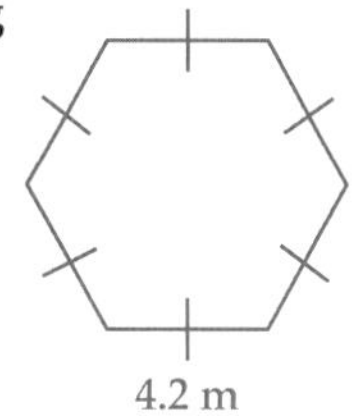

h

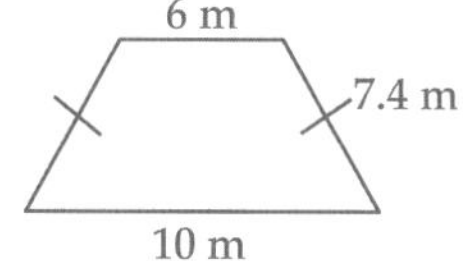

i

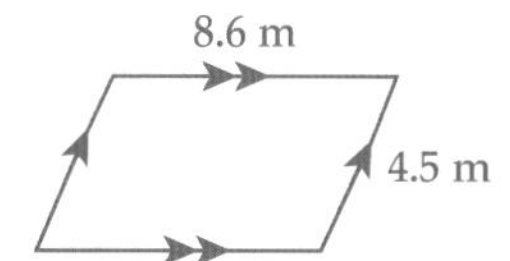

EXERCISE 9-02

4 Find the perimeter of each shape described below.

a a rectangle with length 12.4 m and width 8.7 m

b a square with side lengths 5.75 cm

c a parallelogram with its longer sides 8.6 m and its shorter sides 6.3 m

d a kite with two adjacent sides 46 mm each and the other two sides 16 mm each

e an equilateral triangle with side lengths 6 cm

f an isosceles triangle with equal sides of 14 m each and the other side 8 m

5 If the perimeter of a rectangle is 84 m, what could its length and width be? Give two possible answers.

6 Find the perimeter of each shape.

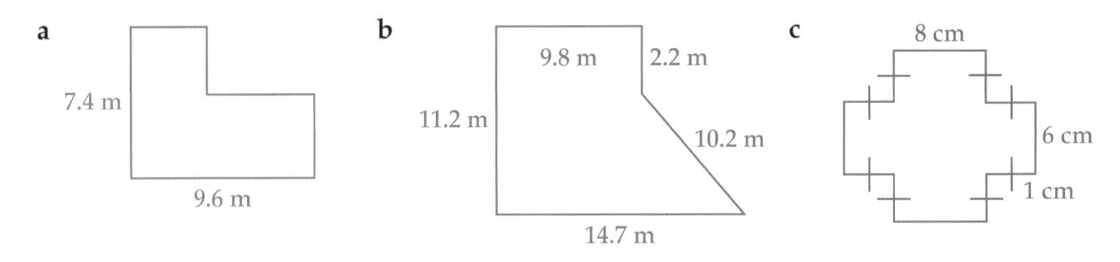

Shutterstock.com/T photography

ISBN 9780170350990

9-03 Parts of a circle

A circle is a completely round shape. Every point on a circle is the same distance from its centre, marked O in the table below. The parts of a circle have special names.

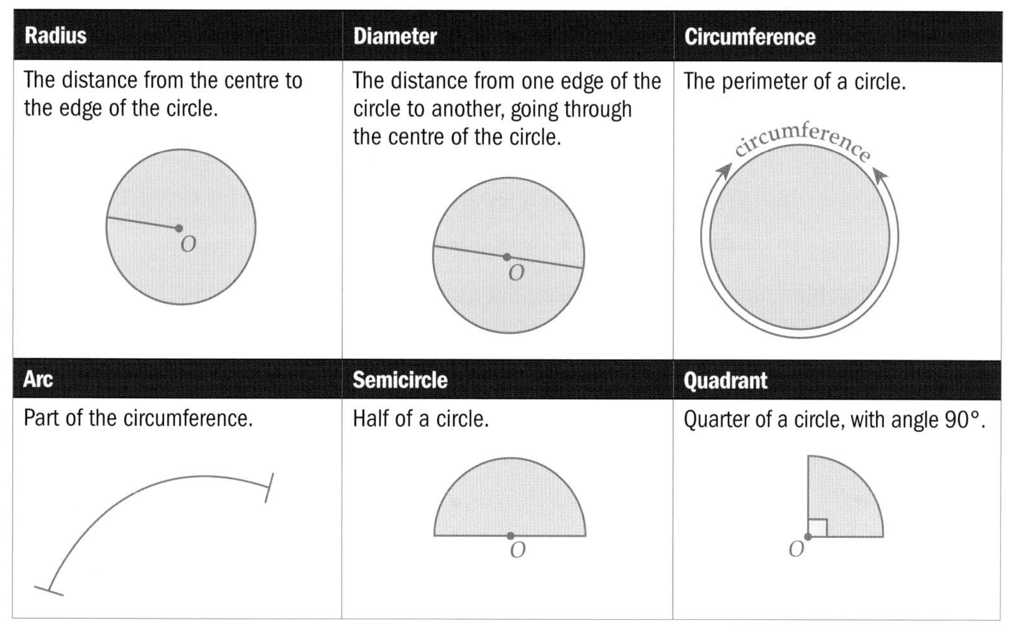

Radius	Diameter	Circumference
The distance from the centre to the edge of the circle.	The distance from one edge of the circle to another, going through the centre of the circle.	The perimeter of a circle.
O	O	circumference
Arc	**Semicircle**	**Quadrant**
Part of the circumference.	Half of a circle.	Quarter of a circle, with angle 90°.
	O	O

Also, a **chord** is an interval joining any two points on the edge of a circle.

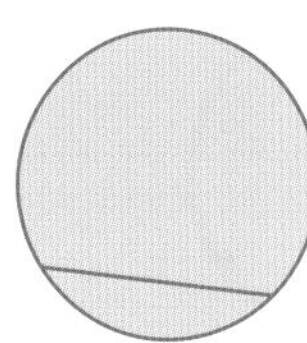

EXAMPLE 4

Draw a circle and mark on it:

a a radius **b** a chord **c** a diameter.

SOLUTION

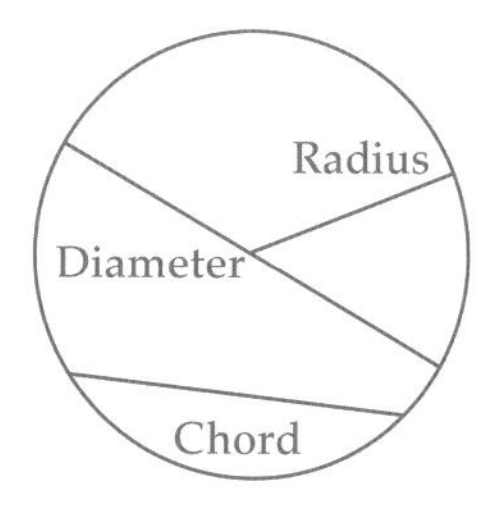

EXERCISE 9-03

1 What is the perimeter of a circle called? Select the correct answer **A, B, C** or **D**.

A radius **B** diameter **C** arc **D** circumference

2 What is part of the edge of a circle called? Select **A, B, C** or **D**.

A radius **B** diameter **C** arc **D** chord

3 Use a ruler and compasses to construct a circle with a radius of 3 cm, then mark and label on the circle:

a the centre **b** a radius **c** a chord

d an arc **e** a diameter **f** a quadrant.

4 **a** Measure the diameter of the circle drawn in Question **3**.

b What is the relationship between the radius and the diameter of the circle?

5 Name each of the following highlighted parts of a circle.

a

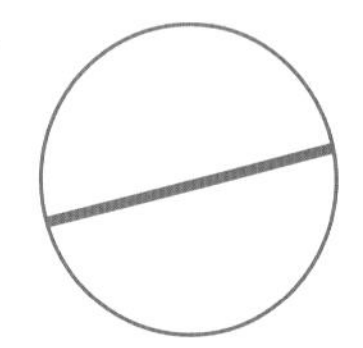

b

c

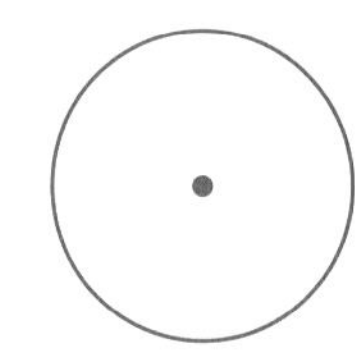

d

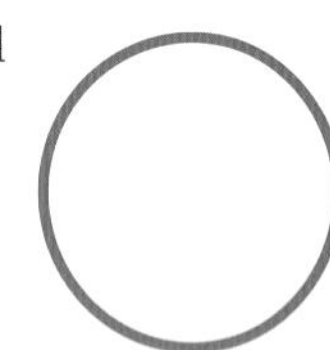

e

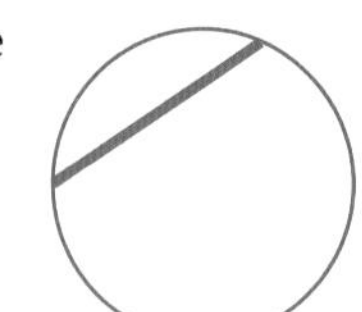

f

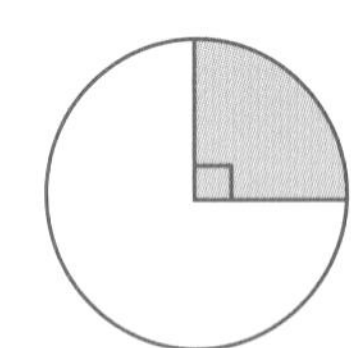

6 How many axes of symmetry has a circle? Select **A, B, C** or **D**.

A 1 **B** 2 **C** 4 **D** an infinite number

7 What is half of a circle called?

8 Use compasses, a ruler and pencil to copy two or more of these circular designs.

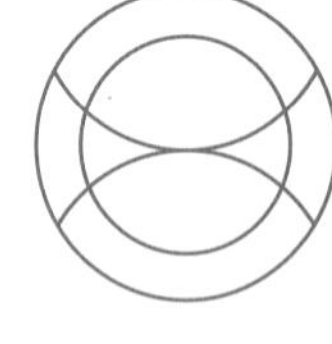

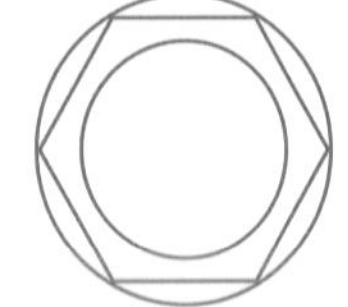
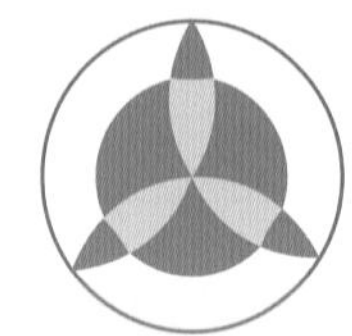

 ISBN 9780170350990

9-04 Circumference of a circle

WORDBANK

pi (π) The special number 3.14159... represented by the Greek letter π, which as a decimal has digits that run endlessly, without repeating or stopping.

The **circumference** of a circle is related to the **diameter** of the circle by a simple formula involving a special number called **pi** (pronounced 'pie'), represented by the Greek letter π.

$$C = \pi \times \text{diameter}$$

where π = 3.14159 …, which can be found on your calculator by pressing the [π] key (you may need to press [SHIFT] first).

Use string, a ruler and a tape measure to measure and record the diameter and circumference of some circular objects such as cans, round cake tins, pipes, coins and bottles. Divide the circumference by the diameter to show that for any circle:

$$\frac{\text{circumference}}{\text{diameter}} = \pi = 3.14\ldots$$

CIRCUMFERENCE OF A CIRCLE

$C = \pi \times \text{diameter}$

$C = \pi d$

Because the diameter of a circle is double its radius, another formula for the circumference of a circle is:

$$C = \pi \times 2 \times \text{radius} = 2\pi r$$

CIRCUMFERENCE OF A CIRCLE

$C = 2 \times \pi \times \text{radius}$

$C = 2\pi r$

EXAMPLE 5

Find correct to two decimal places the circumference of each circle.

a

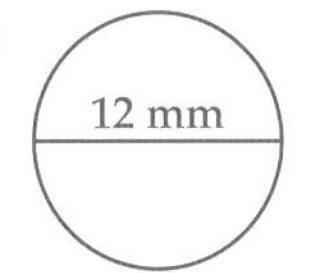

b

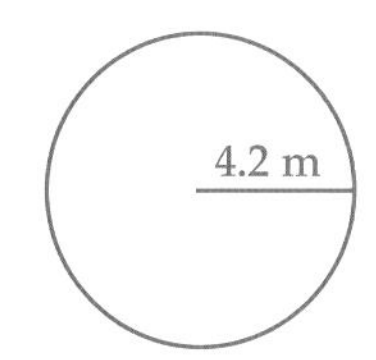

SOLUTION

a $C = \pi d$ ← $d = 12$

$= \pi \times 12$ ← [π] [×] 12 [=] on calculator

$= 37.6991\ldots$

≈ 37.70 mm

b $C = 2\pi r$ ← $r = 4.2$

$= 2 \times \pi \times 4.2$ ← 2 [×] [π] [×] 4.2

$= 26.3893\ldots$

≈ 26.39 m

9–04 Circumference of a circle

EXAMPLE 6

Find correct to one decimal place the perimeter of each shape.

a

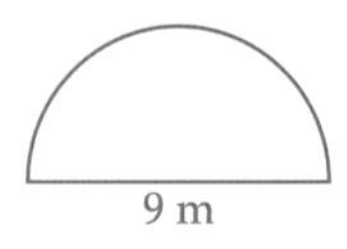

b

3.6 cm

SOLUTION

a Perimeter $= \frac{1}{2} \times \text{circumference} + 9$

$= \frac{1}{2} \times \pi \times 9 + 9$

$= 23.1371...$

≈ 23.1 m

b Perimeter $= \frac{1}{4} \times \text{circumference} + 3.6 + 3.6$

$= \frac{1}{4} \times 2 \times \pi \times 3.6 + 3.6 + 3.6$

$= 12.8548...$

≈ 12.9 cm

✱ quadrant

EXERCISE 9–04

1 Find the circumference of a circle with diameter 5 m. Select the correct answer **A**, **B**, **C** or **D**.

A 15.71 m **B** 78.54 m **C** 31.42 m **D** 7.85 m

2 Find the circumference of a circle with radius 4 cm. Select **A**, **B**, **C** or **D**.

A 50.27 cm **B** 25.13 cm **C** 50.27 cm **D** 6.28 cm

3 **a** Use a ruler and compasses to draw a circle with radius 4.5 cm.

b Measure the diameter of the circle.

c How are the diameter and radius of the circle related to each other?

d Calculate correct to two decimal places the circumference of the circle.

4 Find the circumference of each circle correct to two decimal places.

a

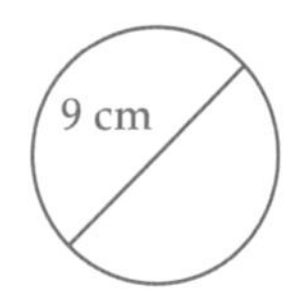

b

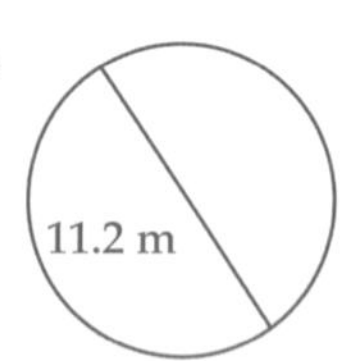

c

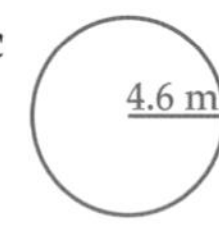

d

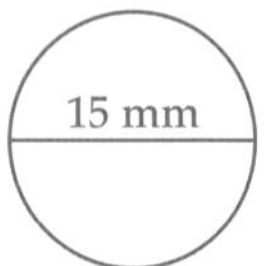

e

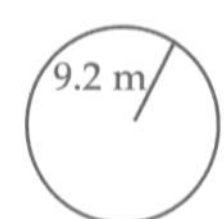

f

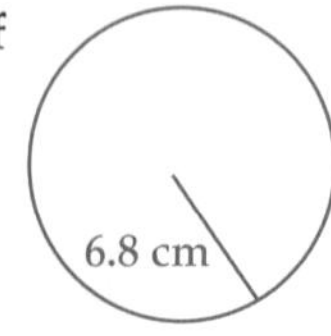

ISBN 9780170350990

EXERCISE 9-04

5 a Describe in words how you could calculate the perimeter of the shape below:

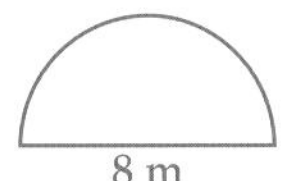

b Calculate the perimeter correct to one decimal place.

6 Find correct to one decimal place the perimeter of each shape.

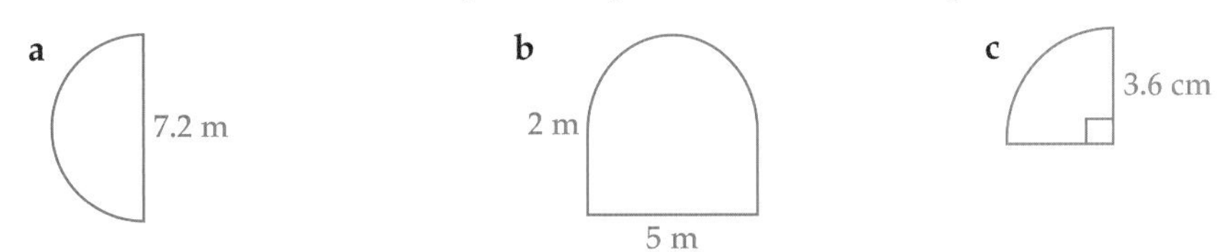

iStockphoto/Claudiad

9-05 24-hour time

WORDBANK

24-hour time Uses four digits to describe the time of day and does not require a.m. or p.m. Instead, the hours of a day are numbered from 0 to 23.

12-hour time	24-hour time	12-hour time	24-hour time
1:00 a.m.	0100	1:00 p.m.	1300
2:00 a.m.	0200	2:00 p.m.	1400
3:00 a.m.	0300	3:00 p.m.	1500
4:00 a.m.	0400	4:00 p.m.	1600
5:00 a.m.	0500	5:00 p.m.	1700
6:00 a.m.	0600	6:00 p.m.	1800
7:00 a.m.	0700	7:00 p.m.	1900
8:00 a.m.	0800	8:00 p.m.	2000
9:00 a.m.	0900	9:00 p.m.	2100
10:00 a.m.	1000	10:00 p.m.	2200
11:00 a.m.	1100	11:00 p.m.	2300
12:00 midday	1200	12:00 midnight	0000

EXAMPLE 7

Write each time in 24-hour time.

a 5:15 a.m. **b** 4:30 p.m. **c** 12:10 a.m. **d** 11:46 p.m.

SOLUTION

a 5:15 a.m. = 0515 ← After 1 a.m., insert a 0 in front to make four digits.
b 4:30 p.m. = 1630 ← After 1 p.m., add 12 to the hour: 4 + 12 = 16.
c 12:10 a.m. = 0010 ← 12 midnight is 00 for the first two digits.
d 11:46 p.m. = 2346 ← After 1 p.m., add 12 to the hour: 11 + 12 = 23.

EXAMPLE 8

Write each time in 12-hour time.

a 1250 **b** 1915 **c** 1047 **d** 0030

SOLUTION

a 1250 = 12:50 p.m. ← 12 is 12 p.m. (midday).
b 1915 = 7:15 p.m. ← After 1300, it is p.m., so subtract 12 from the hour: 19 – 12 = 7.
c 1047 = 10:47 a.m. ← Before 1200, it is a.m. time.
d 0030 = 12:30 a.m. ← 00 is 12 a.m. (midnight).

ISBN 9780170350990

EXERCISE 9–05

1 Write 1245 in 12-hour time. Select the correct answer **A**, **B**, **C** or **D**.

A 2:45 a.m. **B** 12:45 a.m. **C** 2:45 p.m. **D** 12:45 p.m.

2 What is 5:20 p.m. in 24-hour time? Select **A**, **B**, **C** or **D**.

A 0520 **B** 5020 **C** 1520 **D** 1720

3 **a** The Indian Pacific train arrives in Kalgoorlie at 2230. Does it arrive in the afternoon or at night?

b The same train arrives in Adelaide at 0720. What is this in 12-hour time?

c A military operation will begin at 0940 hours. Write this in 12-hour time.

d A bus travels from Batemans Bay to Sydney in 5 hours 45 minutes. If it leaves Batemans Bay at 0620, what time will it arrive in Sydney? Write the answer in 12-hour time and in 24-hour time.

4 Write each time in 24-hour time.

a 3:00 a.m. **b** 6:00 p.m. **c** 4:20 a.m. **d** 5:25 p.m.

e 7:42 a.m. **f** 10:48 p.m. **g** 12 noon **h** 8:54 p.m.

i 4:00 a.m. **j** 2:15 p.m. **k** 9:30 p.m. **l** 8:50 a.m.

m twenty past 3 in the morning **n** half past 10 at night

o quarter to 6 in the morning **p** quarter past 8 at night

5 Write each time in 12-hour time.

a 0520 **b** 1440 **c** 2315 **d** 0605

e 2256 **f** 0338 **g** 1650 **h** 1146

i 0221 **j** 0045 **k** 1212 **l** 0148

6 Joel is catching a flight at 1350 from Sydney to Singapore.

a He has to be at Sydney airport 1 hour 30 minutes before the flight departs. What is this time in 24-hour time?

b If it takes Joel 40 minutes to travel from home to the airport, what time should he leave home? Answer in 12-hour time.

c If the flight to Singapore takes 8 hours 15 minutes, what time should Joel arrive in Singapore (Sydney time)? Answer in 12-hour time and in 24-hour time.

9-06 Time calculations

To round time to the nearest minute:
- look at the number of seconds
- if it is 30 seconds or more, round up
- if it is less than 30 seconds, round down.

To round time to the nearest hour:
- look at the number of minutes
- if it is 30 minutes or more, round up
- if it is less than 30 minutes, round down.

EXAMPLE 9

Round:

a 3 h 38 min to the nearest hour **b** 22 min 27 s to the nearest minute.

SOLUTION

a 3 h 38 min ≈ 4 h ← round up

38 min is more than 30 min (half an hour)

b 22 min 27 s ≈ 22 min ← round down

27 s is less than 30 s (half a minute)

EXAMPLE 10

Calculate the time difference from 3:45 a.m. to 7:25 p.m.

SOLUTION

Use a number line and 'build bridges' like we did in Chapter 1 for mental subtraction.

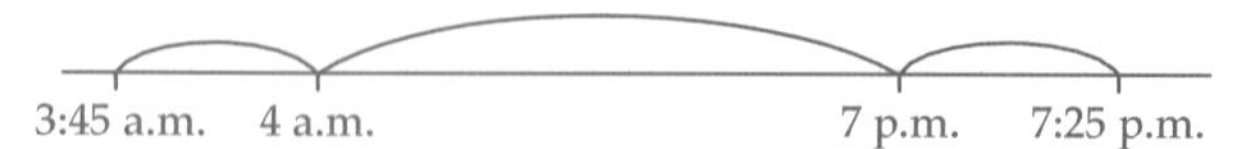

Time difference = 15 min + 15 h + 25 min
= 15 h 40 min

So 15 hours 40 minutes is the time difference.

OR: Convert to 24-hour time and use the calculator's [° ' "] or [DMS] keys:

3:45 a.m. = 0345, 7:25 p.m. = 1925, so enter 19 [° ' "] 25 [° ' "] [−] 3 [° ' "] 45 [° ' "] [=]

So 15 hours 40 minutes is the time difference.

EXAMPLE 11

Convert 215 minutes to hours and minutes.

SOLUTION

215 minutes = (215 ÷ 60) h ← 1 h = 60 min
= 3.583333… h
= 3 h 35 min

Enter [° ' "] or [2ndF] [DMS] on calculator or calculate 0.583333... × 60 for minutes

ISBN 9780170350990

9-06 Time calculations

EXAMPLE 12

How many days is it from 12 May until 23 July?

SOLUTION

12 May to 31 May: 31 – 12 = 19 days ← May has 31 days.
All of June: 30 days ← June has 30 days.
1 July to 23 July: 23 days
Number of days = 19 + 30 + 23
= 72

EXERCISE 9-06

1 Round 15 minutes 32 seconds to the nearest minute. Select the correct answer **A**, **B**, **C** or **D**.

A 14 min **B** 51 min **C** 15 min **D** 16 min

2 What is the time difference from 3:30 p.m. to 8:15 p.m.? Select **A**, **B**, **C** or **D**.

A 5 h 45 min **B** 5h 15 min **C** 4 h 45 min **D** 4 h 15 min

3 Write each time correct to the nearest hour.

a 5 h 23 min **b** 13 h 48 min **c** 11 h 12 min
d 6 h 33 min **e** 14 h 30.4 min **f** 23 h 29.8 min

4 Round each time to the nearest minute.

a 23 min 12 s **b** 48 min 36 s **c** 52 min 48 s

5 How many days is it from:

a 18 January to 23 April **b** 4 May to 22 May the next year?

6 Calculate each time difference.

a 4:25 a.m. to 8:50 a.m. **b** 6:25 p.m. to 11:55 p.m. **c** 2:20 a.m. to 5:40 p.m.
d 11:25 a.m. to 7:10 p.m. **e** 6:40 a.m. to 9:20 p.m. **f** 10:45 p.m. to 11:20 p.m.
g 12:09 p.m. to 3:55 p.m. **h** 8:54 a.m. to 9:23 p.m. **i** 1340 to 2255
j 0428 to 1652 **k** 1245 to 2010 **l** 0345 to 1500

7 Convert each time to hours and minutes.

a 200 min **b** 450 min **c** 325 min

8 Convert each time to minutes and seconds.

a 98 s **b** 500 s **c** 154 s

9 A movie runs for 112 minutes.

a What is this time in hours and minutes?
b If the movie starts at 3:48 p.m., at what time will it finish?
c If the movie finished at 9:56 p.m., at what time did it start?

10 Find the sum of 3 h 40 min, 5 h 12 min and $8\frac{1}{2}$ hours.

9-07 Timetables

EXAMPLE 13

This is a section of a train timetable from Strathfield to Town Hall.

Station	a.m.	a.m.	a.m.	a.m.
Strathfield	8:45	8:58	9:12	9:25
Ashfield	8:58		9:18	
Redfern	9:08	9:12	9:28	9:39
Central	9:15	9:18	9:35	
Town Hall	9:20		9:40	9:49

a How long does it take the 8:45 a.m. train from Strathfield to arrive at Redfern?

b If Josie needs to be at Central at 9:20 a.m., which train should she catch from Strathfield?

c Why are there some blank cells in the timetable?

d Which train is the fastest to go from Redfern to Town Hall?

e If Paul missed the 8:45 a.m. train from Strathfield, what is the earliest he can get to Town Hall?

SOLUTION

a Time difference from 8:45 a.m. to 9:08 a.m. = 15 + 8 = 23 min.

b Latest train to arrive at Central before 9:20 is the one that arrives at 9:18 a.m.
So, Josie should catch the 8:58 a.m. train from Strathfield.

c The blank cells mean that the train does not stop at those stations.

d The fastest is the 9:39 a.m. train from Redfern, which takes 10 minutes. The others take 12 minutes.

e The earliest time is 9:40 a.m.

Shutterstock.com/PomInOz

ISBN 9780170350990

EXERCISE 9-07

Use the train timetable from Example 13 to answer Questions 1 to 8.

1 Which train arrives at Town Hall before 9:30 a.m.? Select the correct answer **A**, **B**, **C** or **D**.

A Strathfield 8:58 **B** Ashfield 9:18 **C** Redfern 9:08 **D** Central 9:18

2 How long does the 9:18 train take to go from Ashfield to Central? Select **A**, **B**, **C** or **D**.

A 16 min **B** 15 min **C** 18 min **D** 17 min

3 How long is each of the following train trips?

a 8:45 from Strathfield to Redfern

b 9:39 from Redfern to Town Hall

c 9:12 from Strathfield to Central

d 8:58 from Ashfield to Town Hall

4 If Tess catches the 8:58 train from Strathfield, what time should she arrive at Central?

5 How long does it take an 'all-stations train' to go from Central to Town Hall?

6 Which train should Khalid catch if he needs to be at Redfern at 9:15?

7 What is the fastest time for a train to travel from Strathfield to Town Hall? Could I catch this train if I had to be at Town Hall by 9:45?

8 Liam catches the 8:58 train from Strathfield to go to Town Hall, but it doesn't stop there. What should he do?

9 A section of a city bus timetable is shown.

Market St	7:45	8:05	8:48
George St	7:56		9:01
York St	8:10	8:20	9:15
Sussex St	8:22		9:27

a How long does the first bus take to travel from George St to Sussex St?

b Do the other two buses take the same time?

c If Gina wants to be at York St by 8:30, which bus should she catch from Market St?

d If Jacob missed the 7:56 bus from George St, what is the earliest time he could arrive at Sussex St?

e How long does the 8:48 bus take to travel from Market St to George St?

f Why could this be different from the time the 7:45 bus takes?

10 Find a train or bus timetable on the Internet and write five questions about it.

9-08 International time zones

The world is divided into 24 different time zones, each one representing a 1 hour time difference. World times are measured in relation to the Greenwich Observatory in London, either ahead or behind UTC (Coordinated Universal Time), also known as GMT (Greenwich Mean Time).

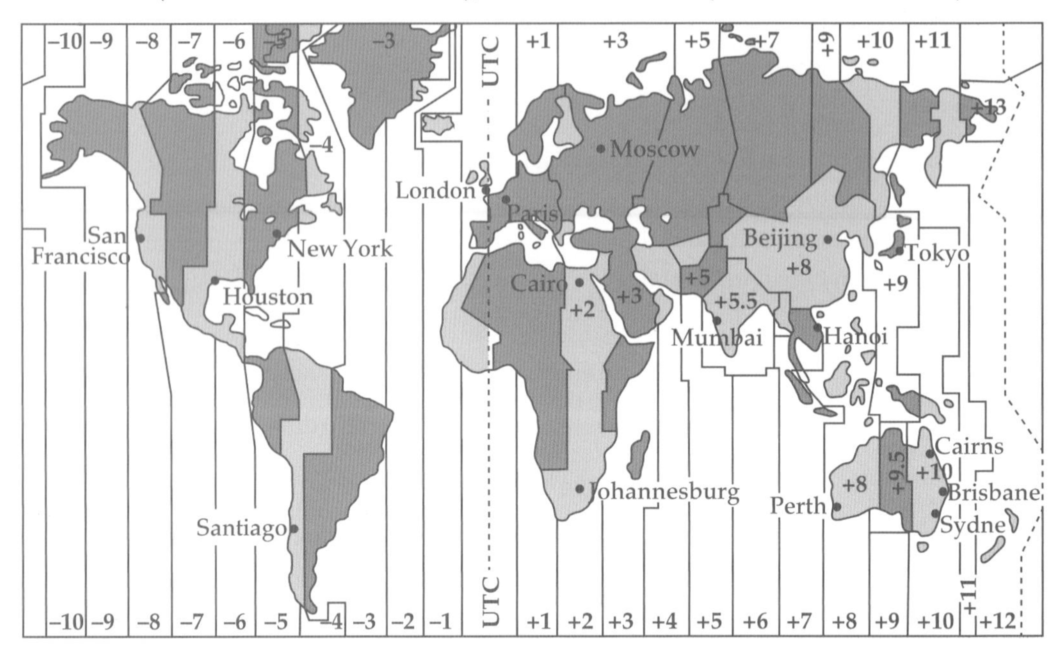

EXAMPLE 14

If it is 6 a.m. in London, what is the time in:

a Melbourne **b** Moscow **c** San Francisco?

SOLUTION

a Melbourne is 10 hours ahead of UTC, so its time is 6 a.m. + 10 h = 4 p.m.

b Moscow is 3 hours ahead of UTC, so its time is 6 a.m. + 3 h = 9 a.m.

c San Francisco is 8 hours behind UTC, so its time is 6 a.m. – 8 h = 10 p.m. the previous day.

Shutterstock.com/Reidl

 ISBN 9780170350990

9-08 International time zones

Australian time zones

Australia has three time zones: Western (UTC+8), Central (UTC+9.5) and Eastern (UTC+10)

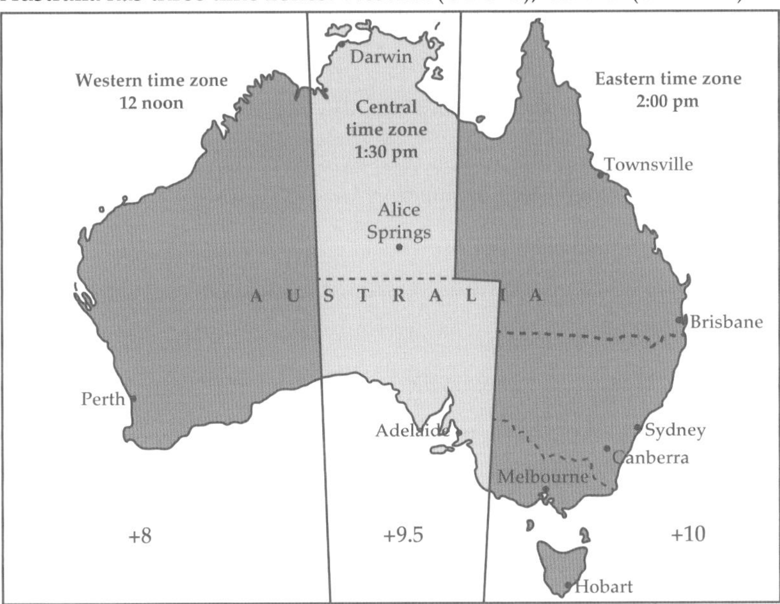

EXAMPLE 15

If it is 9:30 a.m. in Alice Springs, what time is it in:

a Perth **b** Sydney **c** Adelaide?

SOLUTION

a Perth is 1.5 hours behind the time in Alice Springs, so its time is 9:30 a.m. – 1.5 h = 8 a.m.

b Sydney is half an hour ahead of the time in Alice Springs, so its time is 9:30 a.m. + 0.5 h = 10 a.m.

c Adelaide is in the same time zone as Alice Springs, so its time is 9:30 a.m.

EXERCISE 9-08

1 If it is 7:30 a.m. in London, what time will it be in Johannesburg, South Africa? Select the correct answer **A**, **B**, **C** or **D**.

A 8:30 a.m. **B** 9:30 a.m. **C** 5:30 a.m. **D** 6:30 a.m.

2 If it is 6:30 p.m. in Mumbai, India, what time will it be in London? Select **A**, **B**, **C** or **D**.

A 1:30 p.m. **B** 12:30 p.m. **C** 12 midday **D** 1 p.m.

3 Given that it is 8 a.m. in London, find the time in each city.

a Moscow, Russia **b** Santiago, Chile **c** Mumbai, India

d Tokyo, Japan **e** Houston, USA **f** Melbourne, Australia

EXERCISE 9–08

4 Given that it is 1 p.m. in Johannesburg, find the time in each city.

a London **b** Paris **c** New York
d Perth **e** Beijing **f** Brisbane

5 Melissa caught a direct flight from Sydney to London. She left Sydney at 9:20 a.m. on Thursday and the flight was 28 hours long. What time and day did she arrive in London?

6 Mei Lin lives in Canberra and calls her grandmother who lives in Beijing, China, at 4.30 p.m. What time is it in Beijing when Mei Lin rings?

7 If it is 8 a.m. in Darwin, find the time in each city.

a Brisbane **b** Hobart **c** Perth **d** Adelaide

8 If it is 7 p.m. in Perth, find the time in each city.

a Canberra **b** Alice Springs **c** Townsville **d** Melbourne

9 Ania lives in Adelaide and rings her father who is in Sydney on business. If he answers the phone at 3 p.m in Sydney, what time is it in Adelaide?

10 Peter is keen to watch the Wimbledon final live on TV at his home in Brisbane. The match is played in London at 6 p.m. and lasts 3 hours 20 minutes. Between what times does Peter watch the match in Brisbane?

Shutterstock.com/Alison Henley

ISBN 9780170350990

EXERCISE 9–08

11 Most Australian states change to daylight saving time from October to March each year to take advantage of the longer hours of daylight. Clocks are turned forward one hour during this period; for example, from 2 a.m. to 3 a.m., ahead of standard time. During February, what time is it in:

a Victoria when it is 9 a.m. in Western Australia, where daylight saving does not operate

b Queensland, where daylight saving does not operate, when it is 5 p.m. in New South Wales and the ACT?

12 Georgia left Sydney at 9:30 p.m. daylight saving time to travel to Brisbane on a flight that took $1\frac{1}{2}$ hours. What was the local time when she arrived?

Shutterstock.com/wang song

LANGUAGE ACTIVITY

CODE PUZZLE

Use this table to decode the words used in this chapter.

1	2	3	4	5	6	7	8	9	10	11	12	13
A	B	C	D	E	F	G	H	I	J	K	L	M

14	15	16	17	18	19	20	21	22	23	24	25	26
N	O	P	Q	R	S	T	U	V	W	X	Y	Z

1 12 – 5 – 14 – 7 – 20 – 8
2 13 – 5 – 20 – 18 – 9 – 3
3 13 – 1 – 19 – 19
4 20 – 9 – 13 – 5
5 16 – 5 – 18 – 9 – 13 – 5 – 20 – 5 – 18
6 3 – 9 – 18 – 3 – 12 – 5
7 18 – 5 – 3 – 20 – 1 – 14 – 7 – 12 – 5
8 11 – 9 – 20 – 5
9 20 – 18 – 9 – 1 – 14 – 7 – 12 – 5
10 16 – 1 – 18 – 1 – 12 – 12 – 5 – 12 – 15 – 7 – 18 – 1 – 13
11 3 – 9 – 18 – 3 – 21 – 13 – 6 – 5 – 18 – 5 – 14 – 3 – 5
12 17 – 21 – 1 – 4 – 18 – 1 – 14 – 20
13 18 – 1 – 4 – 9 – 21 – 19
14 4 – 9 – 1 – 13 – 5 – 20 – 5 – 18
15 20 – 9 – 13 – 5 – 20 – 1 – 2 – 12 – 5
16 20 – 9 – 13 – 5 – 26 – 15 – 14 – 5

ISBN 9780170350990

PRACTICE TEST 9

Part A General topics

Calculators are not allowed.

1 Find 25% of $160.

2 If $y = -4$, evaluate $6 - 7y$.

3 Find the average of –4, 8, 12, 8 and 6.

4 Copy this diagram and mark two supplementary angles.

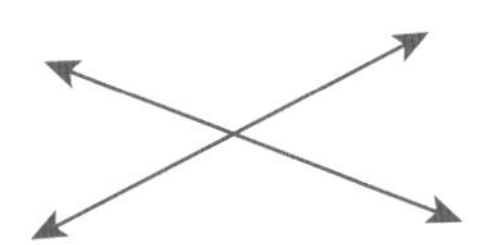

5 Evaluate $\frac{3}{5} \times \frac{15}{12}$.

6 Name the quadrilateral with one pair of parallel sides.

7 Evaluate 8^3.

8 How many axes of symmetry has an isosceles triangle?

9 Round 126.4829 to two decimal places.

10 What is the probability of rolling a prime number on a die?

Part B Length and time

Calculators are allowed.

9-01 The metric system

11 What unit is used to measure the capacity of a cup? Select the correct answer **A**, **B**, **C** or **D**.

A cm **B** cm^3 **C** mL **D** L

12 What is the capacity of a bucket of water closest to? Select **A**, **B**, **C** or **D**.

A 20 L **B** 250 mL **C** 8 L **D** 180 mL

9-02 Perimeter

13 Find the perimeter of this shape. Select **A**, **B**, **C** or **D**.

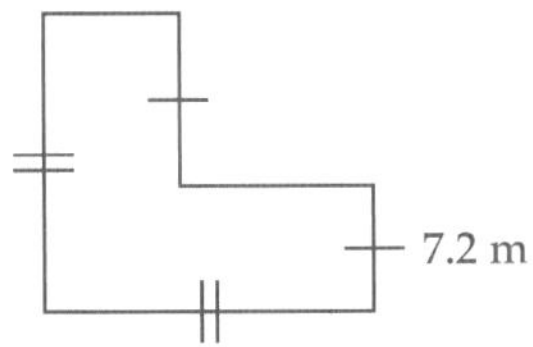

A 14.4 m **B** 28.8 m **C** 21.6 m **D** 57.6 m

9-03 Parts of a circle

14 Describe in words a diameter.

15 Name each part of a circle shown.

a

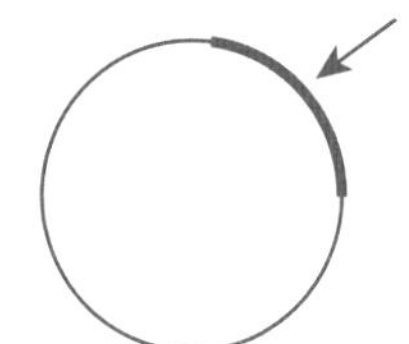

b

ISBN 9780170350990

9-04 Circumference of a circle

16 Find correct to two decimal places the circumference of a circle with:

a diameter 8 cm

b radius 2.6 m.

17 Find correct to one decimal place the perimeter of this quadrant.

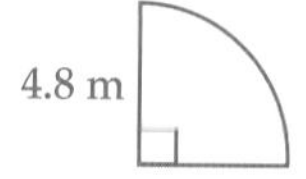

9-05 24-hour time

18 a Write 6:55 p.m. in 24-hour time.

b Write 1542 in 12-hour time.

9-06 Time calculations

19 a If a movie ended at 4:14 p.m. and was 2 h 20 min long, what time did it start?

b How many minutes are there from 1:25 a.m. to 2:08 a.m.?

9-07 Timetables

20 According to this train timetable, what time would I need to catch a train from Strathfield to get to Town Hall by 9:45 a.m.?

Station	a.m.	a.m.	a.m.	a.m.
Strathfield	8:45	8:58	9:12	9:25
Ashfield	8:58		9:18	
Redfern	9:08	9:12	9:28	9:39
Central	9:15	9:18	9:35	
Town Hall	9:20		9:40	9:49

9-08 International time zones

21 When it is 6 a.m. in London, what time is it in eastern Australia if Australian Eastern Standard Time is UTC+10?

ISBN 9780170350990

AREA AND VOLUME

10

WHAT'S IN CHAPTER 10?

IN THIS CHAPTER YOU WILL:

- convert between metric units for area
- find the area of a square, rectangle, triangle and parallelogram
- find the area of composite shapes
- find the area of a trapezium, kite and rhombus
- find the areas of circles and circular shapes
- convert between metric units for volume and capacity
- identify the cross-section of a prism and draw prisms and other solids from different views
- find the volume of a prism
- find the volume of a cylinder

10-01 Metric units for area

WORDBANK

area The amount of surface space inside a flat shape, measured in square units such as mm^2, cm^2, m^2 or km^2.

A **square millimetre** (mm^2) is the area of a square of length 1 mm, about the size of a grain of raw sugar or rock salt.

Actual size →■

A **square centimetre** (cm^2) is the area of a square of length 1 cm, about the size of a face of a die.

$1 \text{ cm} = 10 \text{ mm}$

$1 \text{ cm}^2 = 10 \text{ mm} \times 10 \text{ mm} = 100 \text{ mm}^2$

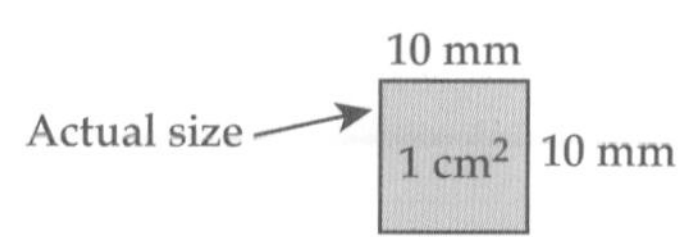

A **square metre** (m^2) is the area of a square of length 1 m, about the size of the base of a shower floor.

$1 \text{ m} = 100 \text{ cm} = 1000 \text{ mm}$

$1 \text{ m}^2 = 100 \text{ cm} \times 100 \text{ cm} = 10\,000 \text{ cm}^2$

$1 \text{ m}^2 = 1000 \text{ mm} \times 1000 \text{ mm} = 1\,000\,000 \text{ mm}^2$

100 cm

1 m²

100 cm

A **hectare** (ha) is the area of a square of length 100 m, about the size of two football fields.

$1 \text{ ha} = 100 \text{ m} \times 100 \text{ m} = 10\,000 \text{ m}^2$

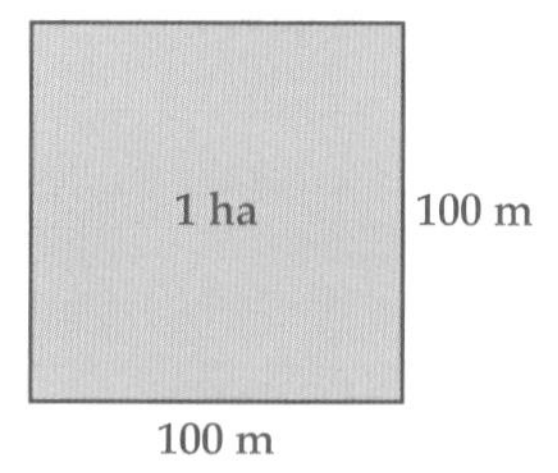

A **square kilometre** (km^2) is the area of a square of length 1 km, about the size of a theme park.

$1 \text{ km} = 1000 \text{ m}$

$1 \text{ km}^2 = 1000 \text{ m} \times 1000 \text{ m} = 1\,000\,000 \text{ m}^2$

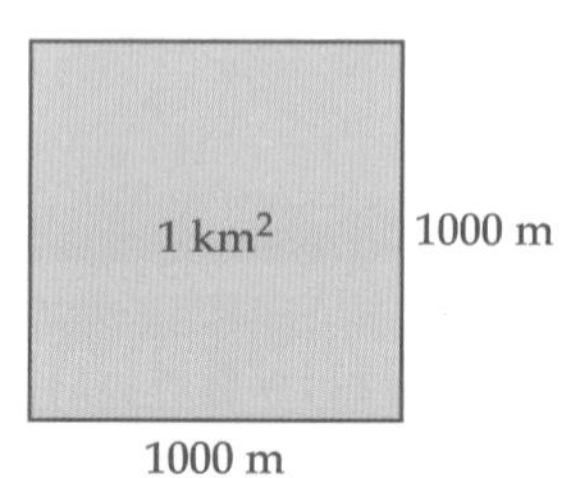

$1 \text{ cm}^2 = 100 \text{ mm}^2$

$1 \text{ m}^2 = 10\,000 \text{ cm}^2 = 1\,000\,000 \text{ mm}^2$

$1 \text{ ha} = 10\,000 \text{ m}^2$

$1 \text{ km}^2 = 1\,000\,000 \text{ m}^2$

EXAMPLE 1

Which unit of area would you use to measure the size of:

a a playground? **b** Canberra? **c** your desktop?

SOLUTION

a m^2 **b** km^2 **c** cm^2

ISBN 9780170350990

10-01 Metric units for area

EXAMPLE 2

Convert:

a $4 \text{ km}^2 =$ ____ m^2

b $0.8 \text{ m}^2 =$ ____ cm^2

c $220 \text{ m}^2 =$ ____ ha

SOLUTION

a $4 \text{ km}^2 = 4 \times 1\,000\,000 \text{ m}^2$ ← LOSM: Large Over to Small Multiply: $1 \text{ km}^2 = 1\,000\,000 \text{ m}^2$

$= 4\,000\,000 \text{ m}^2$

b $0.8 \text{ m}^2 = 0.8 \times 10\,000 \text{ cm}^2$ ← LOSM: Large Over to Small Multiply: $1 \text{ m}^2 = 10\,000 \text{ cm}^2$

$= 8000 \text{ cm}^2$

c $220 \text{ m}^2 = 220 \div 10\,000$ ha ← SOLD: Small Over to Large Divide: $1 \text{ ha} = 10\,000 \text{ m}^2$

$= 0.022$ ha

EXERCISE 10-01

1 Which measurement unit is used for area? Select the correct answer **A**, **B**, **C** or **D**.

A mL **B** cm^3 **C** cm^2 **D** cm

2 Which unit would you use to measure the area of an apartment? Select **A**, **B**, **C** or **D**.

A mm^2 **B** m^2 **C** cm^2 **D** ha

3 Write the unit that would be the most suitable for measuring the area of:

a your backyard
b a sheet of newspaper
c Melbourne
d a large farm
e your fingernail
f your bathroom

4 Convert:

a $250\,000 \text{ cm}^2 =$ ___ m^2
b $28 \text{ m}^2 =$ ___ cm^2
c $56\,000 \text{ m}^2 =$ ___ ha
d $75 \text{ ha} =$ ___ m^2
e $2.3 \text{ m}^2 =$ ___ mm^2
f $180\,000 \text{ mm}^2 =$ ___ m^2
g $9.6 \text{ km}^2 =$ ___ m^2
h $855 \text{ mm}^2 =$ ___ cm^2
i $34\,000\,000 \text{ m}^2 =$ ___ km^2

5 Convert:

a 1 cm to mm
b 1 cm^2 to mm^2
c 1 m to cm
d 1 m^2 to cm^2
e 1 m to mm
f 1 m^2 to mm^2
g 1 km to m
h 1 km^2 to m^2
i 50 mm to cm
j 50 mm^2 to cm^2
k 8 cm to m
l 8 cm^2 to m^2
m 120 000 mm to m
n $120\,000 \text{ mm}^2$ to m^2
o 28 km to m
p 28 km^2 to m^2
q 6500 m to cm
r 6500 m^2 to cm^2

6 Find the sum of the measurements below by converting them all to m^2 first.

28.6 m^2, 4.9 ha, 8.4 km^2, $54\,000 \text{ cm}^2$, 452 m^2

10–02 Areas of rectangles, triangles and parallelograms

Square

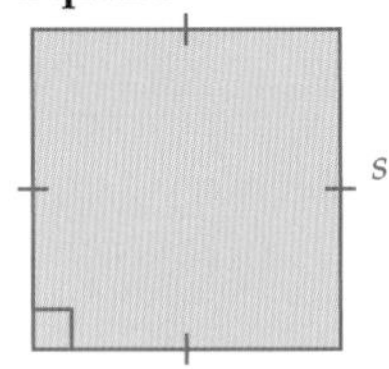

A = (side length)^2
$A = s^2$

Rectangle

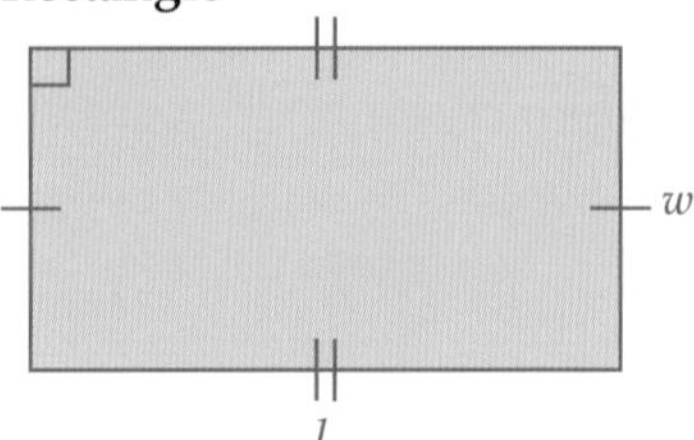

A = length × width
$A = lw$

Triangle

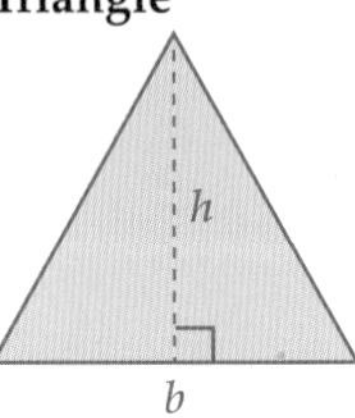

$A = \frac{1}{2}$ × base × height

$A = \frac{1}{2}bh$

Parallelogram

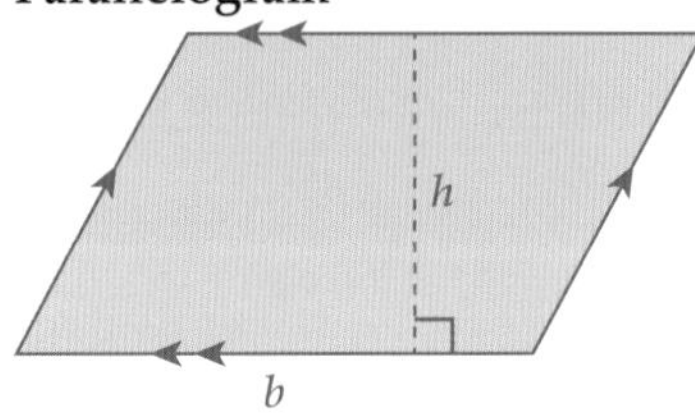

A = base × height
$A = bh$

EXAMPLE 3

Find the area of each shape.

a

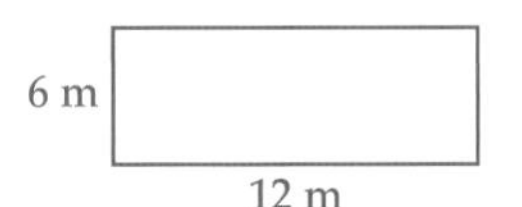

b

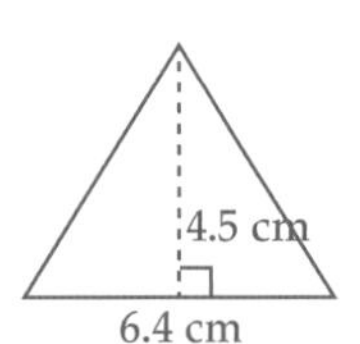

c

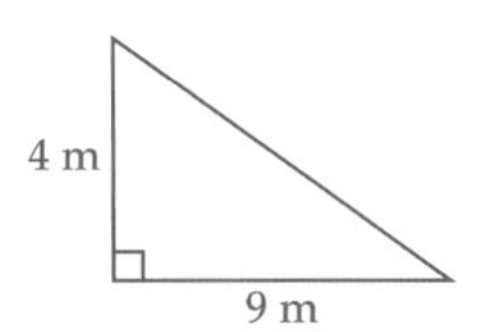

d

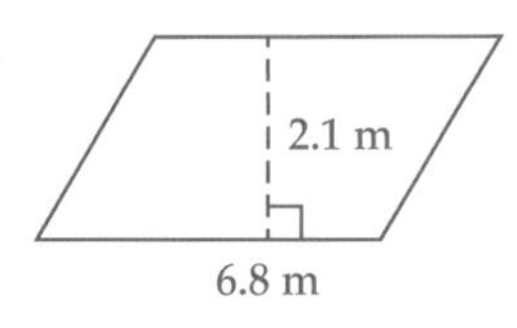

SOLUTION

a $A = l \times w$ ⟵ rectangle

$= 12 \times 6$

$= 72\text{ m}^2$

b $A = \frac{1}{2}bh$ ⟵ triangle

$= \frac{1}{2} \times 6.4 \times 4.5$

$= 14.4\text{ cm}^2$

c $A = \frac{1}{2}bh$ ⟵ triangle

$= \frac{1}{2} \times 9 \times 4$

$= 18\text{ m}^2$

d $A = bh$ ⟵ parallelogram

$= 6.8 \times 2.1$

$= 14.28\text{ m}^2$

ISBN 9780170350990

EXERCISE 10–02

1 Find the area of a rectangle of length 9 cm and width 6 cm. Select the correct answer **A**, **B**, **C** or **D**.

A 63 cm² **B** 56 cm² **C** 72 cm² **D** 54 cm²

2 Find the area of a triangle with base 14 cm and height 12 cm. Select **A**, **B**, **C** or **D**.

A 168 cm² **B** 42 cm² **C** 84 cm² **D** 336 cm²

3 Find the area of a parallelogram with base 8 cm and height 4.2 cm. Select **A**, **B**, **C** or **D**.

A 67.2 cm² **B** 33.6 cm² **C** 336 cm² **D** 16.8 cm²

4 Find the area of each shape.

a

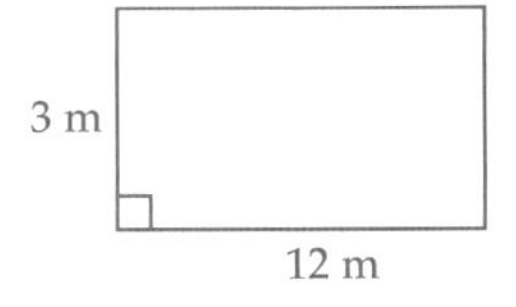

b

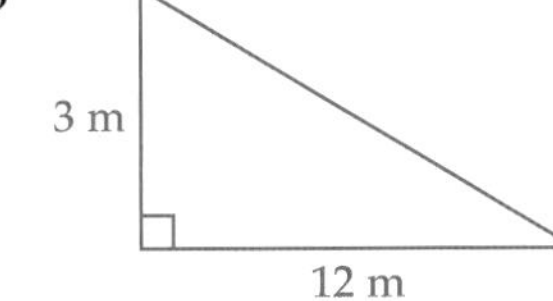

c

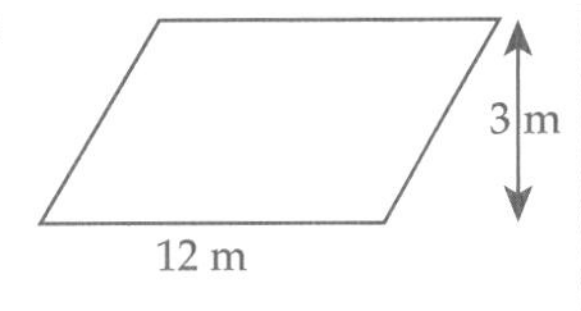

5 **a** What is the same about the rectangle, triangle and parallelogram in Question **4**?

b Compare the areas of the rectangle, triangle and parallelogram in Question **4**.

6 Find the area of each shape.

a

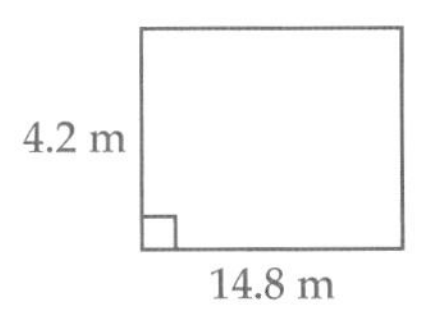

b

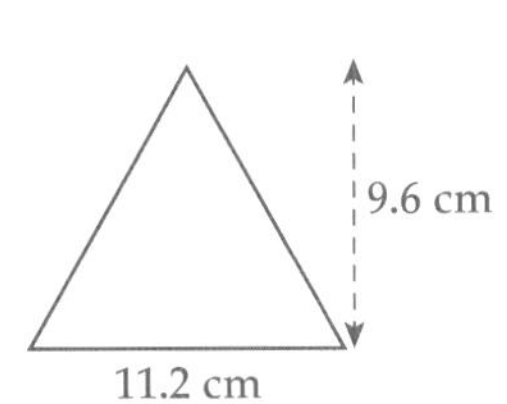

c

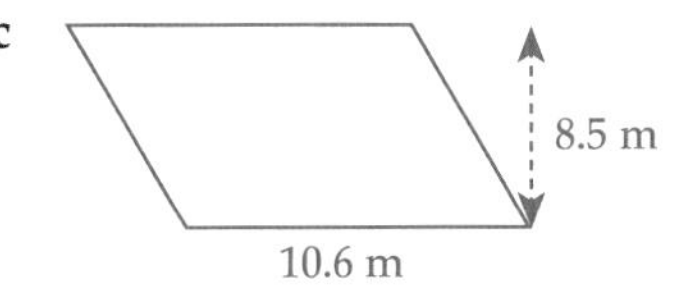

d

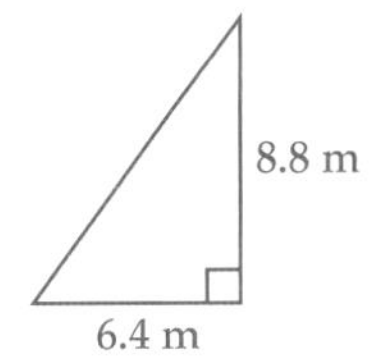

e

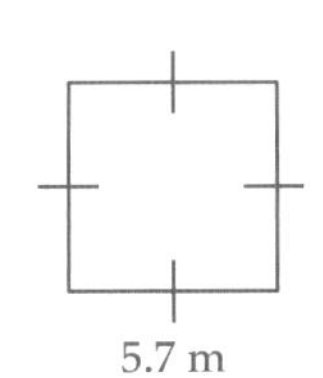

f

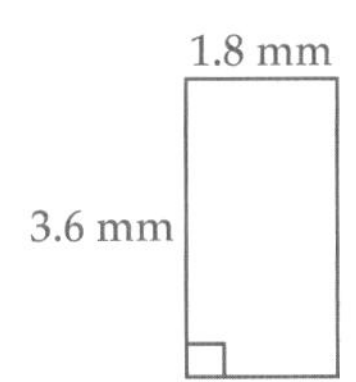

7 Find the area of each of the following shapes.

a A rectangle of length 12.4 m and width 7.2 m.

b A parallelogram with base 23.4 cm and height 15.8 cm.

c A triangle with base 5.4 m and height 8.6 m.

8 Find the area of the garden shown below.

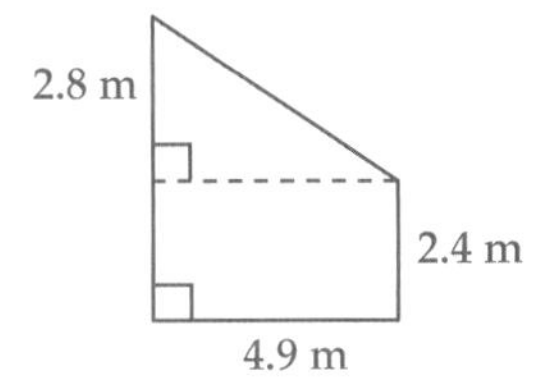

EXAMPLE 4

Find the area of each composite shape.

a

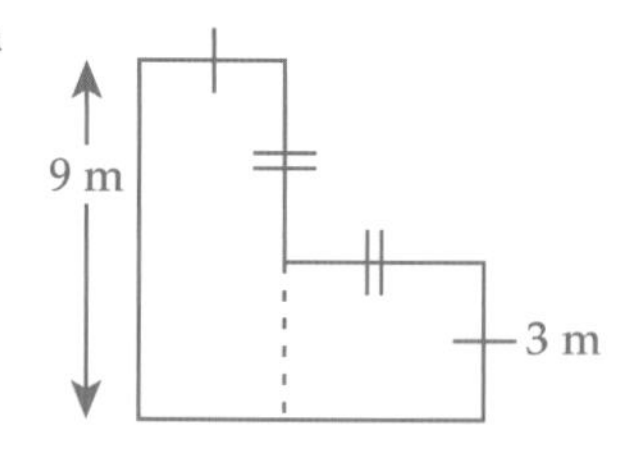

b

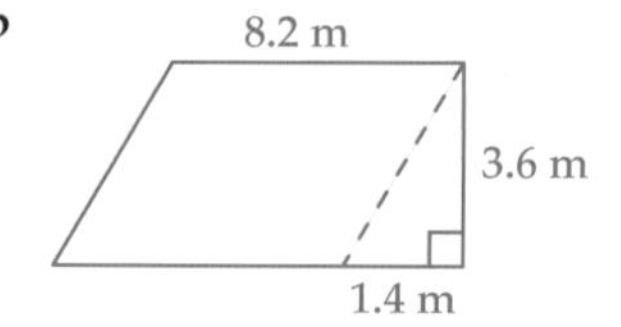

SOLUTION

a Area = Left rectangle + Right rectangle

$= l \times w + l \times w$

$= 3 \times 9 + 6 \times 3$ ⟵ (9 – 3 = 6)

$= 45 \text{ m}^2$

b Area = Parallelogram + Triangle

$= b \times h + \frac{1}{2} \times b \times h$

$= 8.2 \times 3.6 + \frac{1}{2} \times 1.4 \times 3.6$

$= 32.04 \text{ m}^2$

EXAMPLE 5

Find the shaded area.

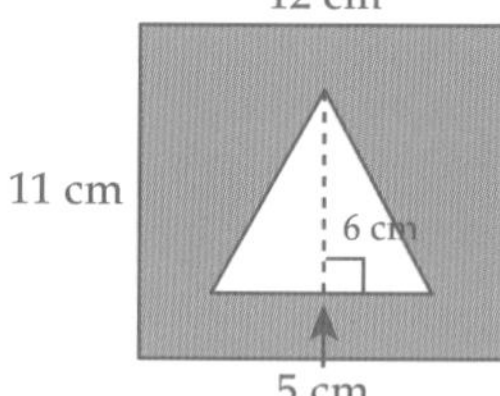

SOLUTION

Shaded area = Rectangle – Triangle

$= 12 \times 11 - \frac{1}{2} \times 5 \times 6$

$= 117 \text{ cm}^2$

Alamy/Chris Hellier

ISBN 9780170350990

EXERCISE 10–03

1 Find the area of each composite shape by adding the areas of the smaller shapes.

a
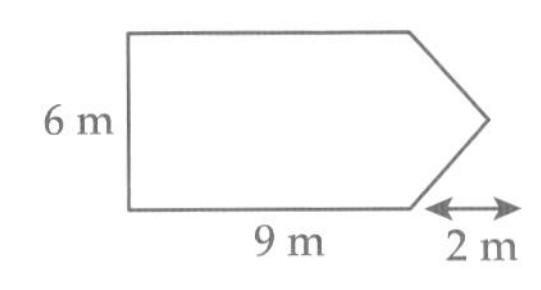

b
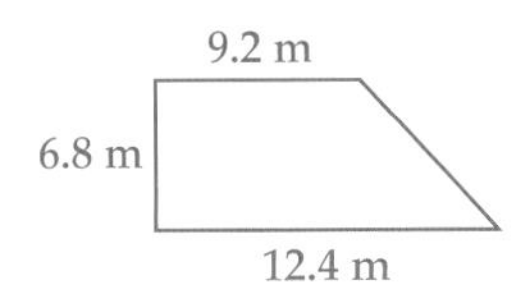

c
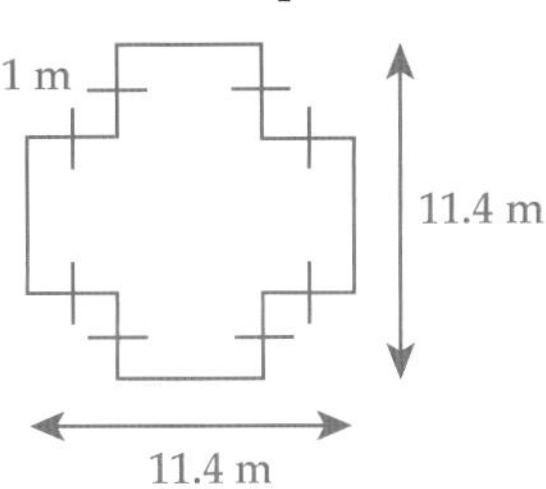

d
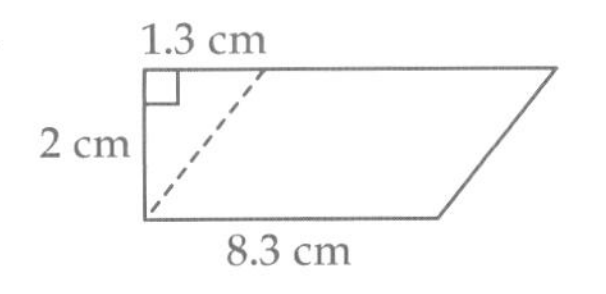

e
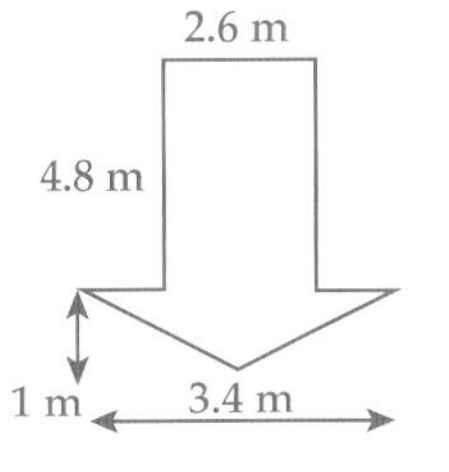

f
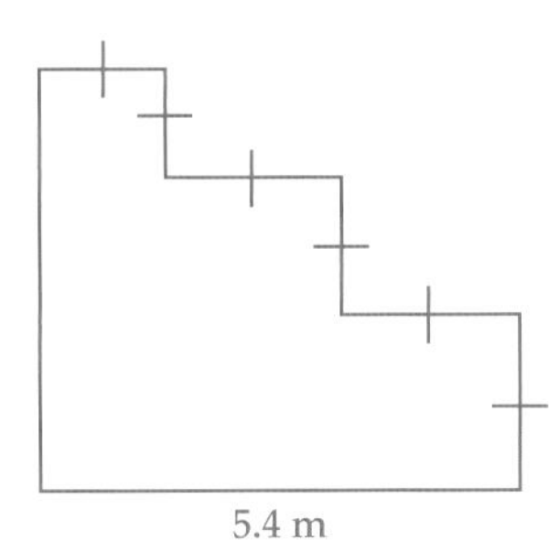

2 Another way of finding the answers to Questions **1c** and **1d** is to subtract areas. Show how you will get the same answer by subtracting areas in Questions **1c** and **1d**.

3 Find the shaded area in each shape.

a
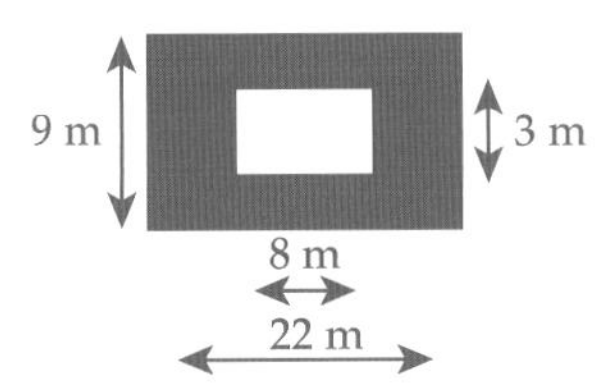

b
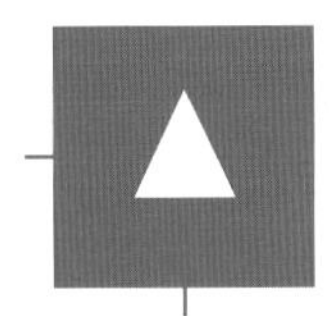

Base of triangle = 4.6 m
Height of triangle = 6.2 m

4 a Calculate the area of this shape:

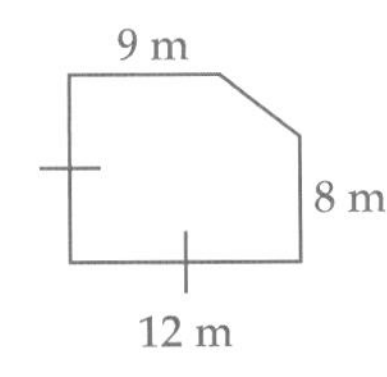

i by adding areas

ii by subtracting areas

b Which method was quicker?

5 Find the area of a garden path 1.2 m wide surrounding a rectangular garden of length 5.66 m and width 2.4 m.

6 Draw a T-shape and show how you could find its area in two different ways.

10–04 Area of a trapezium

This trapezium has parallel sides of length a and b and a perpendicular height of h.
To find its area, we can cut it into pieces along the red dotted lines and rearrange the pieces to make a rectangle.

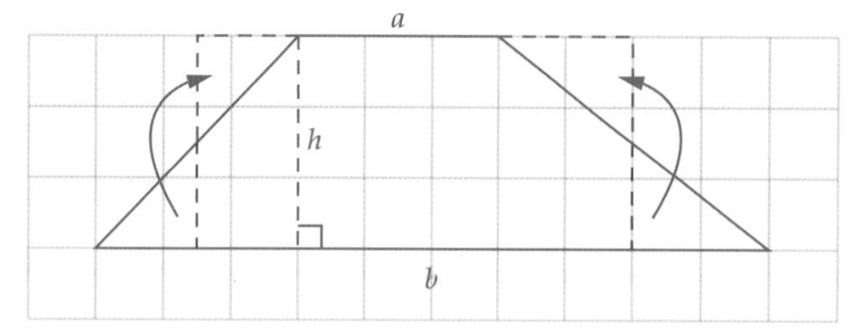

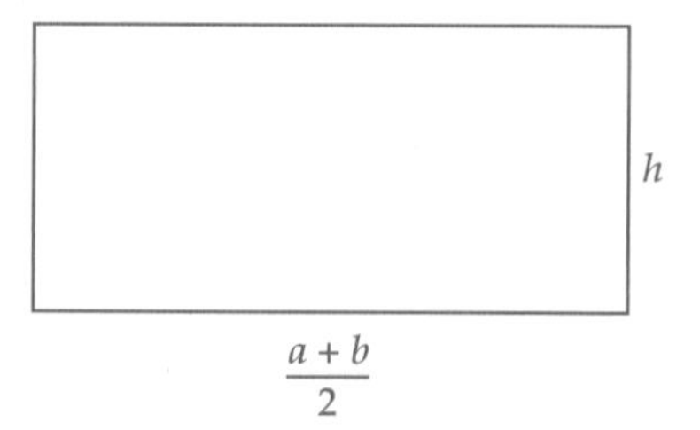

The length of this rectangle is the average of a and b, which is $\frac{a+b}{2}$.
The width of this rectangle is h.

So the area of the trapezium $= \frac{a+b}{2} \times h$ ← $l \times w$

$$= \frac{1}{2} \times (a + b) \times h$$

$$= \frac{1}{2} h(a + b)$$

AREA OF A TRAPEZIUM

$A = \frac{1}{2} \times$ height $\times$ (sum of parallel sides)

$A = \frac{1}{2} h(a + b)$

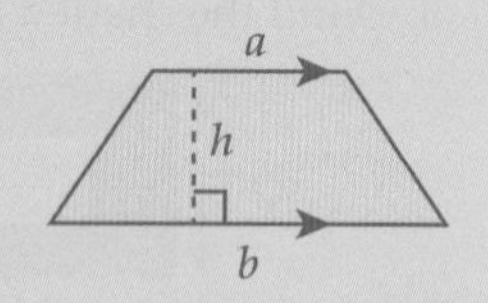

EXAMPLE 6

Find the area of each trapezium.

a

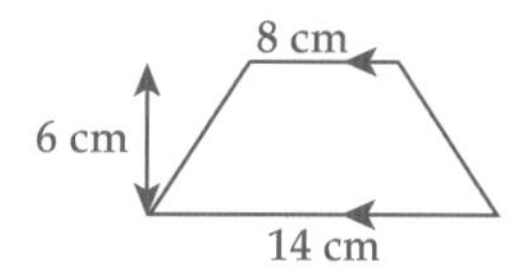

b

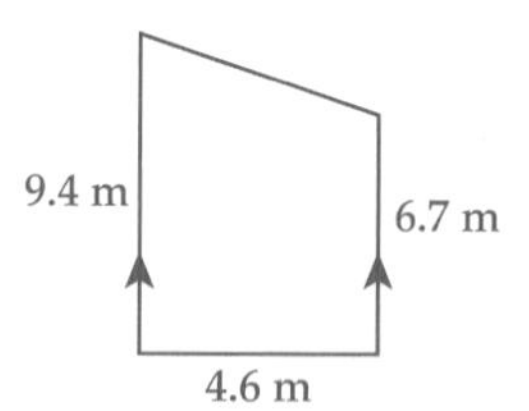

SOLUTION

a $A = \frac{1}{2} h(a + b)$

$= \frac{1}{2} \times 6 \times (8 + 14)$

$= 66 \text{ cm}^2$

b $A = \frac{1}{2} h(a + b)$

$= \frac{1}{2} \times 4.6 \times (9.4 + 6.7)$

$= 37.03 \text{ m}^2$

ISBN 9780170350990

EXERCISE 10–04

1 Find the area of a trapezium with height 6 m and parallel sides 4 m and 5 m. Select the correct answer **A**, **B**, **C** or **D**.

A 27 m² **B** 54 m² **C** 24 m² **D** 30 m²

2 Find the area of each trapezium.

a

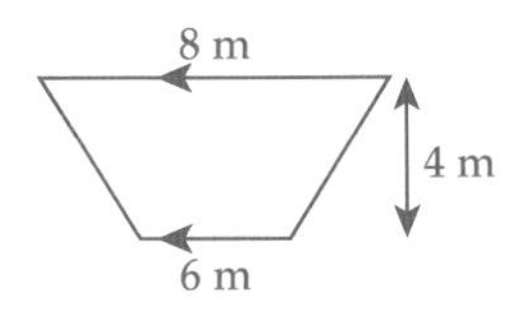

b

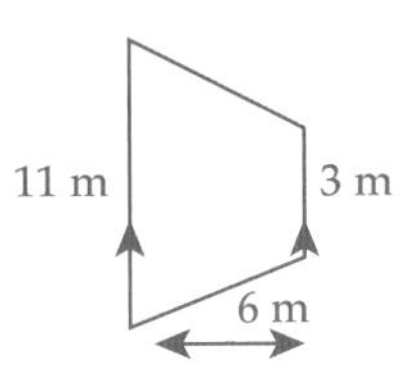

c

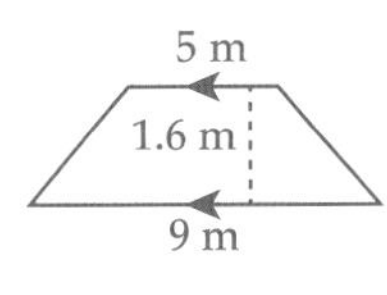

d

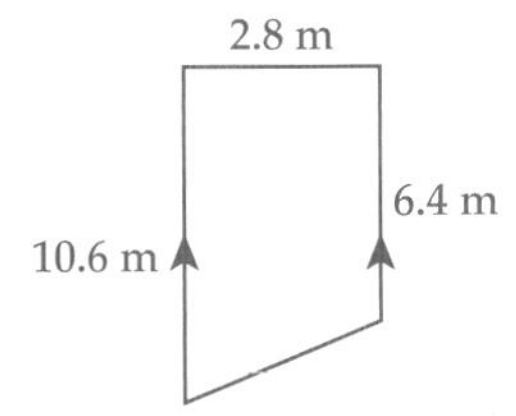

e

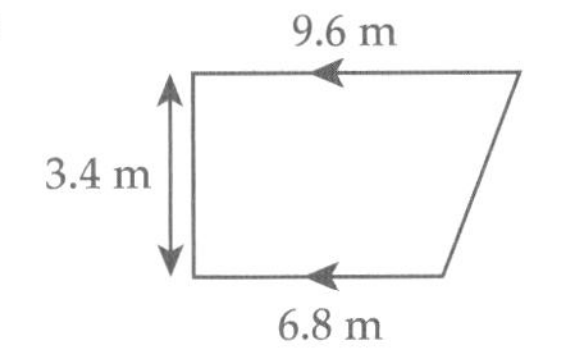

f

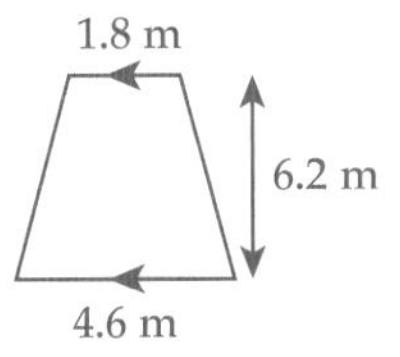

3 Draw each trapezium described and find its area.

a Height of 6 cm and parallel sides of 3 cm and 7 cm.

b Vertical parallel sides of 4 m and 6.2 m and a distance of 3.8 m between the sides.

4 Find the area of each shape by first dividing it into smaller shapes.

a

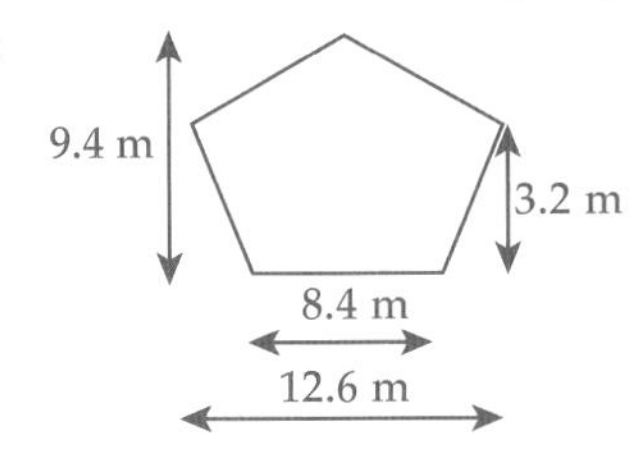

b

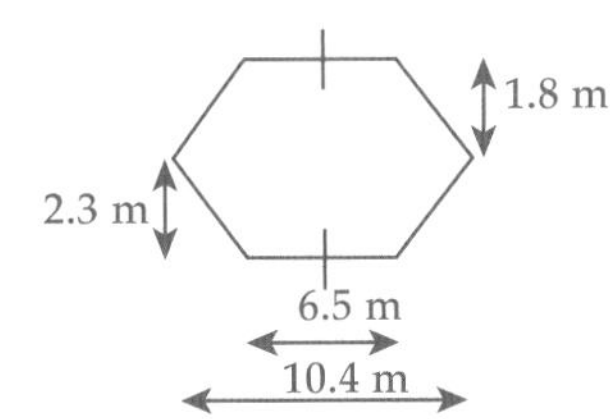

c

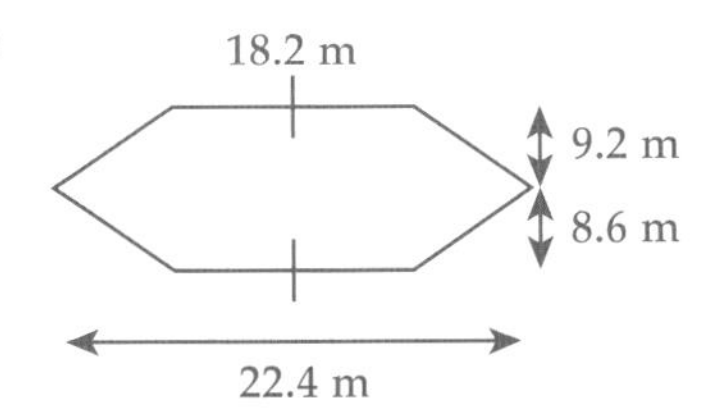

5 A garden tile has the shape of a trapezium with parallel sides 18.5 cm and 12.7 cm and a perpendicular height of 10.8 cm. What is its area?

10–05 Areas of kites and rhombuses

To find the area of a kite or rhombus, we can cut each shape into four triangles and rearrange the triangles to make a rectangle.
This kite and rhombus each have diagonals of length x and y and the y-diagonal cuts the x-diagonal in half. Both can be converted to rectangles of length y and width $\frac{1}{2}x$.

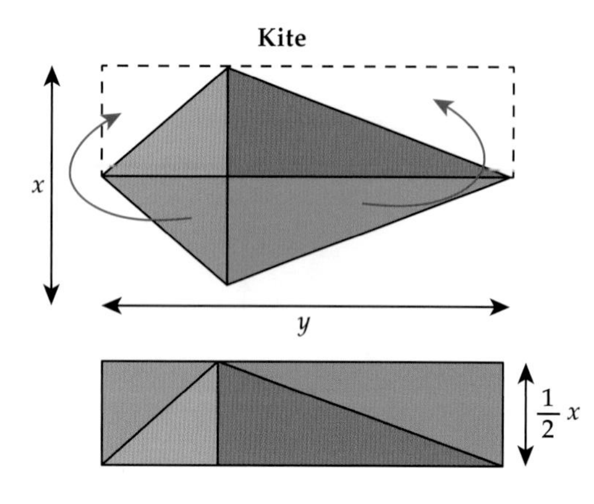

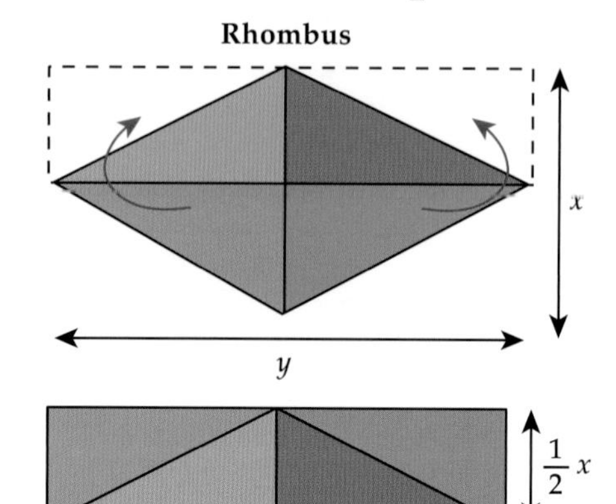

So the area of the kite and the rhombus $= y \times \frac{1}{2}x$ ← $l \times w$

$= \frac{1}{2}xy$

AREA OF A KITE AND RHOMBUS

$A = \frac{1}{2} \times \text{diagonal 1} \times \text{diagonal 2}$

$A = \frac{1}{2}xy$

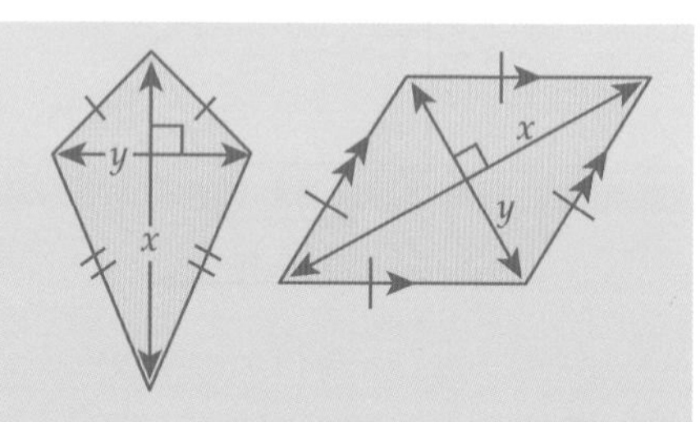

EXAMPLE 7

Find the area of each shape.

a

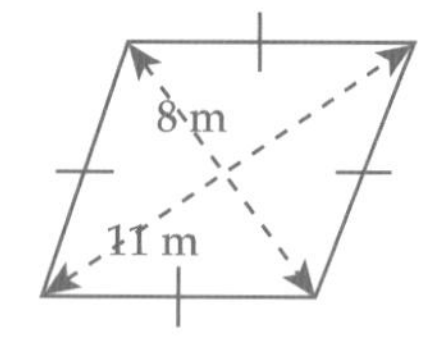

b

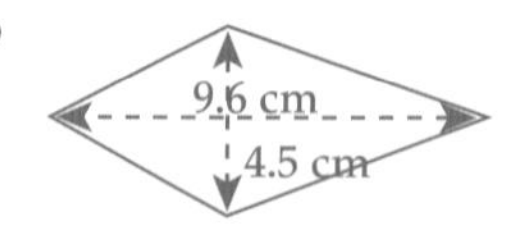

SOLUTION

a $A = \frac{1}{2}xy$

$= \frac{1}{2} \times 8 \times 11$

$= 44 \text{ m}^2$

b $A = \frac{1}{2}xy$

$= \frac{1}{2} \times 4.5 \times 9.6$

$= 21.6 \text{ cm}^2$

ISBN 9780170350990

EXERCISE 10-05

1 If the diagonals of a rhombus are 11 cm and 14 cm, what is its area? Select the correct answer **A, B, C** or **D**.

A 154 cm^2 **B** 77 cm^2 **C** 38.5 cm^2 **D** 308 cm^2

2 Find the area of a kite with diagonals 12 cm and 13 cm. Select **A, B, C** or **D**.

A 39 cm^2 **B** 156 cm^2 **C** 78 cm^2 **D** 312 cm^2

3 Copy and complete this working for the area of a kite.

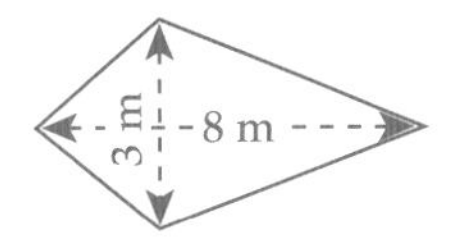

$A = \frac{1}{2} \times \_\_\_ \times 8 = \_\_\_ \text{ m}^2$

4 Find the area of each of the following shapes.

a

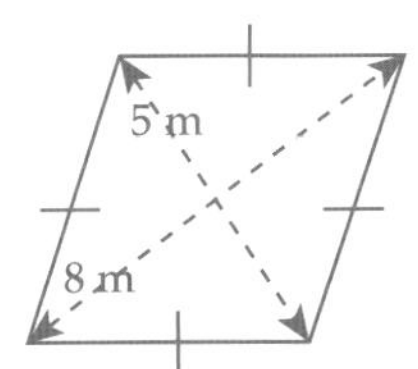

b

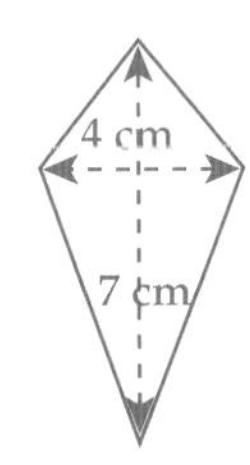

c

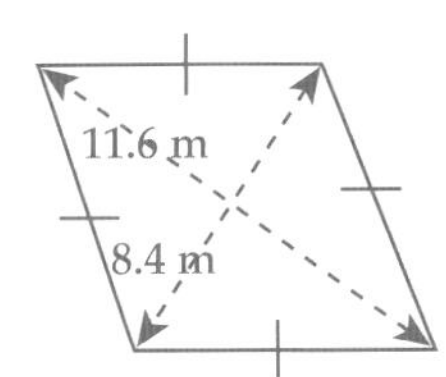

d

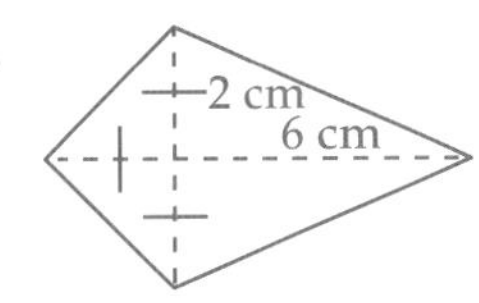

e

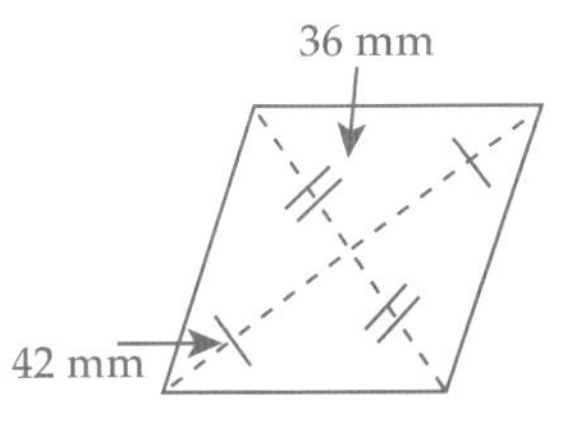

f

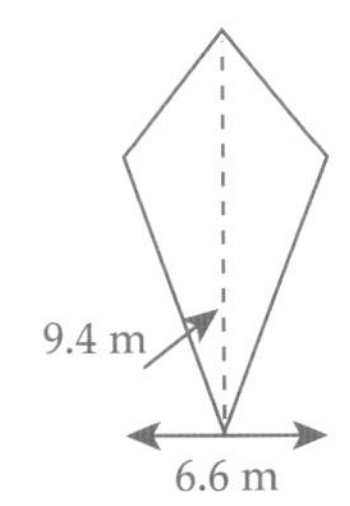

5 Find the area of a playground which is in the shape of a rhombus with one diagonal 6.4 m long and the other 310 cm long. Answer in m^2.

6 Canvas costs \$12.80/m^2 and string costs \$4.90/m when used to build a kite with diagonals 1.4 m and 1.8 m long. The kite requires 20 m of string and the cost of its frame is \$28.50.

a Find the area of the kite and the cost of the canvas used.

b Find the total cost of building this kite.

10–06 Area of a circle

To find the area of a circle, we can cut it into small sectors and rearrange them to approximate a rectangle.

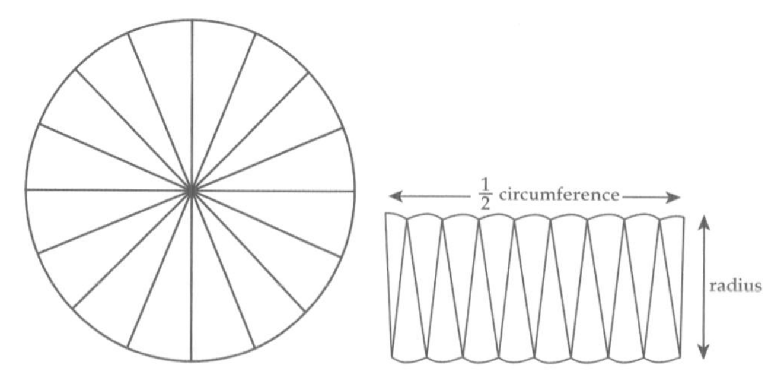

This circle with radius r has a circumference of $2\pi r$. It can be converted to a rectangle of length $\frac{1}{2} \times 2\pi r$ and width r.

So the area of the circle $= \frac{1}{2} \times 2\pi r \times r$ ⟵ $l \times w$

$= \pi r^2$

AREA OF A CIRCLE

$A = \pi \times \text{radius}^2$

$A = \pi r^2$

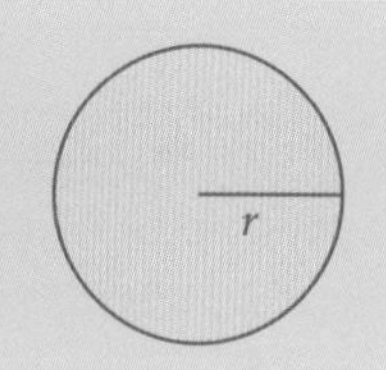

If you are given the diameter, halve it to find the radius first.

EXAMPLE 8

Find correct to two decimal places the area of each circle.

a

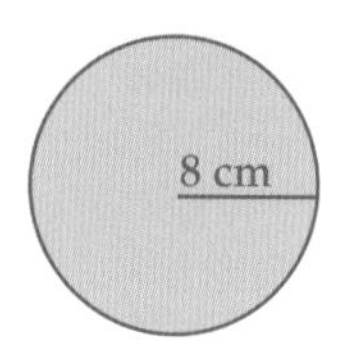

b

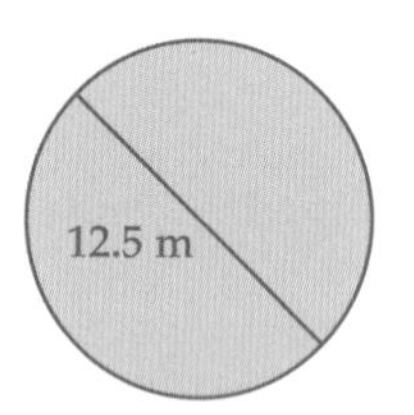

SOLUTION

a $A = \pi r^2$ ⟵ $r = 8$

$= \pi \times 8^2$

$= 201.0619\ldots$

$\approx 201.06 \text{ cm}^2$

b $A = \pi r^2$ ⟵ $r = \frac{1}{2} \times 12.5 = 6.25$

$= \pi \times 6.25^2$

$= 122.71846\ldots$

$= 122.72 \text{ m}^2$

ISBN 9780170350990

10–06 Area of a circle

EXAMPLE 9

Find correct to one decimal place the area of each shape.

a

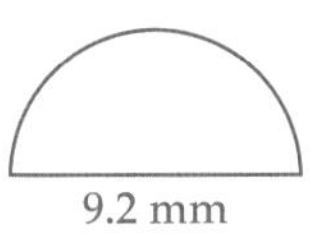

b

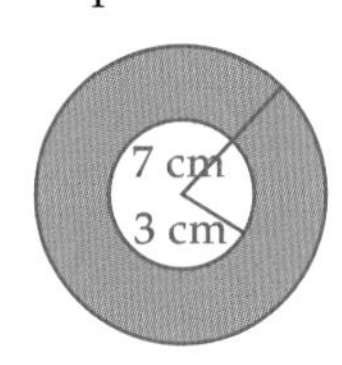

SOLUTION

a $A = \frac{1}{2} \times \text{circle}$ ← semicircle

$= \frac{1}{2}\pi r^2$

$= \frac{1}{2} \times \pi \times 4.6^2$ ← $r = \frac{1}{2} \times 9.2 = 4.6$

$= 33.2380$

$\approx 33.2 \text{ mm}^2$

b $A = \text{large circle} - \text{small circle}$ ← ring shape

$= \pi \times 7^2 - \pi \times 3^2$

$= 125.6637\ldots$

$\approx 125.7 \text{ cm}^2$

EXERCISE 10–06

1 Find the area of a circle with radius 6 cm. Select the correct answer **A**, **B**, **C** or **D**.

A 37.70 cm^2 **B** 113.10 cm^2 **C** 226.19 cm^2 **D** 75.40 cm^2

2 Find the area of a circle with diameter 9.8 m. Select **A**, **B**, **C** or **D**.

A 75.4 m^2 **B** 301.7 m^2 **C** 30.8 m^2 **D** 56.5 m^2

3 Copy and complete the working to find the area of this circle correct to two decimal places.

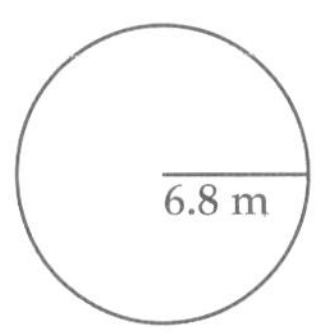

$A = \pi r^2$

$= \pi \times \_\_\_^2$

$= \_\_\_\_$

$\approx \_\_\_\_ \text{ m}^2$

4 Find the area of each circle correct to one decimal place.

a

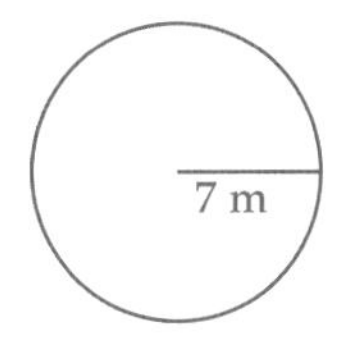

b

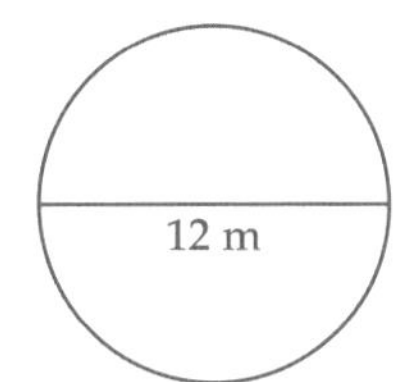

c

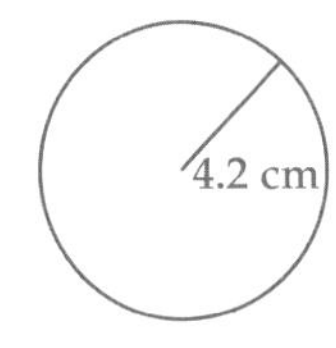

d

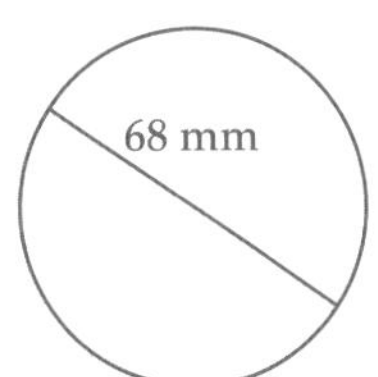

e

f

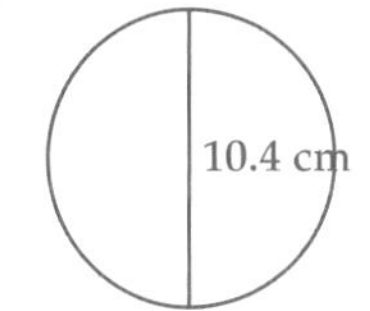

5 Find correct to two decimal places the shaded area of each shape.

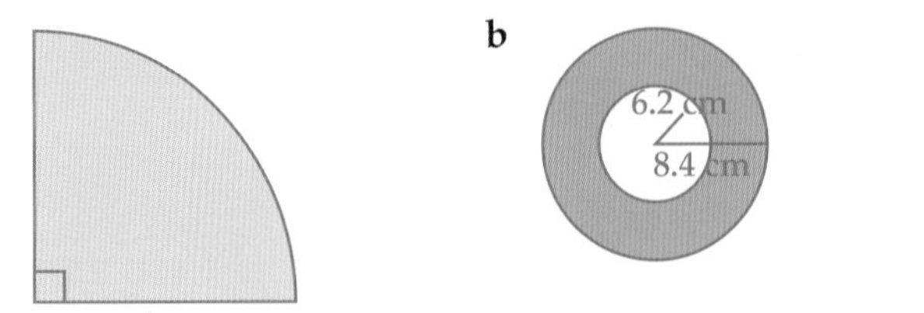

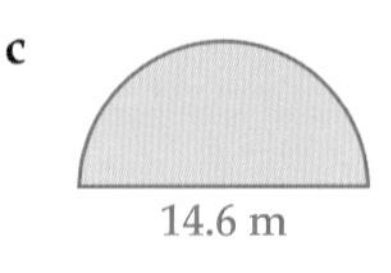

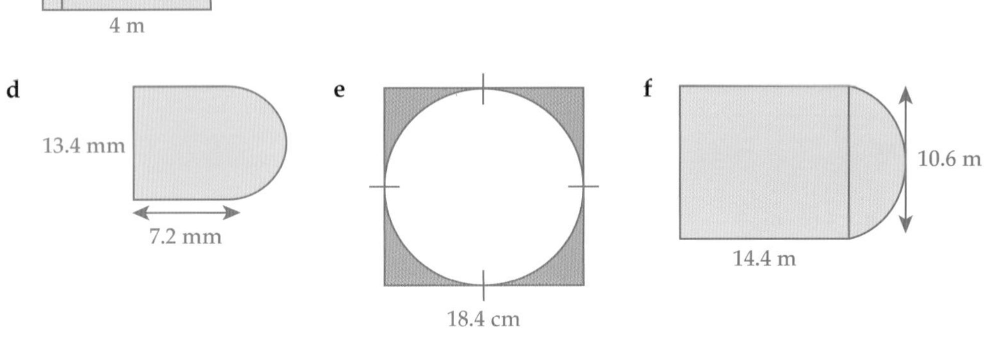

iStockphoto/viviamo

ISBN 9780170350990

10–07 Metric units for volume

WORDBANK

volume The amount of space inside a solid shape, measured in cubic units such as mm^3, cm^3 or m^3.

capacity The amount of liquid or material that a container can hold, measured in mL, L and kL.

A **cubic millimetre** (mm^3) is the volume of a cube of length 1 mm, about the size of a grain of raw sugar or rock salt.
A **cubic centimetre** (cm^3) is the volume of a cube of length 1 cm, about the size of a face of a die.
1 cm = 10 mm
$1\ cm^3 = 10\ mm \times 10\ mm \times 10\ mm = 1000\ mm^3$

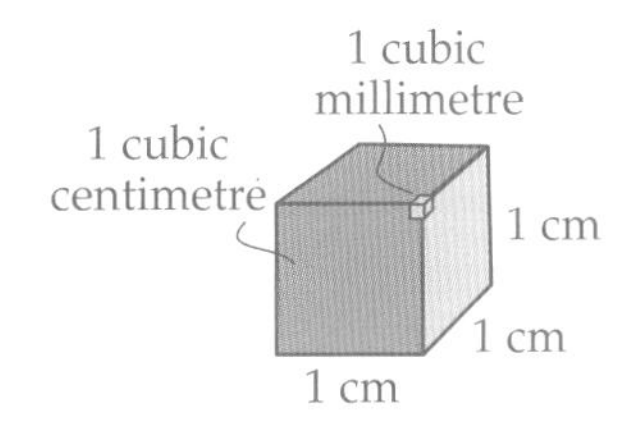

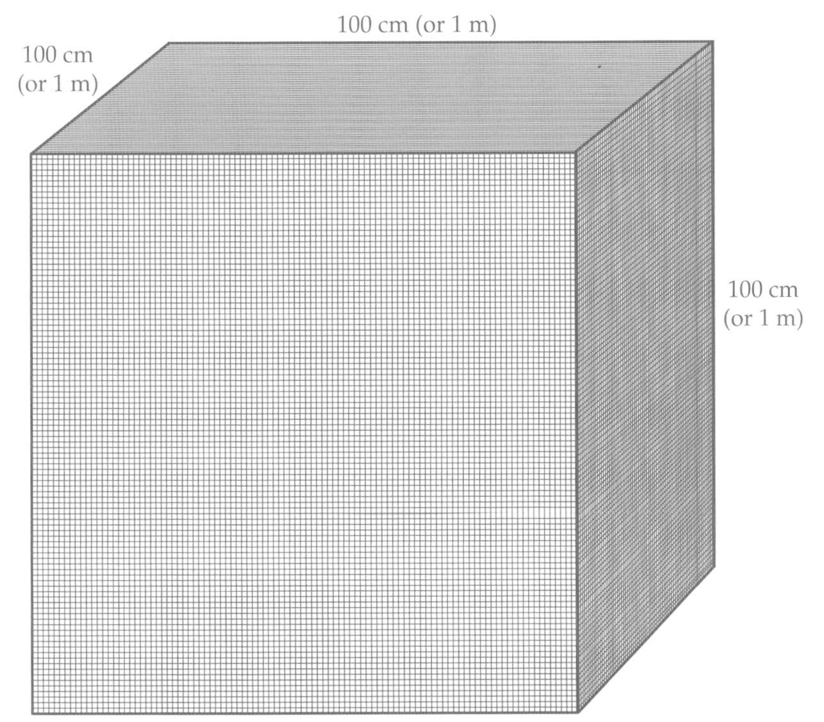

A **cubic metre** (m^3) is the volume of a cube of length 1 m, about the size of two washing machines.
1 m = 100 cm = 1000 mm
$1\ m^3 = 100\ cm \times 100\ cm \times 100\ cm = 1\ 000\ 000\ cm^3$
$1\ m^3 = 1000\ mm \times 1000\ mm \times 1000\ mm = 1\ 000\ 000\ 000\ mm^3$

$1\ cm^3 = 1000\ mm^3$
$1\ m^3 = 1\ 000\ 000\ cm^3 = 1\ 000\ 000\ 000\ mm^3$

ISBN 9780170350990

EXAMPLE 10

Convert:

a $4.5\ m^3 = \_\_\_\_ cm^3$ **b** $5\,200\,000\ mm^3 = \_\_\_\_ m^3$

SOLUTION

a $4.5\ m^3 = 4.5 \times 1\,000\,000\ cm^3$ ← LOSM: Large Over to Small Multiply: $1\ m^3 = 1\,000\,000\ cm^3$

$= 4\,500\,000\ cm^3$ ← LOSM: $4.5 \times 1\,000\,000$

b $5\,200\,000\ mm^3 = 5\,200\,000 \div 1\,000\,000\,000\ m^3$ ← SOLD: Small Over to Large Divide: $1\ m^3 = 1\,000\,000\,000\ mm^3$

$= 0.0052\ m^3$ ← SOLD: $5\,200\,000 \div 1\,000\,000\,000$

VOLUME AND CAPACITY

1 cm^3 contains 1 mL

1 m^3 contains 1 kL or 1000 L

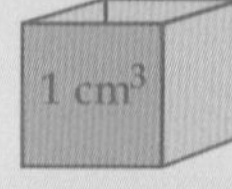

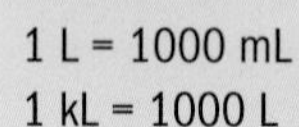

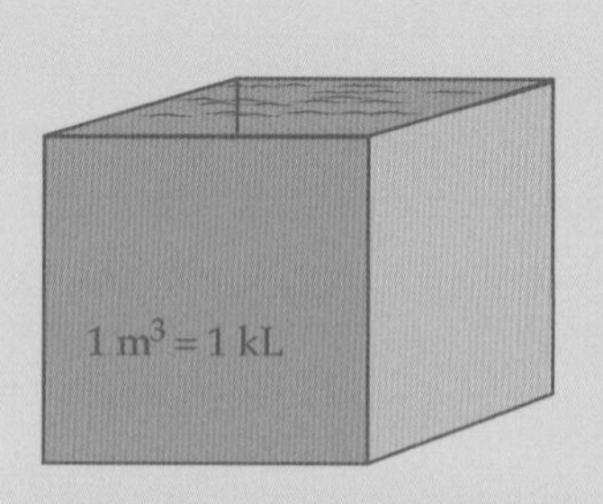

1 L = 1000 mL

1 kL = 1000 L

iStockphoto/Dantesattic

 ISBN 9780170350990

EXERCISE 10-07

1 Which measurement unit is used for volume? Select the correct answer **A**, **B**, **C** or **D**.

A cm^3 **B** mm^2 **C** kL^2 **D** mL^3

2 What unit would you use to measure the volume of a storage shed? Select **A**, **B**, **C** or **D**.

A mL **B** mm^3 **C** cm^3 **D** m^3

3 Write the metric unit that would be the most suitable for measuring:

a the volume of a balloon
b the area of a kitchen bench
c the length of a room
d the area of a playground
e the volume of a bookcase
f the volume of an insect
g your height
h the volume of a truck

4 Convert:

a $550\,000\,000\ cm^3 = \_\_\_\ m^3$
b $47\ m^3 = \_\_\_\ cm^3$
c $56\ cm^3 = \_\_\_\ mm^3$
d $7.5\ m^3 = \_\_\_\ mm^3$
e $0.043\ mm^3 = \_\_\_\ cm^3$
f $280\,000\,000\ mm^3 = \_\_\_\ m^3$
g $9.26\ m^3 = \_\_\_\ cm^3$
h $855\,000\ cm^3 = \_\_\_\ m^3$

5 Write the metric unit that would be most suitable for measuring the capacity of:

a a cup
b a lake
c a water bottle
d a can of soft drink
e a swimming pool
f a dose of cough medicine

6 Copy and complete:

a $4\ cm^3 = \_\_\_\ mL$
b $222\ mL = \_\_\_\ cm^3$
c $7500\ cm^3 = \_\_\_\ L$
d $10.4\ L = \_\_\_\ cm^3$
e $8504\ mL = \_\_\_\ L$
f $67\ L = \_\_\_\ mL$
g $680\ L = \_\_\_\ m^3$
h $2.56\ kL = \_\_\_\ m^3$

7 Convert:

a 2 m to cm
b $2\ m^2$ to cm^2
c $2\ m^3$ to cm^3
d 50 m to cm
e $50\ m^2$ to cm^2
f $50\ m^3$ to cm^3
g 800 m to cm
h $800\ m^2$ to cm^2
i $800\ m^3$ to cm^3

10–08 Drawing prisms

WORDBANK

cross-section A 'slice' of a solid shape.

prism A solid shape that has identical cross-sections with straight sides (not rounded).

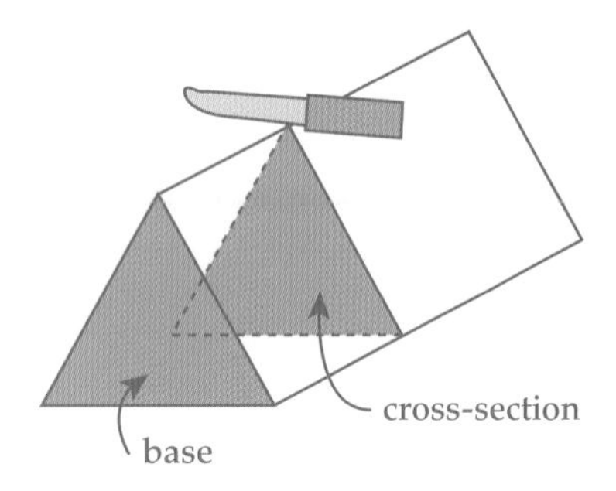

This triangular prism is a **prism** because its cross-sections are identical triangles.

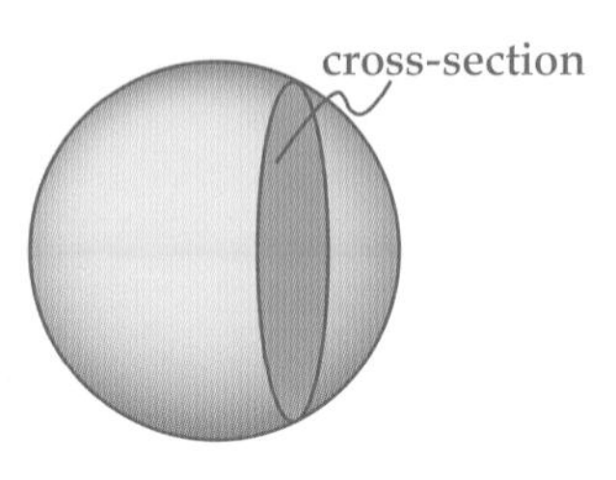

This sphere is **not a prism** because its cross-sections are circles (round) of different sizes.

EXAMPLE 11

a Draw a cross-section of the prism shown.

b What shape is the prism's cross-section?

c What is the name of this prism?

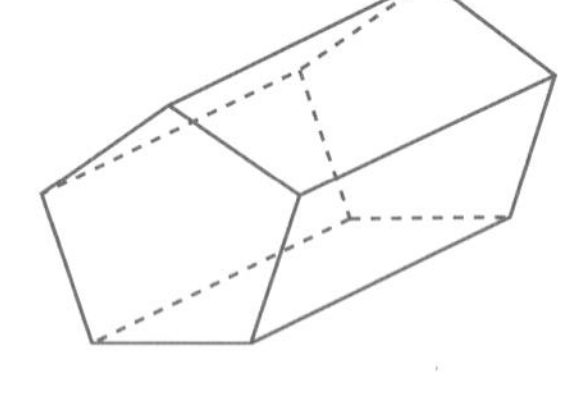

SOLUTION

a

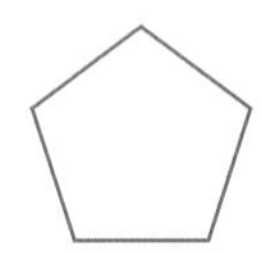

b The base is a pentagon.

c It is a pentagonal prism.

EXAMPLE 12

For the solid shown, draw:

a the front view

b the left view

c the top view.

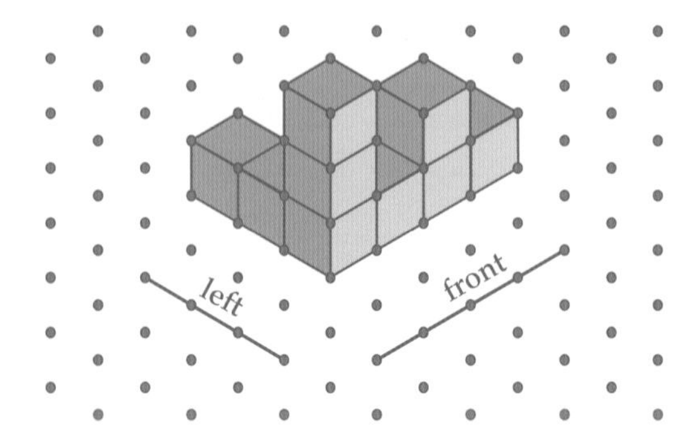

ISBN 9780170350990

SOLUTION

a front view

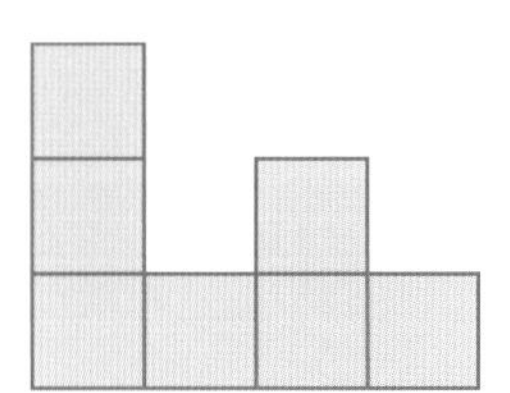

b left view

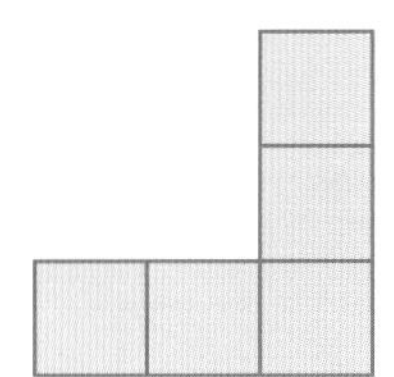

c top view

EXERCISE 10–08

1 Which of these solids are prisms?

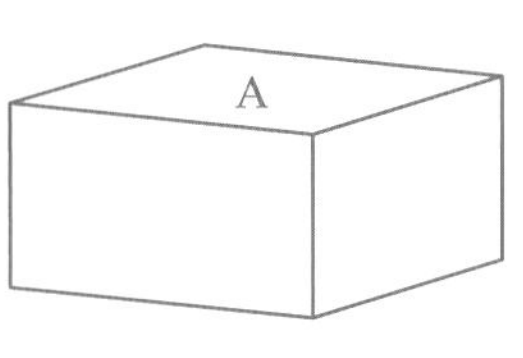

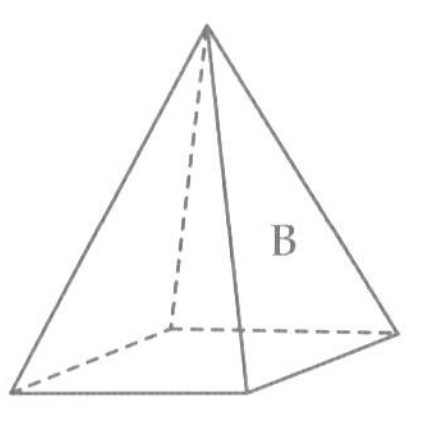

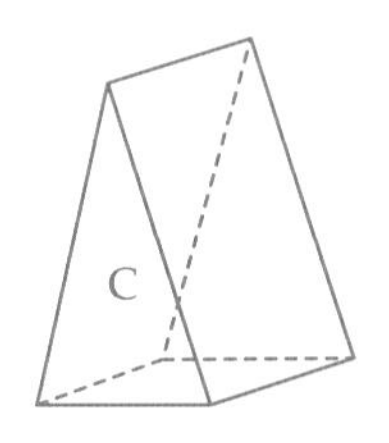

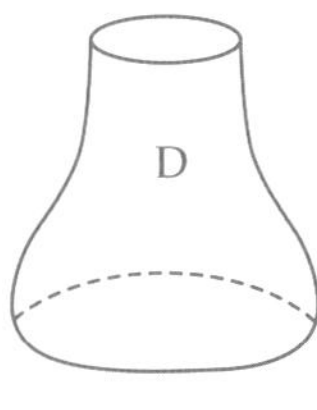

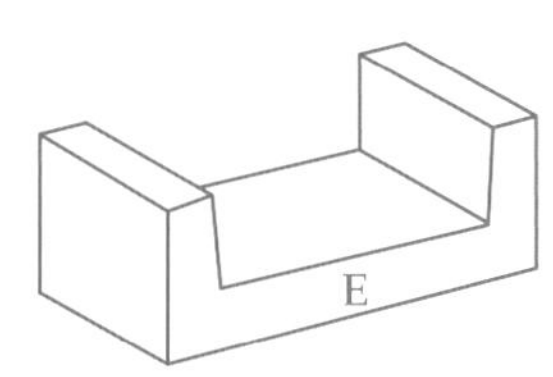

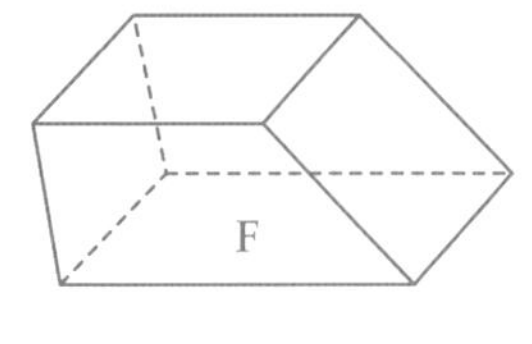

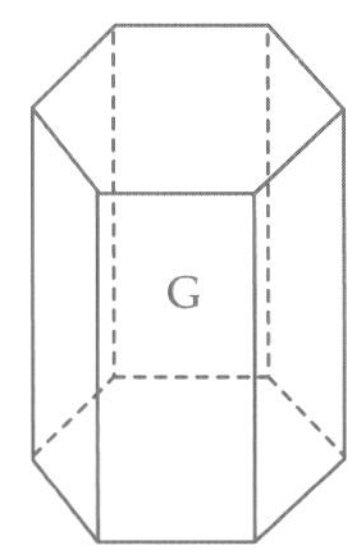

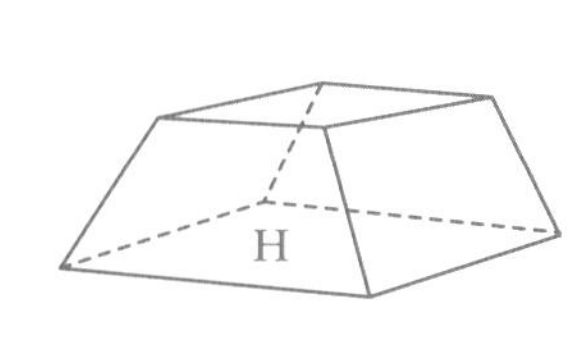

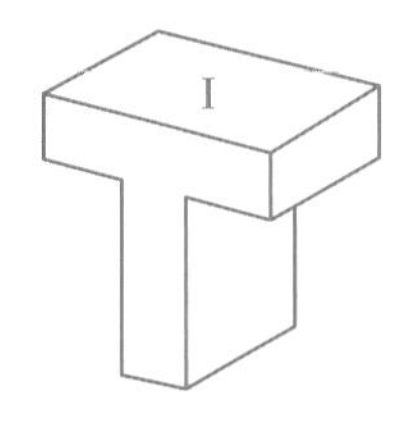

2 Match each name to the correct prism in Question **1**.

a 'T-prism'

b trapezoidal prism (two answers)

c rectangular prism

d hexagonal prism

e 'U-prism'

f triangular prism

3 Draw the cross-section of each prism.

a

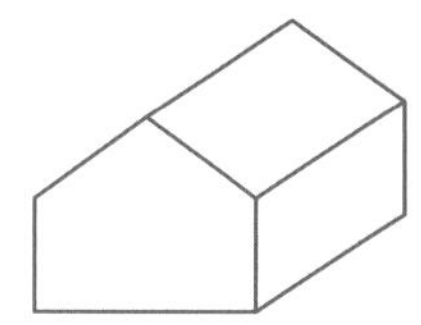

b

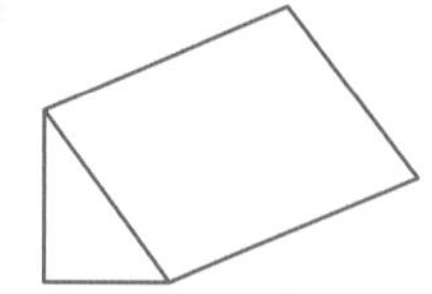

c

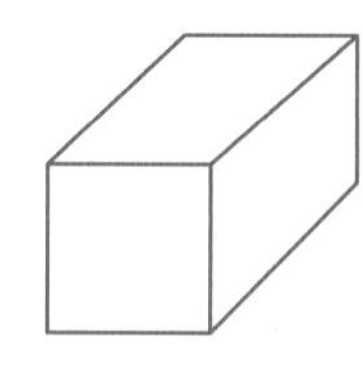

d

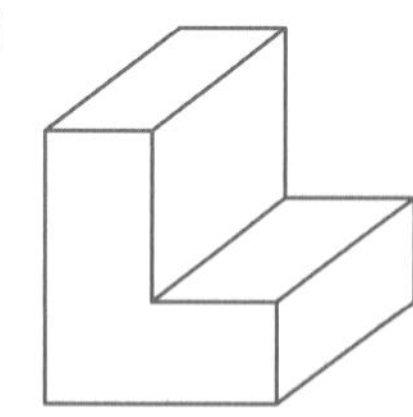

e

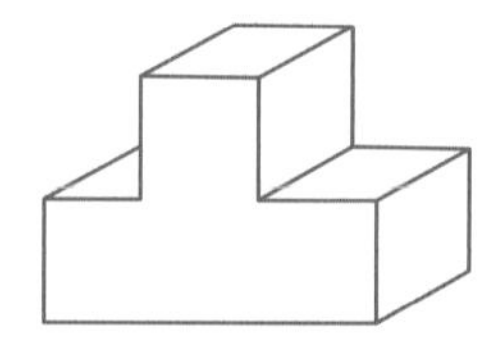

f

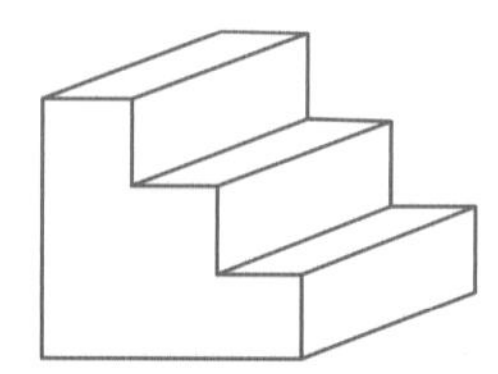

4 Write the name of each prism in Question **3 a**, **b**, **c** above.

5 Draw the following prism on isometric dot paper, then draw the views of this prism from A, from B, and from C.

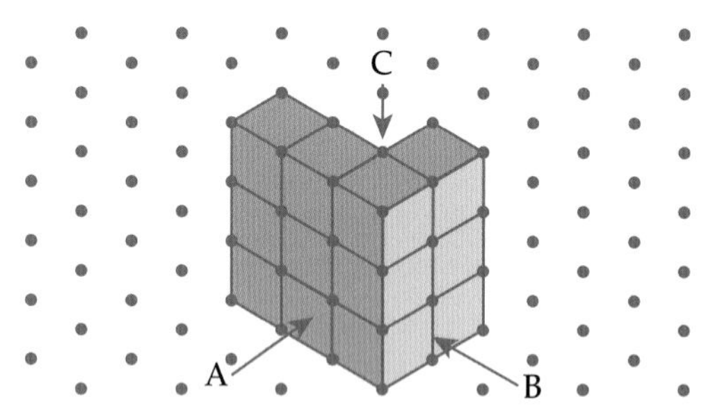

6 What is the top view of this prism? Select the correct answer **A**, **B**, **C** or **D**.

A

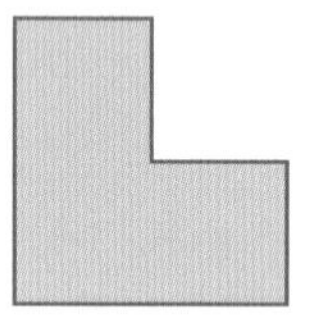

B

C

D

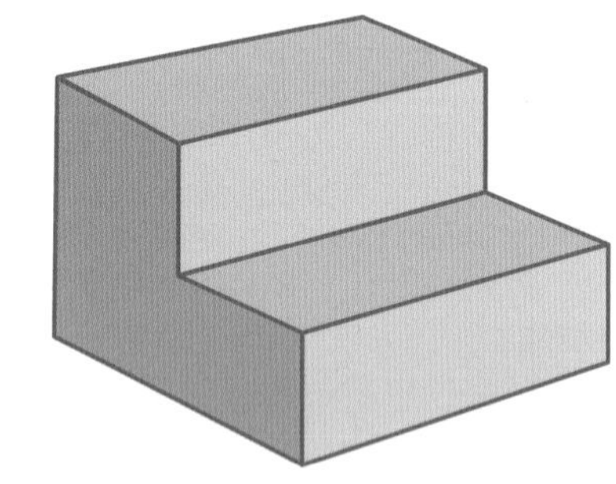

ISBN 9780170350990

7 For each solid, draw each view requested.

a

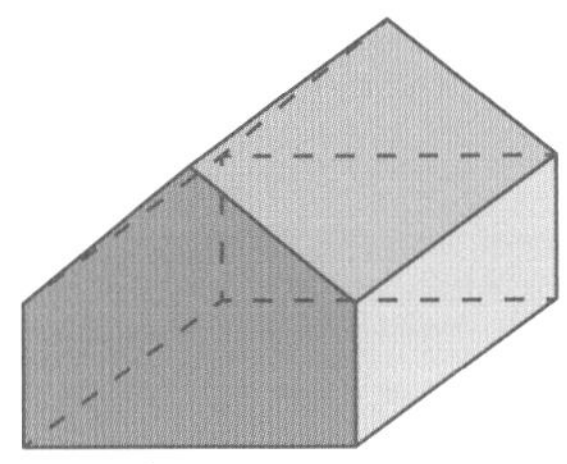

i front view
ii right view

b

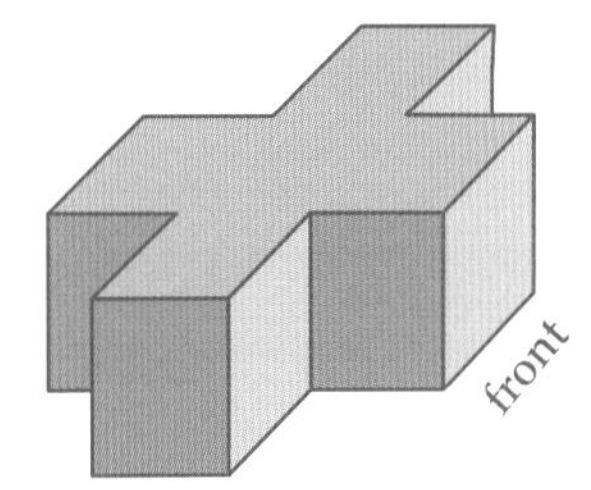

i left view
ii top view
iii front view

c

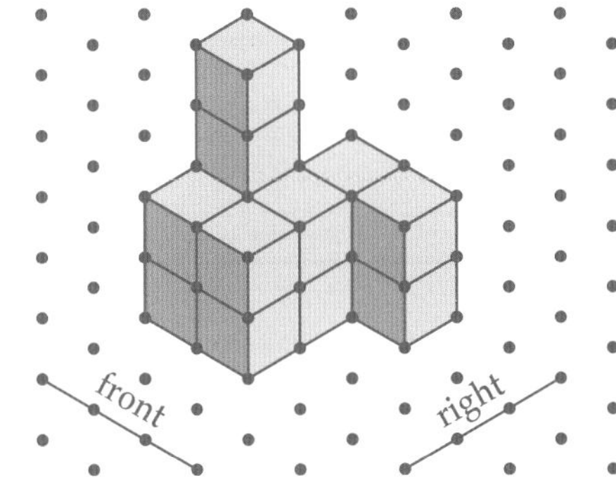

i front view
ii right view
iii top view

d

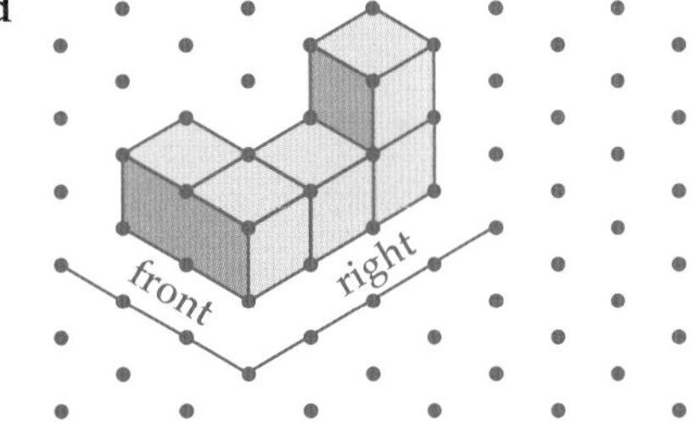

i front view
ii left view
iii top view

10–09 Volume of a prism

VOLUME OF A RECTANGULAR PRISM

V = length × width × height

$V = lwh$

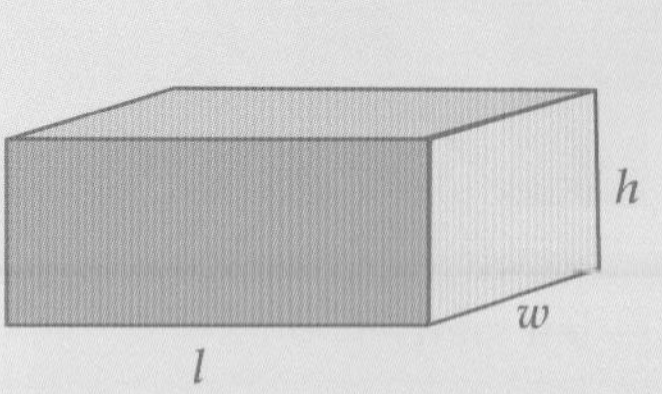

VOLUME OF ANY PRISM

V = area of base or cross-section × height

$V = Ah$

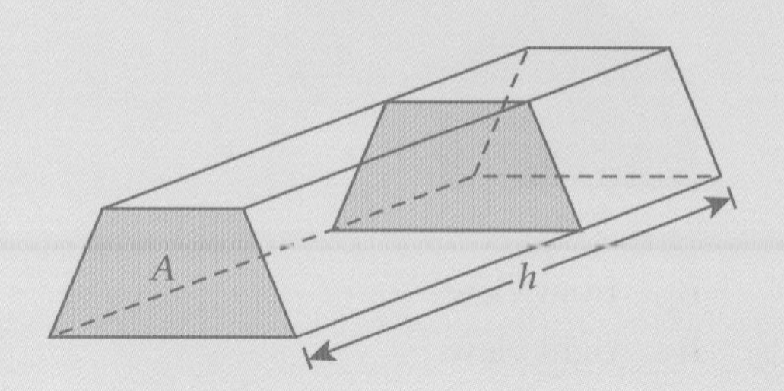

EXAMPLE 13

Find the volume of each prism.

a

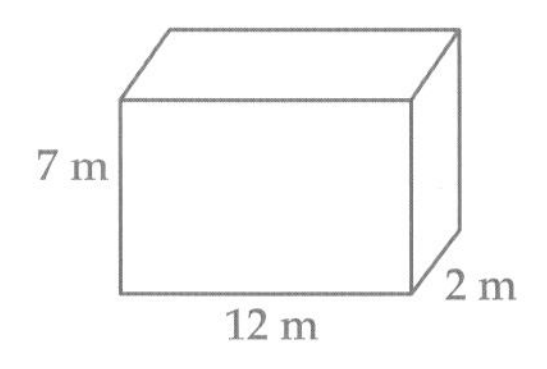

b

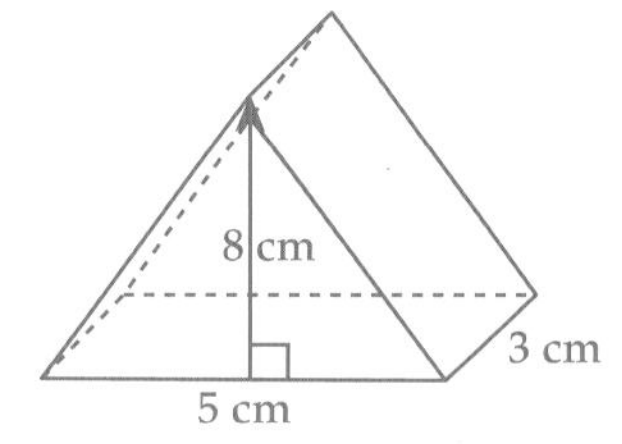

c

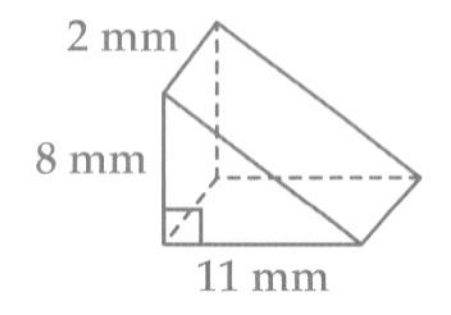

d

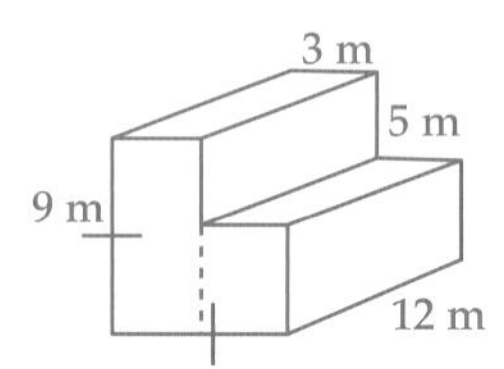

SOLUTION

a $V = 12 \times 2 \times 7$ ← $V = lbh$

$= 168\text{ m}^3$

b First, find A, the area of the base or cross-section.

$A = \frac{1}{2} \times 5 \times 8$ ← Area of a triangle $A = \frac{1}{2}bh$.

$= 20\text{ cm}^2$

$V = Ah$

$= 20 \times 3$ ← $h = 3$

$= 60\text{ cm}^3$

 ISBN 9780170350990

c $A = \frac{1}{2} \times 11 \times 8$ ← Area of a triangle.

$= 44\ \text{mm}^2$

$V = Ah$

$= 44 \times 2$ ← $h = 2$

$= 88\ \text{mm}^3$

d $A = 3 \times 9 + 6 \times 4$ ← left rectangle + right rectangle

✱ length of right rectangle = 9 − 3 = 6, width = 9 − 5 = 4

$= 51\ \text{m}^2$

$V = Ah$

$= 51 \times 12$ ← $h = 12$

$= 612\ \text{m}^3$

EXERCISE 10-09

1 Find the volume of a rectangular prism with length 11 m, width 6 m and height 8 m. Select the correct answer **A**, **B**, **C** or **D**.

A 576 m^3 **B** 1056 m^3 **C** 528 m^3 **D** 264 m^3

2 Find the volume of a triangular prism with base length 12 m, base height 6 m and prism height 8 m. Select **A**, **B**, **C** or **D**.

A 144 m^3 **B** 288 m^3 **C** 576 m^3 **D** 1152 m^3

Alamy/Ian M Butterfield (Ireland)

ISBN 9780170350990

EXERCISE 10–09

3 Find the volume of each prism.

a
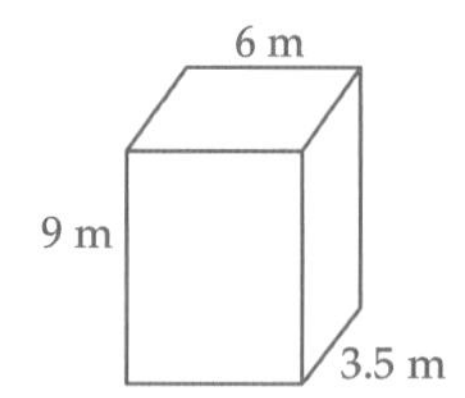

b
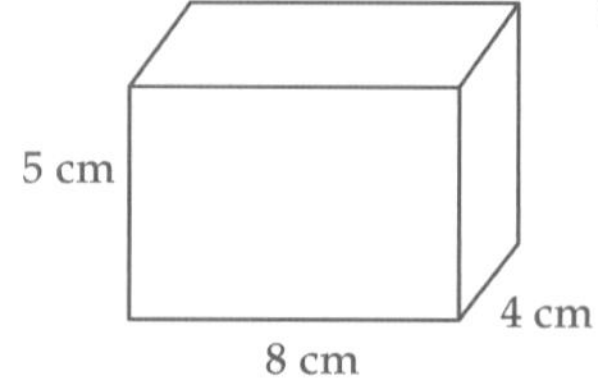

c
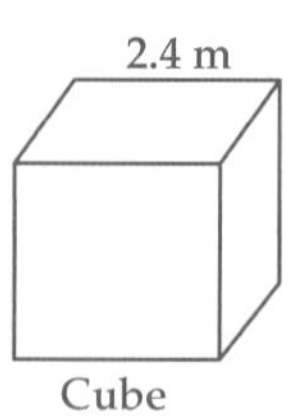

d
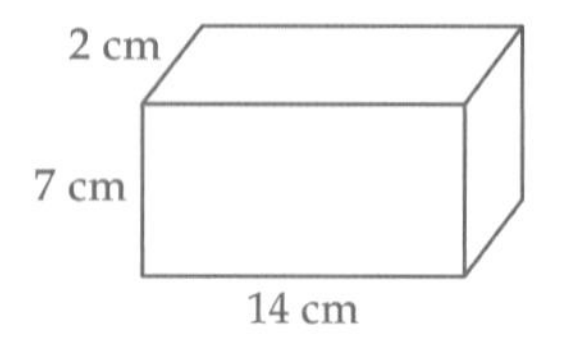

e
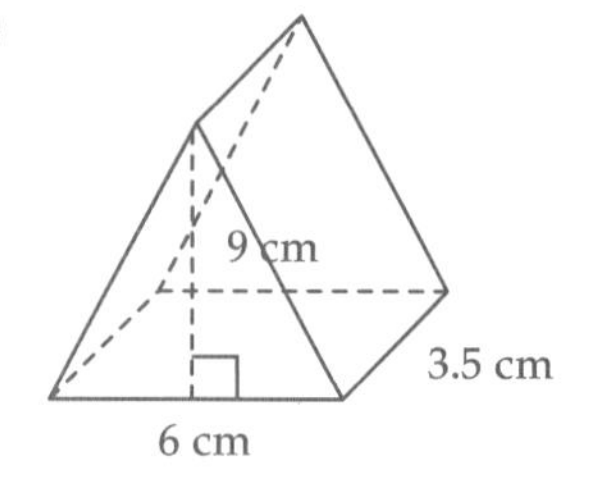

f
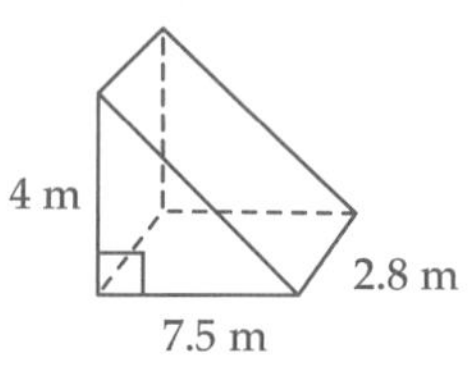

g
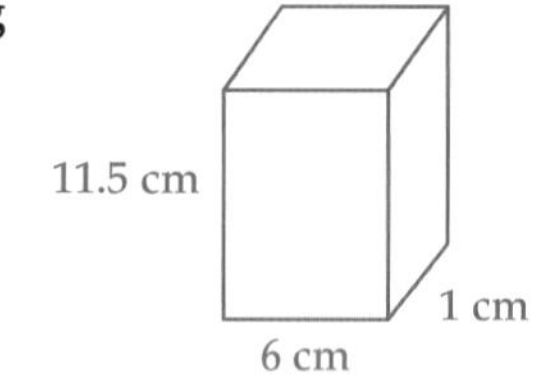

h
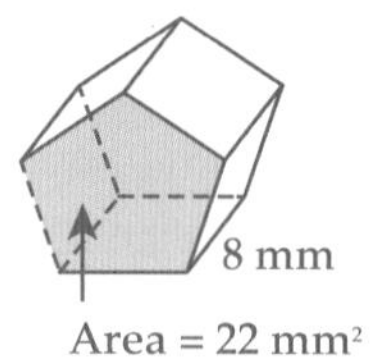

i
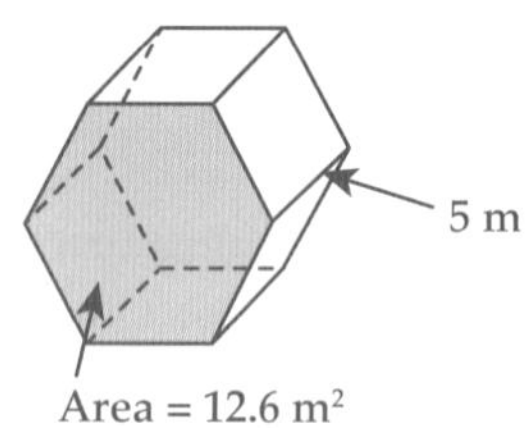

j
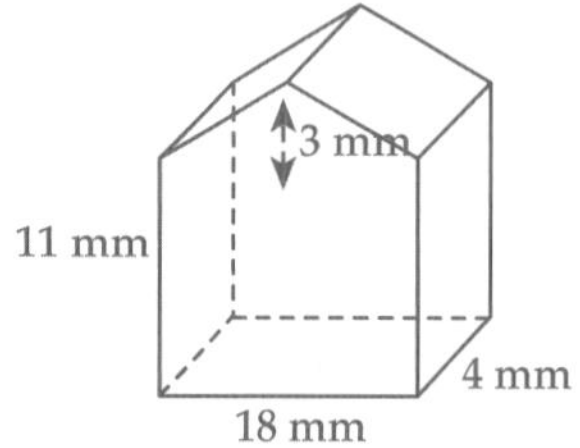

k
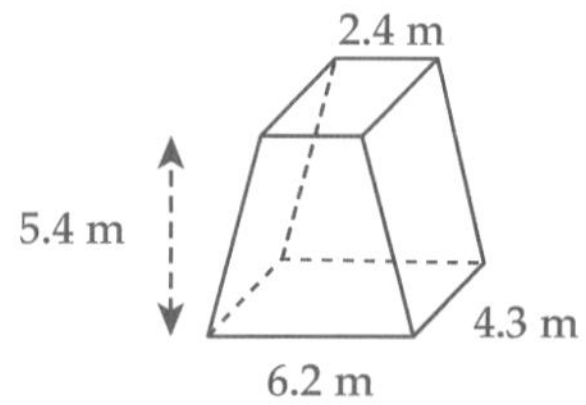

l
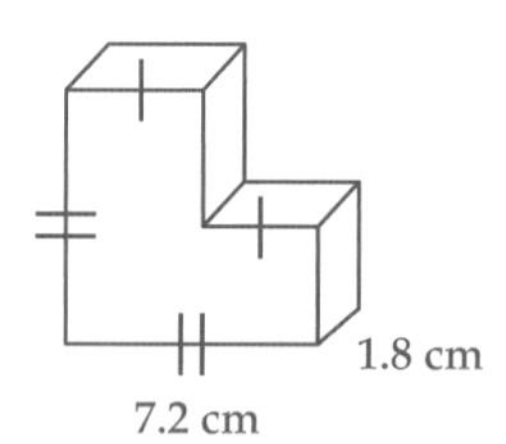

 ISBN 9780170350990

10–10 Volume of a cylinder

A **cylinder** is like a 'circular prism' because its cross-sections are all identical circles. Therefore, we can use the formula $V = Ah$ to find its volume. For a circle, $A = \pi r^2$, so:

$$\begin{aligned}\text{Volume of a cylinder} &= Ah \\ &= \pi r^2 \times h \\ &= \pi r^2 h\end{aligned}$$

VOLUME OF A CYLINDER

$V = \pi \times \text{radius}^2 \times \text{height}$

$V = \pi r^2 h$

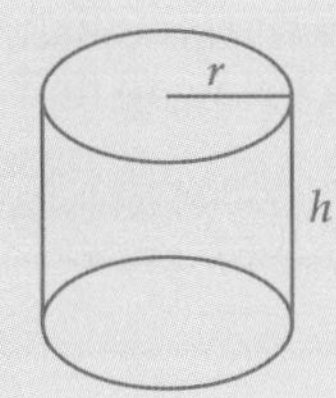

EXAMPLE 14

Find correct to one decimal place the volume of each cylinder.

a

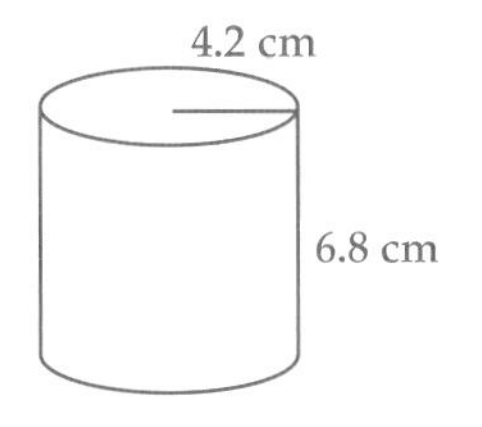

b

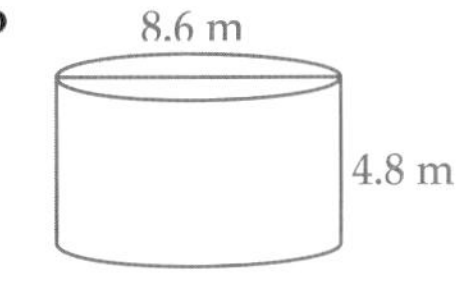

SOLUTION

a $V = \pi r^2 h$

$= \pi \times 4.2^2 \times 6.8$

$= 376.8403\ldots$

$\approx 376.8 \text{ cm}^3$

b $V = \pi r^2 h$

$= \pi \times 4.3^2 \times 4.8$ ← $r = \frac{1}{2} \times 8.6 = 4.3$

$= 278.8226\ldots$

$\approx 278.8 \text{ m}^3$

The **capacity** of a cylinder is the amount of material the cylinder can hold when full. Capacity can be measured in cubic units or mL or L if the cylinder contains liquid. You will need to know that **1 cm³ = 1 mL** and **1 m³ = 1000 L**.

EXAMPLE 15

Find the capacity of this fuel tank correct to the nearest litre.

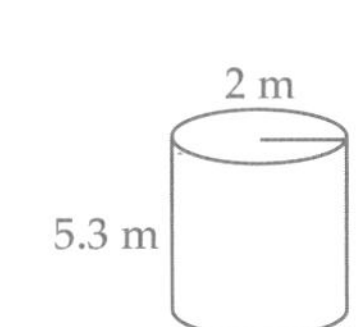

SOLUTION

$V = \pi r^2 h$

$= \pi \times 2^2 \times 5.3$

$= 66.6017\ldots \text{ m}^3$

$= 66.6017\ldots \times 1000 \text{ L}$ ← $1 \text{ m}^3 = 1 \text{ kL} = 1000 \text{ L}$

$= 66\,601.7\ldots \text{ L}$

$\approx 66\,602 \text{ L}$

EXERCISE 10–10

1 Find the volume of a cylinder with radius 6 m and height 9 m. Select the correct answer **A**, **B**, **C** or **D**.

A 1017.9 m^3 **B** 169.6 m^3 **C** 339.3 m^3 **D** 254.5 m^3

2 Find the volume of a cylinder with diameter 7.4 m and height 8.2 m. Select **A**, **B**, **C** or **D**.

A 1410.7 m^3 **B** 190.6 m^3 **C** 2821.4 m^3 **D** 352.7 m^3

3 Copy and complete the working below to find the volume of this cylinder.

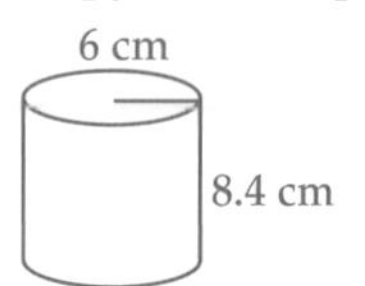

$V = \pi r^2 h$

$= \pi \times \_\_\_ \times 8.4$

$\approx \_\_\_\_$ cm^3

4 Find correct to one decimal place the volume of each cylinder below.

a

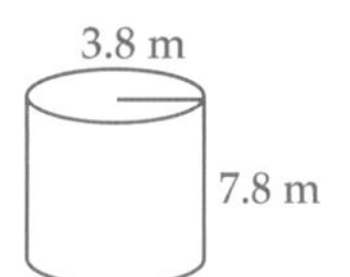

b

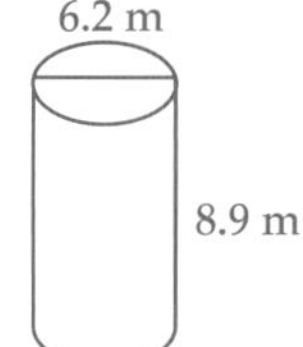

c

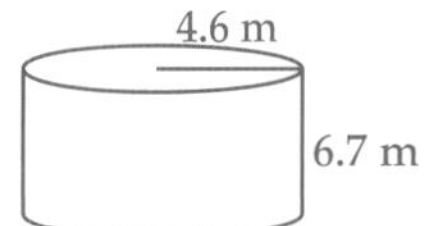

d

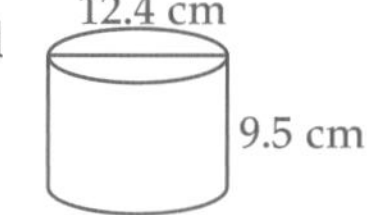

5 **a** Describe in words how you would find the volume of this solid.

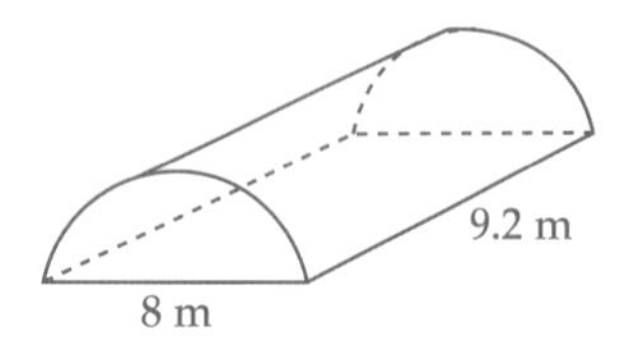

b Find correct to one decimal place the volume of this solid.

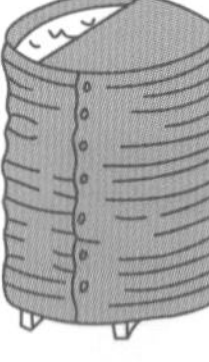

6 A water tank has a diameter of 2.4 m and a height of 6.4 m.

a What volume (correct to two decimal places) of water can it hold in cubic metres?

b What is the volume correct to the nearest litre?

c What is the volume correct to the nearest kilolitre?

7 A $1 coin has a radius of 12 mm and a thickness of 3 mm. What is its volume correct to the nearest cubic millimetre?

8 A can of soft drink has a diameter of 7 cm and a height of 8.5 cm.

a What is the volume of the can correct to one decimal place?

b What is its capacity correct to the nearest mL?

ISBN 9780170350990

CROSSWORD PUZZLE

Make a copy of this puzzle, then complete the crossword using words from the list below.

AREA	CIRCUMFERENCE	COMPOSITE	CIRCLE
CONVERT	DIAMETER	FIGURE	PARALLEL
PARALLELOGRAM	PI	RADIUS	RECTANGLE

ISBN 9780170350990

PRACTICE TEST 10

Part A General topics

Calculators are not allowed.

1 If $x = 8$ and $y = -4$, evaluate $x^2 + 2y$.

2 What are the possible outcomes when a die is tossed?

3 Complete this pattern: 81, 27, 9, 3, ______

4 Evaluate $-9 + 12 \times 6$.

5 Evaluate $-8 - (-3) + (-2)$.

6 What is the value of $a + b$ for the following diagram?

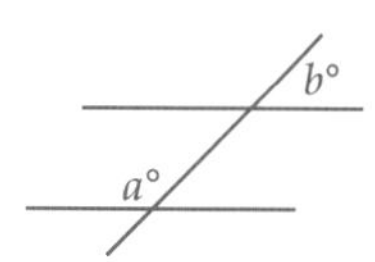

7 A plane flight started at 1925 and lasted 3 hours 15 minutes. What time did it finish?

8 Copy and complete: 268 cm = ______ m.

9 Convert 0.6 to a simple fraction.

10 Use a factor tree to write 72 as a product of its prime factors.

Part B Area and volume

Calculators are allowed.

10–01 Metric units for area

11 What unit would you use to measure the area of a national park? Select the correct answer **A**, **B**, **C** or **D**.

A mm^2 **B** m^2 **C** cm^2 **D** ha

12 Copy and complete:

a $9.2\ m^2$ = ______ cm^2

b $77\ 000\ m^3$ = ___ ha.

10–02 Areas of rectangles, triangles and parallelograms

13 What is the area of a parallelogram with base 34 mm and height 4.6 cm? Select **A**, **B**, **C** or **D**.

A 15.64 cm^2 **B** 156.4 mm^2 **C** 156.4 cm^2 **D** 15.64 mm^2

14 Find the area of each shape.

a

4.2 cm

12.3 cm

b

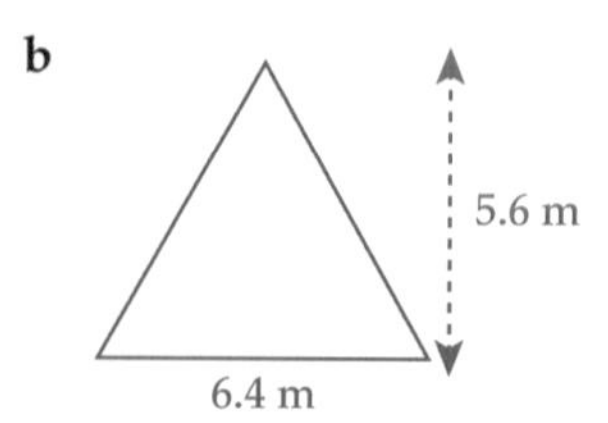

ISBN 9780170350990

10-03 Areas of composite shapes

15 Find the area of this shape.

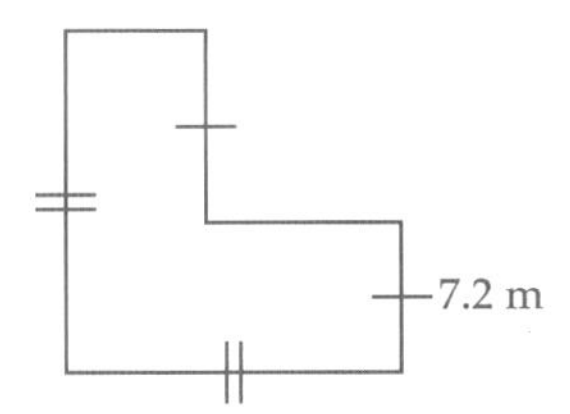

10-04 Area of a trapezium

16 Find the area of this trapezium.

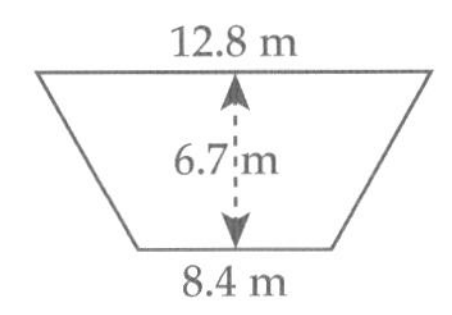

10-05 Areas of kites and rhombuses

17 Find the area of each shape.

a

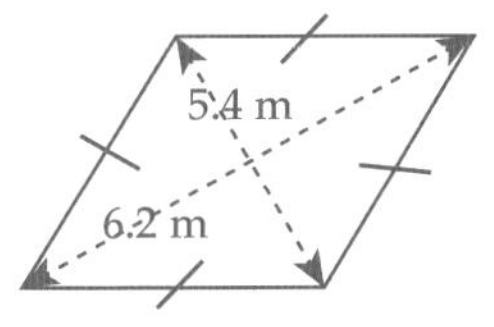

b

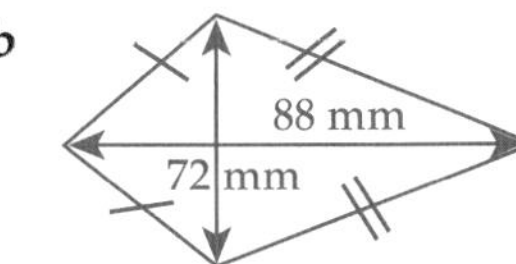

10-06 Area of a circle

18 Find correct to one decimal place the area of this semicircle.

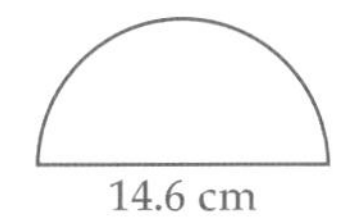

10-07 Metric units for volume

19 Copy and complete:

a 42 356 cm^3 = ______ m^3

b 56.8 cm^3 = ___ mL

10-08 Drawing prisms

20 For this solid shape, draw its:

a back view

b left view

c top view

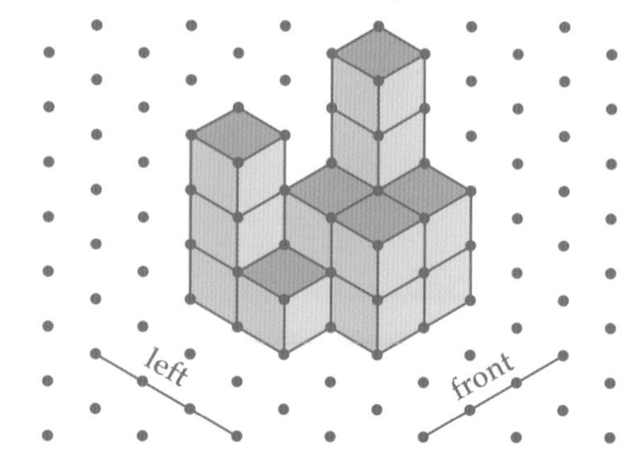

10-09 Volume of a prism

21 Find the volume of this prism.

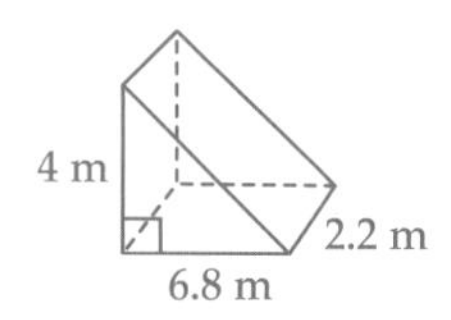

10-10 Volume of a cylinder

22 Find the volume of each cylinder correct to two decimal places.

a

b

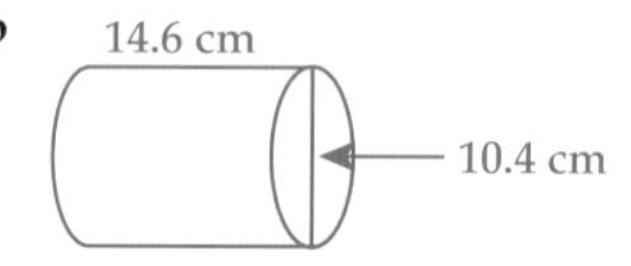

ISBN 9780170350990

FRACTIONS

11

WHAT'S IN CHAPTER 11?

IN THIS CHAPTER YOU WILL:

- find equivalent fractions
- simplify fractions
- convert between improper fractions and mixed numerals
- order fractions, including on a number line
- add and subtract fractions, including mixed numerals
- find a fraction of a number or amount
- multiply fractions, including mixed numerals
- find the reciprocal of a fraction
- divide fractions, including mixed numerals

ISBN 9780170350990

11-01 Simplifying fractions

WORDBANK

numerator The number at the top in a fraction.

denominator The number at the bottom in a fraction.

$\frac{2}{5}$ ← numerator ← denominator

equivalent fractions Fractions that are the same size. They have the same value.

$\frac{1}{3}$, $\frac{2}{6}$ and $\frac{3}{9}$ are equivalent fractions.

simplify a fraction To make the numerator and the denominator of a fraction as small as possible by dividing by the same factor.

To find an equivalent fraction:

- multiply the numerator and the denominator by the same number, or
- divide the numerator and the denominator by the same number.

EXAMPLE 1

Complete this pair of equivalent fractions: $\frac{4}{5} = \frac{24}{?}$.

SOLUTION:

To find the missing denominator, look at the two numerators: 4 and 24.

4 is multiplied by **6** to give 24, so do the same thing to the denominator 5.

$$\frac{4}{5} = \frac{4 \times 6}{5 \times 6} = \frac{24}{30}$$

To simplify a fraction, divide the numerator and the denominator by the same number, preferably a large number such as their highest common factor (HCF), until the fraction is in lowest form.

EXAMPLE 2

Simplify each fraction.

a $\frac{20}{25}$ **b** $\frac{28}{49}$

SOLUTION

a $\frac{20 \div 5}{25 \div 5} = \frac{4}{5}$ ← dividing numerator and denominator by 5, the HCF of 20 and 25, or on calculator, enter 20 a b/c 25 =

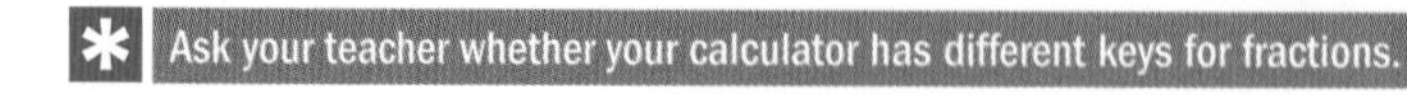

b $\frac{28 \div 7}{49 \div 7} = \frac{4}{7}$

ISBN 9780170350990

EXERCISE 11-01

1 Which fraction is equivalent to $\frac{4}{5}$? Select the correct answer **A**, **B**, **C** or **D**.

A $\frac{6}{10}$ **B** $\frac{12}{10}$ **C** $\frac{12}{15}$ **D** $\frac{16}{25}$

2 Simplify $\frac{16}{36}$. Select **A**, **B**, **C** or **D**.

A $\frac{2}{3}$ **B** $\frac{4}{9}$ **C** $\frac{4}{12}$ **D** $\frac{2}{9}$

3 Copy and complete:

To form an equivalent fraction we multiply both the ________ and the ________ by the same number, or we can divide both the ________ and the ________ by the same number.

To simplify a fraction, ________ both the numerator and the ________ by the same number until the fraction is in its ________ form.

4 Which fractions below are equivalent?

$\frac{12}{15}$ $\frac{6}{8}$ $\frac{4}{5}$ $\frac{15}{20}$ $\frac{24}{30}$ $\frac{3}{4}$

5 Complete each pair of equivalent fractions.

a $\frac{3}{5} = \frac{}{15}$ **b** $\frac{5}{8} = \frac{}{24}$ **c** $\frac{4}{9} = \frac{}{45}$

d $\frac{2}{5} = \frac{12}{}$ **e** $\frac{18}{30} = \frac{6}{}$ **f** $\frac{15}{25} = \frac{3}{}$

g $\frac{7}{10} = \frac{}{80}$ **h** $\frac{3}{8} = \frac{12}{}$ **i** $\frac{3}{4} = \frac{}{100}$

6 Simplify each fraction.

a $\frac{8}{12}$ **b** $\frac{15}{45}$ **c** $\frac{12}{20}$ **d** $\frac{16}{18}$

e $\frac{16}{24}$ **f** $\frac{25}{75}$ **g** $\frac{42}{56}$ **h** $\frac{32}{40}$

i $\frac{55}{85}$ **j** $\frac{30}{400}$ **k** $\frac{50}{280}$ **l** $\frac{36}{81}$

7 Is each equation true or false?

a $\frac{18}{27} = \frac{2}{9}$ **b** $\frac{3}{5} = \frac{45}{75}$ **c** $\frac{56}{49} = \frac{7}{8}$ **d** $\frac{120}{600} = \frac{2}{5}$

11–02 Improper fractions and mixed numerals

WORDBANK

proper fraction A fraction such as $\frac{4}{10}$ where the numerator is smaller than the denominator.

improper fraction A fraction such as $\frac{7}{3}$ where the numerator is larger than or equal to the denominator.

mixed numeral A number such as $3\frac{2}{5}$, made up of a whole number and a fraction.

If a fraction's numerator is larger than its denominator, then the value of the fraction is greater than 1. For example, the improper fraction $\frac{3}{2}$ is represented by the diagram below:

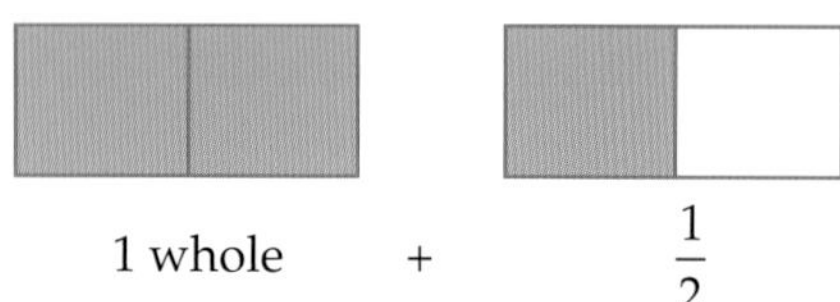

1 whole + $\frac{1}{2}$

Three parts are shaded and there are two parts in each whole, so $\frac{3}{2} = 1\frac{1}{2}$, a mixed numeral.

To convert an improper fraction to a mixed numeral, divide the numerator by the denominator and write the remainder as a proper fraction.

EXAMPLE 3

Convert each improper fraction to a mixed numeral.

a $\frac{5}{4}$ **b** $\frac{13}{5}$

SOLUTION

a $\frac{5}{4} = 5 \div 4$

$= 1$ remainder 1

$= 1\frac{1}{4}$

b $\frac{13}{5} = 13 \div 5$

$= 2$ remainder 3

$= 2\frac{3}{5}$

* Write the remainder in the numerator of the fraction.

EXAMPLE 4

Convert each mixed numeral to an improper fraction.

a $2\frac{1}{4}$ **b** $3\frac{2}{5}$

SOLUTION

a $2\frac{1}{4} = 2 + \frac{1}{4}$

$= \frac{2 \times 4}{4} + \frac{1}{4}$

$= \frac{8}{4} + \frac{1}{4}$

$= \frac{9}{4}$

b $3\frac{2}{5} = 3 + \frac{2}{5}$

$= \frac{3 \times 5}{5} + \frac{2}{5}$

$= \frac{15}{5} + \frac{2}{5}$

$= \frac{17}{5}$

ISBN 9780170350990

11–02 Improper fractions and mixed numerals

A quick shortcut is to work clockwise from the denominator, multiply it by the whole number and add the numerator.

$$3\frac{2}{5} = \frac{5 \times 3 + 2}{5} = \frac{17}{5}$$

To convert a mixed numeral to an improper fraction, multiply the denominator by the whole number, then add the numerator. Write the total as the new numerator of the fraction.

EXERCISE 11–02

1 Convert $\frac{5}{3}$ to a mixed numeral. Select the correct answer **A, B, C** or **D**.

A $1\frac{5}{3}$ B $1\frac{1}{3}$ C $1\frac{3}{5}$ D $1\frac{2}{3}$

2 Convert $2\frac{3}{4}$ to an improper fraction. Select **A, B, C** or **D**.

A $\frac{11}{4}$ B $\frac{11}{3}$ C $\frac{9}{4}$ D $\frac{7}{4}$

3 Classify each fraction as being proper (P), improper (I) or a mixed numeral (M).

a $1\frac{1}{3}$ b $\frac{3}{50}$ c $3\frac{1}{4}$

d $\frac{1}{4}$ e $\frac{7}{3}$ f $\frac{20}{45}$

g $\frac{15}{25}$ h $\frac{92}{8}$ i $\frac{13}{2}$

j $\frac{3}{8}$ k $4\frac{3}{26}$ l $\frac{100}{7}$

4 Convert each improper fraction to a mixed numeral.

a $\frac{7}{5}$ b $\frac{8}{5}$ c $\frac{9}{4}$ d $\frac{7}{6}$

e $\frac{12}{5}$ f $\frac{6}{5}$ g $\frac{11}{9}$ h $\frac{17}{4}$

EXERCISE 11-02

5 Convert each mixed numeral into an improper fraction.

a $2\frac{1}{3}$ b $3\frac{3}{4}$ c $4\frac{3}{5}$ d $1\frac{7}{8}$

e $3\frac{2}{3}$ f $2\frac{5}{6}$ g $4\frac{4}{5}$ h $6\frac{3}{8}$

6 Is each equation true or false?

a $3\frac{1}{4} = \frac{13}{3}$ b $\frac{15}{7} = 2\frac{1}{7}$ c $4\frac{3}{4} = \frac{19}{4}$

d $\frac{22}{5} = 4\frac{2}{4}$ e $3\frac{2}{3} = \frac{11}{3}$ f $\frac{23}{6} = 3\frac{4}{6}$

7 Raj bought 5 pizzas and cut them into 8 slices each. His friends ate $4\frac{1}{4}$ of the pizzas.

a How many pieces of pizza did they eat?

b How much pizza was left?

iStockphoto/Aleaimage

ISBN 9780170350990

11-03 Ordering fractions

- **To order fractions**, write them with the same **denominator**, then compare their **numerators**.
- A **common denominator** can be found by multiplying the denominators of both fractions together or by using the lowest common multiple (LCM) of the denominators.

EXAMPLE 5

Which fraction is larger: $\frac{2}{3}$ or $\frac{7}{12}$?

SOLUTION

Change both fractions to equivalent fractions with the same denominator.

Method 1

Common denominator = 3 × 12 = 36.

$\frac{2}{3} = \frac{2 \times 12}{3 \times 12} = \frac{24}{36}$ $\quad \frac{7}{12} = \frac{7 \times 3}{12 \times 3} = \frac{21}{36}$

As 24 > 21, $\frac{24}{36}$ is larger, so $\frac{2}{3}$ is larger.

Method 2

The lowest common multiple (LCM) of 3 and 12 is 12.

$\frac{2}{3} = \frac{2 \times 4}{3 \times 4} = \frac{8}{12}$ $\quad \frac{7}{12}$ already has 12 as a denominator.

As 8 > 7, $\frac{8}{12}$ is larger, so $\frac{2}{3}$ is larger.

EXAMPLE 6

Plot the fractions $\frac{2}{3}$, $-\frac{1}{2}$ and $\frac{1}{6}$ on a number line.

SOLUTION

The lowest common multiple (LCM) of 2, 3 and 6 is 6.

$\frac{2}{3} = \frac{2 \times 2}{3 \times 2} = \frac{4}{6}$ $\quad -\frac{1}{2} = -\frac{1 \times 3}{2 \times 3} = -\frac{3}{6}$ $\quad \frac{1}{6} = \frac{1}{6}$

Divide a number line into intervals of $\frac{1}{6}$.

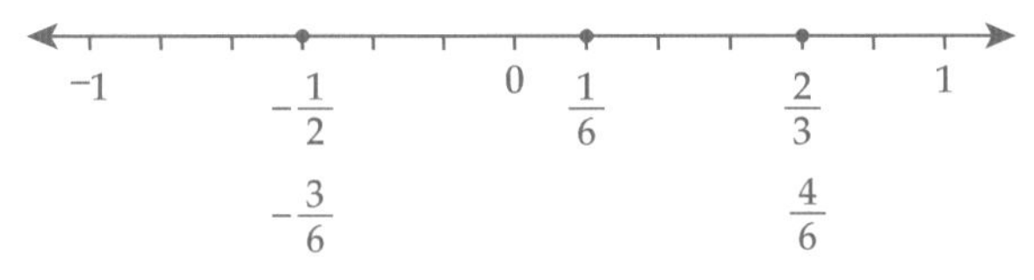

EXERCISE 11-03

1 Which fraction is largest: $\frac{1}{3}, \frac{3}{4}, \frac{7}{10}$ or $\frac{3}{5}$? Select the correct answer **A, B, C** or **D**.

A $\frac{1}{3}$ B $\frac{3}{4}$ C $\frac{7}{10}$ D $\frac{3}{5}$

2 Which fraction is smallest: $\frac{1}{3}, \frac{3}{4}, \frac{7}{10}$ or $\frac{3}{5}$? Select **A, B, C** or **D**.

A $\frac{1}{3}$ B $\frac{3}{4}$ C $\frac{7}{10}$ D $\frac{3}{5}$

3 For each pair of fractions, find the larger fraction.

a $\frac{7}{8}, \frac{3}{8}$ b $\frac{3}{5}, \frac{2}{5}$ c $\frac{3}{5}, \frac{7}{8}$

d $\frac{4}{5}, \frac{5}{6}$ e $2\frac{1}{3}, 2\frac{1}{4}$ f $3\frac{5}{7}, 2\frac{3}{4}$

4 Is each statement true or false?

a $\frac{4}{5} < \frac{9}{10}$ b $\frac{7}{12} > \frac{3}{4}$ c $\frac{5}{6} = \frac{25}{36}$

d $\frac{15}{20} > \frac{3}{4}$ e $\frac{24}{28} < \frac{7}{6}$ f $\frac{30}{45} \leq \frac{2}{3}$

5 Write each set of fractions in ascending order.

a $\frac{3}{12}, \frac{3}{4}, -\frac{7}{12}, \frac{1}{2}, \frac{8}{12}$

b $\frac{1}{4}, \frac{5}{8}, -\frac{3}{8}, \frac{9}{4}, -\frac{1}{2}$

c $\frac{1}{3}, \frac{1}{6}, \frac{3}{4}, \frac{11}{12}, \frac{5}{6}$

6 Plot the fractions from Question **5b** on a number line.

7 Write the fractions below in descending order:

a $\frac{4}{6}, \frac{1}{2}, \frac{9}{6}, \frac{1}{3}, -\frac{3}{12}$

b $-\frac{5}{9}, -\frac{2}{3}, \frac{1}{3}, \frac{13}{9}, \frac{7}{9}$

c $\frac{7}{4}, \frac{2}{3}, \frac{7}{12}, \frac{5}{6}, \frac{9}{6}$

8 Plot the fractions from Question **7b** on a number line.

ISBN 9780170350990

11–04 Adding and subtracting fractions

To add and subtract fractions:

- **with the same denominator**, simply add or subtract the numerators
- **with different denominators**, first convert them to equivalent fractions with the same denominator, then add or subtract numerators.

EXAMPLE 7

Evaluate each expression by first writing both fractions with the same denominator.

a $\frac{1}{6}+\frac{2}{3}$ **b** $\frac{2}{5}+\frac{7}{8}$ **c** $\frac{7}{8}-\frac{2}{3}$

SOLUTION

a $\frac{1}{6}+\frac{2}{3}=\frac{1}{6}+\frac{2\times 2}{3\times 2}$ ← Write with a common denominator of 6.

$=\frac{1}{6}+\frac{4}{6}$

$=\frac{5}{6}$

b $\frac{2}{5}+\frac{7}{8}=\frac{2\times 8}{5\times 8}+\frac{7\times 5}{8\times 5}$ ← Write with a common denominator of 40.

$=\frac{16}{40}+\frac{35}{40}$

$=\frac{51}{40}$

$=1\frac{11}{40}$ ← Change improper fraction to a mixed numeral.

c $\frac{7}{8}-\frac{2}{3}=\frac{7\times 3}{8\times 3}-\frac{2\times 8}{3\times 8}$ ← Write with a common denominator of 24.

$=\frac{21}{24}-\frac{16}{24}$

$=\frac{5}{24}$

11–04 Adding and subtracting fractions

To add and subtract mixed numerals:
- add or subtract the whole numbers first
- then add or subtract the fractions.

EXAMPLE 8

Evaluate each expression.

a $2\frac{1}{4}+1\frac{2}{5}$

b $3\frac{3}{8}-2\frac{1}{4}$

SOLUTION

a
$$\begin{aligned}2\frac{1}{4}+1\frac{2}{5}&=2+1+\frac{1}{4}+\frac{2}{5}\\&=3+\frac{1\times5}{4\times5}+\frac{2\times4}{5\times4}\\&=3\frac{5}{20}+\frac{8}{20}\\&=3\frac{13}{20}\end{aligned}$$

b
$$\begin{aligned}3\frac{3}{8}-2\frac{1}{4}&=3-2+\frac{3}{8}-\frac{1}{4}\\&=1+\frac{3}{8}-\frac{1\times2}{4\times2}\\&=1+\frac{3}{8}-\frac{2}{8}\\&=1\frac{1}{8}\end{aligned}$$

These expressions can also be evaluated on your calculator as follows.

a $2\frac{1}{4}+1\frac{2}{5}=3\frac{13}{20}$ ← On calculator, enter 2 [a b/c] 1 [a b/c] 4 [+] 1 [a b/c] 2 [a b/c] 5 [=]

EXERCISE 11–04

1 Evaluate $\frac{1}{3}+\frac{2}{5}$. Select the correct answer **A, B, C** or **D**.

A $\frac{3}{8}$ **B** $\frac{3}{15}$ **C** $\frac{11}{15}$ **D** $\frac{7}{15}$

2 Evaluate $\frac{3}{4}-\frac{1}{3}$. Select **A, B, C** or **D**.

A 2 **B** $\frac{5}{12}$ **C** $\frac{2}{12}$ **D** $\frac{4}{12}$

3 Evaluate each expression.

a $\frac{3}{5}+\frac{1}{5}$ **b** $\frac{7}{8}-\frac{2}{8}$ **c** $\frac{1}{7}+\frac{5}{7}$ **d** $\frac{4}{5}-\frac{1}{5}$

4 Evaluate each sum.

a $\frac{1}{4}+\frac{2}{5}$ **b** $\frac{1}{6}+\frac{3}{5}$ **c** $\frac{3}{8}+\frac{2}{3}$ **d** $\frac{5}{8}+\frac{3}{5}$

e $\frac{4}{9}+\frac{1}{3}$ **f** $\frac{7}{12}+\frac{2}{3}$ **g** $\frac{11}{15}+\frac{3}{5}$ **h** $\frac{7}{8}+\frac{5}{6}$

ISBN 9780170350990

5 Evaluate each difference.

a $\frac{3}{4}-\frac{2}{5}$ b $\frac{5}{6}-\frac{1}{4}$ c $\frac{7}{8}-\frac{1}{3}$ d $\frac{7}{10}-\frac{2}{5}$

e $\frac{5}{9}-\frac{1}{6}$ f $\frac{7}{12}-\frac{1}{3}$ g $\frac{14}{15}-\frac{3}{5}$ h $\frac{11}{12}-\frac{3}{4}$

6 Is each equation true or false?

a $\frac{7}{9}+\frac{2}{3}=\frac{7}{9}+\frac{4}{9}$ b $\frac{9}{10}-\frac{3}{4}=\frac{18}{20}-\frac{15}{20}$

7 Evaluate each expression.

a $1\frac{1}{4}+1\frac{2}{4}$ b $1\frac{1}{5}+2\frac{1}{4}$ c $3\frac{4}{7}-1\frac{1}{7}$ d $4\frac{4}{5}-2\frac{1}{3}$

e $3\frac{1}{2}-2\frac{3}{4}$ f $2\frac{1}{3}-1\frac{1}{4}$ g $3\frac{1}{5}-1\frac{2}{3}$ h $2\frac{2}{3}-1\frac{3}{4}$

8 Luka bought some berries and ate $\frac{1}{6}$ of them. He gave Lachlan $\frac{1}{4}$ of them and Liam $\frac{1}{3}$ of them. What fraction of the berries are left?

Shutterstock.com/Olga Rosi

11–05 Fraction of a quantity

What is $\frac{2}{3}$ of 12? Here are 12 lollies:

If these lollies were divided into three equal piles, each pile would be $\frac{1}{3}$ of 12.

$\frac{1}{3}$ of 12 = 4 $\qquad$ $\frac{1}{3}$ of 12 = 4 $\qquad$ $\frac{1}{3}$ of 12 = 4

$$\frac{2}{3} \times 12 = 8$$

EXAMPLE 9

Find each quantity.

a $\frac{3}{4}$ of \$280 $\qquad$ **b** $\frac{5}{12}$ of 4 hours (in minutes)

SOLUTION

a $\frac{3}{4}$ of \$280 $= \left(\frac{1}{4} \times \$280\right) \times 3$

$= \$70 \times 3$ ← $280 \div 4 = 70$

$= \$210$

b $\frac{5}{12}$ of 4 hours $= \frac{5}{12}$ of 240 min ← 4×60 min

$= \left(\frac{1}{12} \times 240\right) \times 5$

$= 20 \times 5$ ← $240 \div 12 = 20$

$= 100$ min

 The amount is divided by the denominator of the fraction and then multiplied by the numerator.

ISBN 9780170350990

EXERCISE 11–05

1 Find $\frac{1}{3}$ of \$240. Select the correct answer **A**, **B**, **C** or **D**.

A \$60 **B** \$80 **C** \$160 **D** \$100

2 Find $\frac{2}{3}$ of \$240. Select **A**, **B**, **C** or **D**.

A \$120 **B** \$240 **C** \$160 **D** \$200

3 Find each quantity.

a $\frac{1}{4} \times 28$ **b** $\frac{2}{4} \times 28$ **c** $\frac{3}{4} \times 28$ **d** $\frac{4}{4} \times 28$

e $\frac{1}{8}$ of 56 **f** $\frac{2}{8}$ of 56 **g** $\frac{5}{8}$ of 56 **h** $\frac{7}{8}$ of 56

i $\frac{1}{4}$ of 1 hour (in min) **j** $\frac{3}{5}$ of \$40 **k** $\frac{3}{4}$ of 200 km

l $\frac{5}{8}$ of \$240 **m** $\frac{3}{7}$ of 42 m **n** $\frac{2}{3}$ of 2 hours (in min)

o $\frac{7}{8}$ of 64 pages **p** $\frac{4}{3}$ of \$150 **q** $\frac{7}{12}$ of 96 L

r $\frac{7}{9}$ of 360 mL **s** $\frac{2}{15}$ of 330 min **t** $\frac{6}{5}$ of \$75

4 Is each equation true or false?

a $\frac{1}{5}$ of 6 m = 1.2 m **b** $\frac{2}{3}$ of 3 h = 100 min **c** $\frac{3}{4}$ of \$260 = \$195

d $\frac{3}{8}$ of 12 L = 4 L **e** $\frac{5}{12}$ of \$720 = \$300 **f** $\frac{9}{10}$ of 2 L = 180 mL

5 At a party, 12 people ate $\frac{3}{4}$ pizza each and 8 people ate $\frac{5}{8}$ pizza each. How many pizzas were eaten?

6 Geri saved \$12 000 for her African holiday. After 1 week, she had spent $\frac{1}{10}$ of her savings. The next week she spent $\frac{1}{20}$ of her savings.

a How much did she spend in the first week?

b How much did she spend in the second week?

c What fraction of Geri's savings was left for the rest of her holiday?

7 Samir is writing a novel and wrote 120 pages on Monday. On Tuesday, he only wrote $\frac{5}{6}$ of this amount. After that, he was only able to write $\frac{4}{5}$ of his previous day's effort. How many pages did Samir write on:

a Tuesday?

b Thursday?

11-06 Multiplying fractions

What is $\frac{1}{2}$ of $\frac{1}{3}$?

$\frac{1}{3}$ of this diagram is shaded:

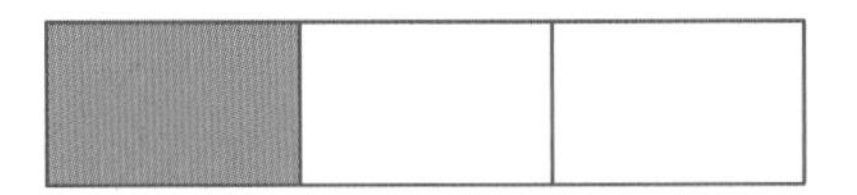

$\frac{1}{2} \times \frac{1}{3} = \frac{1}{2}$ of $\frac{1}{3}$ is shaded:

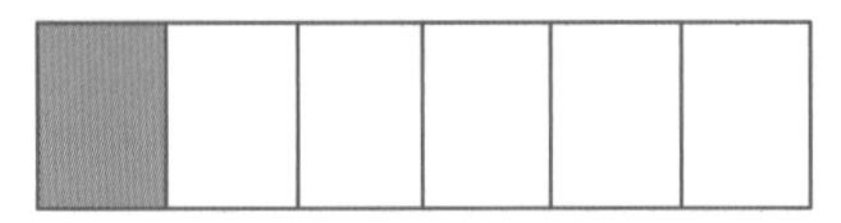

So $\frac{1}{2} \times \frac{1}{3} = \frac{1}{6}$

To **multiply** fractions:

- simplify numerators with denominators (if possible) by dividing by a common factor
- then **multiply numerators** and **multiply denominators** separately.

EXAMPLE 10

Evaluate each product.

a $\frac{3}{8} \times \frac{7}{15}$

b $\frac{5}{9} \times \frac{18}{7}$

SOLUTION

a $\frac{3}{8} \times \frac{7}{15} = \frac{3 \times 7}{8 \times 15}$ or $\frac{3}{8} \times \frac{7}{15} = \frac{\cancel{3}^{1} \times 7}{8 \times \cancel{15}^{5}}$ ← simplify first: divide by 3

$= \frac{21}{120}$ $\qquad$ $= \frac{7}{40}$

$= \frac{7}{40}$

b $\frac{5}{9} \times \frac{18}{7} = \frac{5 \times 18}{9 \times 7}$ or $\frac{5}{9} \times \frac{18}{7} = \frac{5 \times \cancel{18}^{2}}{\cancel{9}^{1} \times 7}$ ← simplify first: divide by 9

$= \frac{90}{63}$ $\qquad$ $= \frac{10}{7}$

$= \frac{10}{7}$ $\qquad$ $= 1\frac{3}{7}$

$= 1\frac{3}{7}$ ← change to a mixed numeral

To **multiply mixed numerals**, first convert them to improper fractions.

EXAMPLE 11

Evaluate each product.

a $1\frac{1}{3} \times 2\frac{1}{2}$

b $1\frac{3}{5} \times 2\frac{1}{4}$

ISBN 9780170350990

SOLUTION

a $1\frac{1}{3}\times 2\frac{1}{2}=\frac{4}{3}\times\frac{5}{2}$

$=\frac{{}^{2}\cancel{4}\times 5}{3\times\cancel{2}_{1}}$

$=\frac{10}{3}$

$=3\frac{1}{3}$

b $1\frac{3}{5}\times 2\frac{1}{4}=\frac{8}{5}\times\frac{9}{4}$

$=\frac{{}^{2}\cancel{8}\times 9}{5\times\cancel{4}_{1}}$

$=\frac{18}{5}$

$=3\frac{3}{5}$

* These products can also be evaluated on your calculator.

EXERCISE 11–06

1 Evaluate $\frac{1}{4}\times\frac{2}{5}$. Select the correct answer **A**, **B**, **C** or **D**.

A $\frac{2}{9}$ **B** $\frac{1}{10}$ **C** $\frac{3}{20}$ **D** $\frac{1}{5}$

2 Evaluate $\frac{5}{8}\times\frac{2}{9}$. Select **A**, **B**, **C** or **D**.

A $\frac{5}{36}$ **B** $\frac{7}{17}$ **C** $\frac{5}{13}$ **D** $\frac{5}{9}$

3 Is each equation true or false?

a $\frac{1}{2}\times\frac{1}{3}=\frac{1}{6}$ **b** $\frac{1}{5}\times\frac{1}{3}=\frac{1}{8}$ **c** $\frac{1}{2}\times\frac{1}{6}=\frac{1}{12}$ **d** $\frac{2}{7}\times\frac{1}{4}=\frac{2}{11}$

4 Copy and complete:

a $\frac{2}{5}\times\frac{3}{8}=\frac{2\times\square}{5\times\square}$

$=\frac{\square}{40}$

$=$ ___

b $\frac{8}{9}\times\frac{3}{4}=\frac{8\times\square}{9\times\square}$

$=\frac{24}{\square}$

$=$ ___

5 Evaluate each product.

a $\frac{2}{5}\times\frac{5}{4}$ **b** $\frac{3}{7}\times\frac{14}{15}$ **c** $\frac{5}{9}\times\frac{18}{10}$ **d** $\frac{4}{9}\times\frac{15}{24}$

e $\frac{6}{7}\times\frac{3}{9}$ **f** $\frac{4}{3}\times\frac{15}{20}$ **g** $\frac{5}{3}\times\frac{6}{10}$ **h** $\frac{5}{8}\times\frac{16}{15}$

6 Copy and complete this sentence.

To multiply mixed numerals, first convert them to ____________ fractions, then multiply the ____________ and multiply the ____________.

7 Evaluate each product.

a $1\frac{1}{2}\times 2\frac{1}{5}$ **b** $1\frac{1}{2}\times 3\frac{2}{3}$ **c** $1\frac{1}{2}\times 1\frac{7}{9}$ **d** $2\frac{1}{3}\times\frac{1}{5}$

e $\frac{3}{4}\times 2\frac{1}{3}$ **f** $1\frac{1}{3}\times 1\frac{2}{5}$ **g** $1\frac{2}{3}\times\frac{1}{5}$ **h** $2\frac{2}{3}\times 1\frac{3}{8}$

i $2\frac{1}{2}\times 1\frac{1}{4}$ **j** $1\frac{2}{3}\times 1\frac{1}{5}$ **k** $3\frac{1}{4}\times 1\frac{1}{3}$ **l** $3\frac{1}{4}\times 4\frac{4}{5}$

11–07 Dividing fractions

The **reciprocal of a fraction** is the fraction 'turned upside down', for example, the reciprocal of $\frac{3}{8}$ is $\frac{8}{3}$.

To **divide by a fraction** $\frac{a}{b}$, multiply by its reciprocal $\frac{b}{a}$.

EXAMPLE 12

Evaluate each quotient.

a $\frac{3}{4} \div \frac{5}{12}$

b $\frac{4}{7} \div \frac{8}{5}$

SOLUTION

a
$$\begin{aligned} \frac{3}{4} \div \frac{5}{12} &= \frac{3}{4} \times \frac{12}{5} \\ &= \frac{3 \times \cancel{12}^{3}}{{}_{1}\cancel{4} \times 5} \\ &= \frac{9}{5} \\ &= 1\frac{4}{5} \end{aligned}$$

b
$$\begin{aligned} \frac{4}{7} \div \frac{8}{5} &= \frac{4}{7} \times \frac{5}{8} \\ &= \frac{\cancel{4}^{1} \times 5}{7 \times \cancel{8}^{2}} \\ &= \frac{5}{14} \end{aligned}$$

To divide mixed numerals, first convert them to improper fractions.

EXAMPLE 13

Evaluate each quotient.

a $1\frac{1}{3} \div \frac{2}{9}$

b $1\frac{1}{4} \div 1\frac{1}{8}$

SOLUTION

a
$$\begin{aligned} 1\frac{1}{3} \div \frac{2}{9} &= \frac{4}{3} \div \frac{2}{9} \\ &= \frac{4}{3} \times \frac{9}{2} \\ &= \frac{{}^{2}\cancel{4} \times \cancel{9}^{3}}{{}^{1}\cancel{3} \times \cancel{2}^{1}} \\ &= \frac{6}{1} \\ &= 6 \end{aligned}$$

b
$$\begin{aligned} 1\frac{1}{4} \div 1\frac{1}{8} &= \frac{5}{4} \div \frac{9}{8} \\ &= \frac{5}{4} \times \frac{8}{9} \\ &= \frac{5 \times \cancel{8}^{2}}{{}_{1}\cancel{4} \times 9} \\ &= \frac{10}{9} \\ &= 1\frac{1}{9} \end{aligned}$$

ISBN 9780170350990

EXERCISE 11-07

1 Which product has the same value as $\frac{2}{5} \div \frac{3}{4}$? Select the correct answer **A**, **B**, **C** or **D**.

A $\frac{5}{2} \times \frac{3}{4}$ **B** $\frac{2}{5} \times \frac{4}{3}$ **C** $\frac{5}{2} \times \frac{4}{3}$ **D** $\frac{2}{5} \times \frac{3}{4}$

2 Evaluate $\frac{2}{5} \div \frac{3}{4}$. Select **A**, **B**, **C** or **D**.

A $\frac{8}{15}$ **B** $3\frac{1}{3}$ **C** $\frac{15}{8}$ **D** $\frac{3}{10}$

3 Find the reciprocal of each fraction or numeral.

a $\frac{3}{4}$ **b** $\frac{1}{5}$ **c** $\frac{5}{9}$ **d** $\frac{1}{12}$

e 5 **f** $\frac{4}{5}$ **g** $\frac{5}{7}$ **h** 8

4 Copy and complete this sentence:

To divide by a fraction, ____________ by its ____________.

5 Is each equation true or false?

a $\frac{5}{8} \div \frac{1}{3} = \frac{5}{24}$ **b** $\frac{3}{5} \div \frac{6}{5} = \frac{1}{2}$ **c** $\frac{1}{12} \div \frac{3}{2} = \frac{1}{18}$ **d** $\frac{2}{3} \div \frac{4}{15} = \frac{2}{5}$

6 Copy and complete:

a $\frac{2}{5} \div \frac{3}{8} = \frac{2}{5} \times \frac{8}{\square}$

$= \frac{\square}{15}$

$=$ ___

b $\frac{8}{9} \div \frac{7}{3} = \frac{8}{9} \times \frac{\square}{7}$

$= \frac{24}{\square}$

$=$ ___

7 Evaluate each quotient.

a $\frac{3}{5} \div \frac{15}{10}$ **b** $\frac{12}{7} \div \frac{6}{14}$ **c** $\frac{3}{8} \div \frac{15}{12}$ **d** $\frac{4}{5} \div \frac{8}{15}$

e $\frac{24}{9} \div \frac{8}{27}$ **f** $\frac{18}{5} \div \frac{6}{20}$ **g** $\frac{6}{7} \div \frac{30}{42}$ **h** $\frac{8}{9} \div \frac{32}{27}$

8 Copy and complete this sentence.

To divide mixed numerals, first convert them to ____________ fractions, then multiply by the ____________ of the second fraction.

9 Evaluate each quotient.

a $\frac{5}{7} \div \frac{10}{21}$ **b** $2\frac{1}{2} \div 4\frac{2}{3}$ **c** $1\frac{1}{3} \div 2\frac{2}{4}$ **d** $\frac{3}{8} \div 1\frac{1}{8}$

e $1\frac{1}{4} \div \frac{5}{2}$ **f** $2\frac{1}{10} \div \frac{8}{3}$ **g** $1\frac{3}{8} \div \frac{2}{5}$ **h** $2\frac{1}{3} \div 3\frac{1}{2}$

10 Max was on a building site and had $10\frac{3}{4}$ m of timber that had to be cut into pieces $1\frac{1}{4}$ m long. How many pieces could he cut?

ISBN 9780170350990

CODE PUZZLE

In order, list the letters for the clues below to spell out a two-word phrase relating to this topic.

The first $\frac{2}{7}$ of FREEWAY

The first $\frac{2}{5}$ of ACTOR

The middle $\frac{1}{7}$ of DAYTIME

The middle $\frac{1}{5}$ of POINT

The last $\frac{2}{11}$ of EXCLAMATION

The first $\frac{2}{9}$ of FRIGHTENS

The last $\frac{2}{6}$ of SPOKEN

The last $\frac{1}{4}$ of JAZZ

The first $\frac{1}{6}$ of YELLOW

ISBN 9780170350990

PRACTICE TEST 11

Part A General topics

Calculators are not allowed.

1 Convert 1745 to 12-hour time.

2 Simplify $\frac{32xy}{16y}$.

3 Evaluate 5^0.

4 Write Pythagoras' theorem for this triangle.

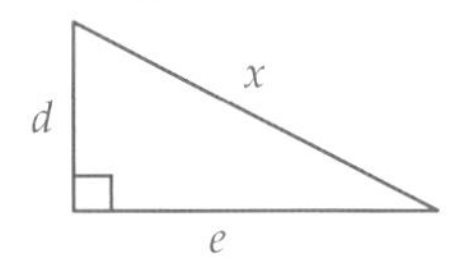

5 Evaluate $(-2)^3$.

6 Write an algebraic expression for the number of hours in d days.

7 Solve $4x - 26 = x + 1$.

8 Evaluate $486 \div 9$.

9 Convert $\frac{1}{3}$ to a decimal.

10 What is the probability of rolling a factor of 4 on a die?

Part B Fractions

Calculators are allowed.

11-01 Simplifying fractions

11 Simplify $\frac{27}{33}$. Select the correct answer **A**, **B**, **C** or **D**.

A $\frac{3}{11}$ **B** $\frac{9}{11}$ **C** $\frac{27}{33}$ **D** $\frac{9}{33}$

12 Which fraction is equivalent to $\frac{5}{12}$? Select **A**, **B**, **C** or **D**.

A $\frac{10}{36}$ **B** $\frac{35}{72}$ **C** $\frac{30}{72}$ **D** $\frac{20}{36}$

11-02 Improper fractions and mixed numerals

13 Convert $3\frac{3}{4}$ to an improper fraction. Select **A**, **B**, **C** or **D**.

A $\frac{36}{4}$ **B** $\frac{15}{3}$ **C** $\frac{9}{4}$ **D** $\frac{15}{4}$

14 Convert each improper fraction to a mixed numeral.

a $\frac{12}{7}$ **b** $\frac{25}{8}$

11-03 Ordering fractions

15 Write these fractions in descending order: $\frac{1}{4}, \frac{1}{3}, \frac{3}{8}, \frac{1}{6}$.

11-04 Adding and subtracting fractions

16 Evaluate each expression.

a $\frac{2}{3} + \frac{3}{4}$

b $\frac{7}{8} - \frac{3}{5}$

11–05 Fraction of a quantity

17 Find each quantity.

a $\frac{3}{4}$ of \$680

b $\frac{2}{3}$ of 2 hours

11–06 Multiplying fractions

18 Evaluate each product.

a $\frac{4}{5} \times \frac{15}{16}$

b $2\frac{1}{2} \times 3\frac{1}{4}$

11–07 Dividing fractions

19 Evaluate each quotient.

a $\frac{7}{9} \div \frac{14}{27}$

b $2\frac{1}{3} \div 3\frac{1}{6}$

ISBN 9780170350990

PERCENTAGES

12

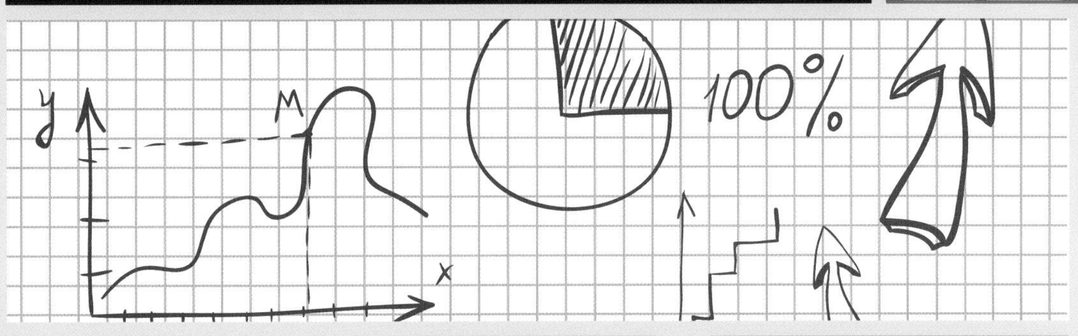

WHAT'S IN CHAPTER 12?

IN THIS CHAPTER YOU WILL:

- convert between percentages, fractions and decimals
- compare percentages, fractions and decimals
- find a percentage of a number or metric quantity
- express quantities as fractions and percentages of a whole
- calculate percentage increases and decreases
- use the unitary method to find a whole amount given a percentage of it
- solve problems involving profit and loss, cost price and selling price
- solve problems involving discounts and GST

* Shutterstock.com/JustMarie

ISBN 9780170350990

12–01 Percentages and fractions

A **percentage** is a fraction with a denominator of 100; for example, 12% means 12 out of 100 or $\frac{12}{100}$.

To convert a percentage to a fraction, write the percentage as a fraction over 100 and simplify if possible.

EXAMPLE 1

Convert each percentage to a fraction.

a 40% **b** 65% **c** 70% **d** $12\frac{1}{2}\%$

SOLUTION

a $40\% = \frac{40}{100} = \frac{2}{5}$

b $65\% = \frac{65}{100} = \frac{13}{20}$

c $70\% = \frac{70}{100} = \frac{7}{10}$

d $12\frac{1}{2}\% = \frac{12\frac{1}{2}}{100} = \frac{25}{200} = \frac{1}{8}$

* Multiply numerator and denominator by 2 to form whole numbers.

To convert a fraction to a percentage, multiply it by 100%. This does not change the amount as 100% = 1.

EXAMPLE 2

Convert each fraction to a percentage.

a $\frac{3}{4}$ **b** $\frac{4}{5}$ **c** $\frac{1}{3}$

SOLUTION

a $\frac{3}{4} = \frac{3}{4} \times 100 = 75\%$

b $\frac{4}{5} = \frac{4}{5} \times 100 = 80\%$

c $\frac{1}{3} = \frac{1}{3} \times 100 = 33\frac{1}{3}\%$

EXAMPLE 3

Write in ascending order: 42%, $\frac{2}{5}$, 40.6%, $\frac{3}{8}$.

SOLUTION

Write all numbers as percentages to compare:

42%, $\frac{2}{5} \times 100\% = 40\%$, 40.6%, $\frac{3}{8} \times 100\% = 37.5\%$

Ascending order is from small to large:

37.5%, 40%, 40.6%, 42%

$\frac{3}{8}$, $\frac{2}{5}$, 40.6% , 42%

ISBN 9780170350990

EXERCISE 12–01

1 Convert 20% to a fraction. Select the correct answer **A**, **B**, **C** or **D**.

A $\frac{1}{50}$ **B** $\frac{2}{5}$ **C** $\frac{1}{20}$ **D** $\frac{1}{5}$

2 Convert each percentage to a fraction.

a 50% **b** 75% **c** 25% **d** 100%

e 70% **f** 44% **g** 30% **h** 65%

i 58% **j** 110% **k** 7% **l** 63%

m 5% **n** 95% **o** 80% **p** 16%

3 Copy and complete:

a $33\frac{1}{3}\% = \dfrac{33\frac{1}{3}}{\square}$

$= \dfrac{33\frac{1}{3} \times \square}{100 \times 3}$

$= \dfrac{\square}{300}$

$= \dfrac{\square}{3}$

b $62\frac{1}{2}\% = \dfrac{\square}{100}$

$= \dfrac{62\frac{1}{2} \times 2}{100 \times \square}$

$= \dfrac{\square}{200}$

$= \dfrac{\square}{8}$

4 Convert each percentage to a fraction.

a $83\frac{1}{3}\%$ **b** $37\frac{1}{2}\%$ **c** $66\frac{2}{3}\%$ **d** $16\frac{1}{2}\%$

5 Convert $\frac{4}{5}$ to a percentage. Select **A**, **B**, **C** or **D**.

A 80% **B** 40% **C** 50% **D** 45%

6 Convert each fraction to a percentage.

a $\frac{1}{10}$ **b** $\frac{3}{4}$ **c** $\frac{1}{5}$ **d** $\frac{2}{9}$

e $\frac{7}{20}$ **f** $\frac{2}{3}$ **g** $\frac{3}{8}$ **h** $\frac{5}{6}$

7 Write in ascending order: 68%, $\frac{3}{5}$, 62.5%, $\frac{65}{100}$.

8 Write in descending order: 88%, $\frac{4}{5}$, 81.5%, $\frac{85}{100}$.

12–02 Percentages and decimals

To convert a percentage to a decimal, divide it by 100: move the decimal point two places left.

EXAMPLE 4

Convert each percentage to a decimal.

a 2% **b** 70% **c** 4.5% **d** 120%

SOLUTION

a $2\% = 2 \div 100$
$= 0.02$

b $70\% = 70 \div 100$
$= 0.7$

c $4.5\% = 4.5 \div 100$
$= 0.045$

d $120\% = 120 \div 100$
$= 1.2$

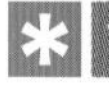
The decimal point moves two places left.

To convert a decimal to a percentage, multiply it by 100%: move the decimal point two places right.

EXAMPLE 5

Convert each decimal to a percentage.

a 0.3 **b** 0.07 **c** 0.35 **d** 0.275

SOLUTION

a $0.3 = 0.3 \times 100\%$
$= 30\%$

b $0.07 = 0.07 \times 100\%$
$= 7\%$

c $0.35 = 0.35 \times 100\%$
$= 35\%$

d $0.275 = 0.275 \times 100\%$
$= 27.5\%$

The decimal point moves two places right.

EXAMPLE 6

Write in descending order: 0.415, 47%, 0.45, 42.6%.

SOLUTION

Write all numbers as percentages to compare:

0.415 = 41.5%, 47%, 0.45 = 45%, 42.6%

Descending order is from large to small:

47%, 45%, 42.6%, 41.5%

47%, 0.45, 42.6%, 0.415

ISBN 9780170350990

EXERCISE 12–02

1 Convert 63% to a decimal. Select the correct answer **A**, **B**, **C** or **D**.

A 6.3 B 0.063 C 6.30 D 0.63

2 Convert 0.4 to a percentage. Select **A**, **B**, **C** or **D**.

A 40% B 4% C 0.4% D 0.04%

3 Is each statement true or false?

a 0.3 = 3% **b** 0.09 = 9% **c** 0.09 = 90%

d 0.52 = 52% **e** 0.6 = 60% **f** 0.08 = 80%

4 Convert each percentage to a decimal.

a 25% **b** 40% **c** 65% **d** 80%

e 12% **f** 71% **g** 100% **h** 120%

5 Convert each decimal to a percentage.

a 0.7 **b** 0.02 **c** 0.28 **d** 0.5

e 0.47 **f** 0.55 **g** 0.9 **h** 0.85

i 0.524 **j** 0.75 **k** 1.2 **l** 2.6

6 Write in ascending order: 53%, 0.51, 50.2%, 0.505

7 Write in descending order: 98%, 0.9, 97.5%, 0.099

8 Copy and complete this table.

Percentage	Fraction	Decimal
10%		
15%		
25%		
30%		
50%		
60%		
75%		
80%		
95%		
100%		

12-03 Percentage of a quantity

This table lists some commonly used percentages.

Percentage	Fraction	Decimal
10%	$\frac{1}{10}$	0.1
12.5%	$\frac{1}{8}$	0.125
25%	$\frac{1}{4}$	0.25
$33\frac{1}{3}\%$	$\frac{1}{3}$	$0.\dot{3}$
50%	$\frac{1}{2}$	0.5
$66\frac{2}{3}\%$	$\frac{2}{3}$	$0.\dot{6}$
75%	$\frac{3}{4}$	0.75
80%	$\frac{4}{5}$	0.8
100%	1	1.0

EXAMPLE 7

Find each quantity.

a 40% of \$60 000 **b** 25% of 75 mL **c** $12\frac{1}{2}\%$ of 2 km

SOLUTION

a $40\% \text{ of } \$60\,000 = \frac{40}{100} \times \$60\,000 \text{ or } 0.4 \times \$60\,000$

* It's faster to enter 0.4 for 40% on a calculator.

$= \$24\,000$

b $25\% \text{ of } 75 \text{ mL} = \frac{25}{100} \times 75 \text{ mL or } 0.25 \times 75 \text{ mL}$

$= 18.75 \text{ mL}$

c $12\frac{1}{2}\% \text{ of } 2 \text{ km} = 12\frac{1}{2}\% \text{ of } 2000 \text{ m}$ ← converting 2 km to 2000 m

$= \frac{12\frac{1}{2}}{100} \times 2000 \text{ m or } 0.125 \times 2000 \text{ m}$

$= 250 \text{ m}$

ISBN 9780170350990

EXERCISE 12–03

1 Find 30% of \$630. Select the correct answer **A**, **B**, **C** or **D**.

A \$63 **B** \$189 **C** \$1890 **D** \$126

2 Find 37.5% of 1600 L. Select **A**, **B**, **C** or **D**.

A 600 L **B** 60 L **C** 16 L **D** 6 L

3 Is each statement true or false?

a 25% = 0.25 **b** 40% = 0.04 **c** 75% = 0.75

d 80% = 0.8 **e** 45% = 0.405 **f** 55% = 0.55

4 Copy and complete:

a 25% of \$848 = 0.____ × \$848
= \$____

b 75% of 96 m = 0.75 × ____
= ____ m

c $33\frac{1}{3}$% of 1 day = $0.\dot{3}$ × ____ hours
= ____ hours

5 Find each quantity.

a 12.5% of 24 m **b** $33\frac{1}{3}$% of 18 cm **c** 37.5% of \$2400

d 8% of \$900 **e** 5% of 600 L **f** $66\frac{2}{3}$% of 72 kg

g 30% of \$420 **h** 50% of 72 km **i** 75% of \$5200

j 20% of 1 hour **k** $12\frac{1}{2}$% of 1 day **l** 60% of \$8000

m 130% of 60 kg **n** 62.5% of 24 km **o** $33\frac{1}{3}$% of 36 mL

p 250% of \$95 **q** $66\frac{2}{3}$% of 99 L **r** 110% of 580 g

6 At the ski camp, 5% of the students were injured. If there were 280 students at the camp, how many were injured?

Shutterstock.com/Phillip Minnis

7 Zac's money box has 250 coins. 20% of them were \$1 and the rest were \$2.

a How many coins were \$1?

b What percentage of the coins were \$2?

c How many \$2 coins were there?

d Altogether, how much money did Zac have in his money box?

12–04 Expressing amounts as fractions and percentages

To express an amount as a fraction of a whole amount, write the fraction as $\frac{\text{amount}}{\text{whole amount}}$ and simplify if possible.

EXAMPLE 8

Express each amount as a fraction.

a 56 marks out of 88

b 250 mL of 8 L

c 780 g of 9.6 kg

SOLUTION

a $56 \text{ marks out of } 88 = \frac{56}{88} = \frac{7}{11}$

b $250 \text{ mL of } 8 \text{ L} = \frac{250 \text{ mL}}{8000 \text{ mL}} = \frac{1}{32}$ ← 8 L = 8000 mL

c $780 \text{ g of } 9.6 \text{ kg} = \frac{780 \text{ g}}{9600 \text{ g}} = \frac{13}{160}$ ← 9.6 kg = 9600 g

To express an amount as a percentage of a whole amount:

calculate $\frac{\text{amount}}{\text{whole amount}} \times 100\%$.

EXAMPLE 9

Express each amount as a percentage.

a 120 marks out of 200 marks

b 350 cm of 5 m

c 84c of $6.40

SOLUTION

a $120 \text{ marks out of } 200 = \frac{120}{200} \times 100\% = 60\%$

b $350 \text{ cm of } 5 \text{ m} = \frac{350}{500} \times 100\% = 70\%$ ← 5 m = 500 cm

c $84\text{c of } \$6.40 = \frac{84\text{c}}{640\text{c}} \times 100\% = 13\frac{1}{8}\%$ ← $6.40 = 640c

ISBN 9780170350990

EXERCISE 12–04

1 Express 30 out of 65 as a fraction. Select the correct answer **A, B, C** or **D**.

A $\frac{5}{13}$ **B** $\frac{6}{15}$ **C** $\frac{6}{13}$ **D** $\frac{3}{13}$

2 Express 35 out of 140 as a percentage. Select **A, B, C** or **D**.

A 25% **B** 20% **C** 15% **D** 35%

3 Copy and complete to express each amount as a fraction.

a 50 cm of 5 m $= \frac{50\text{ cm}}{\_\_\_\_\text{ cm}} = \frac{1}{\square}$

b 36c of \$4 $= \frac{36\text{c}}{\_\_\_\_\text{c}} = \frac{9}{\square}$

4 Express each amount as a fraction.

a 20 mL of 6 L

b 55c of \$15

c 30 mm of 6 cm

d 85 g of 5 kg

e 620 cm of 8 m

f 45 mL of 7.5 L

g 72c of \$12

h 4500 mm of 15 m

i 220 mg of 4 g

5 Copy and complete to express each amount as a percentage.

a 50 cm of 5 m $= \frac{50\text{ cm}}{\_\_\_\_\text{ cm}} \times 100\% = \_\_\_\_\%$

b 36c of \$4 $= \frac{36\text{c}}{\_\_\_\_\text{c}} \times 100\% = \_\_\_\_\%$

6 Express each amount as a percentage.

a 25c of \$5

b 180 mL of 9 L

c 64 cm of 8 m

d 25 mm of 5 cm

e 56 g of 8 kg

f 22c of \$8.80

g 540 mg of 90 g

h 4.5 m of 9 km

i 75c of \$500

7 Fatima bought 18 m^2 of silk and uses 15 m^2 of it to make some scarves. What fraction of the material is used?

8 Harry ordered 22 kg of flour to make some bread. Each loaf of bread needs 550 g of flour.

a What percentage of Harry's flour is used to make one loaf of bread?

b How many loaves of bread can Harry make with this flour?

Shutterstock.com/Scorpp

ISBN 9780170350990

12-05 Percentage increase and decrease

WORDBANK

increase To make larger.

decrease To make smaller.

- **To increase an amount by a percentage**, find the percentage of the amount and add it to the amount.
- **To decrease an amount by a percentage**, find the percentage of the amount and subtract it from the amount.

EXAMPLE 10

a Increase \$500 by 40%

b Decrease 200 m by 25%

SOLUTION

a Increase = 40% of \$500
= \$200

Increased amount = \$500 + \$200
= \$700

b Decrease = 25% of 200 m
= 50 m

Decreased amount = 200 m – 50 m
= 150 m

iStockphoto/123ducu

ISBN 9780170350990

EXERCISE 12–05

1 Increase $1250 by 20%. Select the correct answer **A**, **B**, **C** or **D**.

A $1350 **B** $1450 **C** $1400 **D** $1500

2 Decrease 80 L by 25%. Select **A**, **B**, **C** or **D**.

A 20 L **B** 40 L **C** 60 L **D** 55 L

3 Copy and complete:

a Increase $680 by 10%

10% of $680 = ____ × 680

= $____

Increased amount = $680 + ____

= $____

b Decrease 850 m by 20%

20% of 850 m = 0.2 × ____

= ____ m

Decreased amount = 850 m – ____ m

= ____ m

4 **a** Increase $48 by 25%
b Decrease 600 kg by 20%
c Decrease 50 kg by 15%
d Increase 6500 m by 75%
e Increase $3.20 by 10%
f Decrease 580 mL by 28%
g Increase $3400 by 12%
h Decrease 120 L by 20%
i Decrease 7.8 m by 16%
j Increase 128 kg by 38%

5 Jeremy increased his $2000 savings by 20%.
He then decreased this amount by 20%.

a Is he back to $2000?

b How much money did he end up with?

6 Brock opened a new bike shop and decided to add 20% profit to the $550 price of a bike. After 2 weeks, he did not sell much stock so he decided to decrease the new marked price of the bike by 20%. What is the final price of the bike?

iStockphoto/Izabela Habur

12–06 The unitary method

WORDBANK

unit Unit means one or each. $6 per unit means $6 for one.

unitary method A method to find a unit amount and then use this amount to find the total amount.

To use the unitary method:

- use the given amount to find 1%
- multiply 1% by 100 to find the total amount (100%).

EXAMPLE 11

Mina withdrew $550 from her bank account, which was 20% of her savings. What were Mina's savings?

SOLUTION

20% of savings = $550
1% of savings = $550 ÷ 20 ← first calculating 1%
= $27.50

100% of savings = 27.50 × 100 ← multiplying by 100
= $2750

Mina's savings were $2750.

Check: 20% × $2750 = $550.

EXAMPLE 12

Ewan scored 6 goals during the football season, which was 15% of his team's total goals. How many goals did his team score during the season?

SOLUTION

15% of goals = 6

1% of goals = 6 ÷ 15 = 0.4 ← first calculating 1%

100% of goals = 0.4 × 100 ← multiplying by 100
= 40 goals

The team scored a total of 40 goals.

Check: 15% × 40 = 6.

ISBN 9780170350990

EXERCISE 12-06

1 If 25% of an amount is \$640, what is the amount? Select the correct answer **A**, **B**, **C** or **D**.

A \$1600 B \$2560 C \$160 D \$256

2 If 30% of an amount is \$27, what is the amount? Select **A**, **B**, **C** or **D**.

A \$90 B \$900 C \$30 D \$300

3 Is each statement true or false?

a If 10% = \$350, then 1% = \$35.

b If 25% = \$800, then 1% = \$20.

c If 78% = \$2340, then 1% = \$30.

4 Copy and complete:

If 25% of Dani's loan is \$4800, how much is the loan?

25% of loan = \$____

1% of loan = \$4800 ÷ ___

= ___

100% of loan = ___ × 100

Loan = \$_______.

5 Use the unitary method to find:

a Tina's wage if 40% of it is \$320

b Jim's savings if 70% of it is \$2800

c Ly's salary if 26% of it is \$9000

d Amy's pocket money if 52% of it is \$16

e Toni's bill if 45% of it is \$280

f Rob's score if 38% of it is 600

6 A jeweller charges \$75 to repair a necklace. If this is 2.5% of the value of the necklace, what is the necklace worth?

7 If Poh pays 18% of her weekly wage in tax, and her tax payment is \$282 per week, how much is her weekly wage, correct to the nearest cent?

8 James found that 25% of people in Westvale live in apartments. If 1629 people live in apartments, how many people live in Westvale?

9 Zoe paid 12% deposit for a new house. If the deposit was \$66 000, what is the full price of the house?

12–07 Profit, loss and discounts

WORDBANK

cost price The price at which an item is purchased by a retailer.

selling price The price at which an item is sold by a retailer.

profit To sell an item at a higher price than that at which it was purchased.

loss To sell an item at a lower price than that at which it was purchased.

discount A saving when the selling price of an item is lowered.

GST 10% goods and services tax charged by the Australian Government on most goods and services purchased.

- Profit = selling price – cost price (because selling price is higher).
- Loss = cost price – selling price (because selling price is lower).

EXAMPLE 13

A skateboard was bought by a store for \$60 and sold for \$75.

a What was the cost price?

b What was the profit?

c Find the profit as a percentage of the cost price.

SOLUTION

a Cost price = \$60 ⟵ original cost of the skateboard

b Profit = \$75 – \$60 ⟵ selling price – cost price
= \$15

c Profit as a percentage of the cost price $= \frac{\$15}{\$60} \times 100\%$ ⟵ $\frac{\text{profit}}{\text{cost price}} \times 100\%$

$= 25\%$

EXAMPLE 14

Jenny bought a new washing machine, which cost her \$1385. When she moved house, she sold it for \$850.

a Find the loss.

b Calculate correct to one decimal place the loss as a percentage of the cost price.

SOLUTION

a Loss = \$1385 – \$850 ⟵ cost price – selling price
= \$535

b Profit as a percentage of the cost price $= \frac{\$535}{\$1385} \times 100\%$ ⟵ $\frac{\text{loss}}{\text{cost price}} \times 100\%$

$= 38.6281\ldots\%$

$\approx 38.6\%$

ISBN 9780170350990

12-07 Profit, loss and discounts

EXAMPLE 15

The original price of a dress was $56 but 10% GST is added to the price.

a What is the GST and the selling price?

b If the selling price is discounted by 9%, what is the new price?

SOLUTION

a GST = 10% of $56
= $5.60

Selling price = $56 + $5.60
= $61.60

b Discount = 9% × $61.60
= $5.544

New price = $61.60 – $5.544
= $56.056
≈ $56.06 ← to the nearest cent

EXERCISE 12-07

1 A ring was bought for $250 and sold for $320. What was the profit or loss? Select the correct answer **A**, **B**, **C** or **D**.

A $70 loss **B** $80 profit **C** $70 profit **D** $80 loss

2 A boat was bought for $3600 and sold for $3100. What was the profit or loss? Select **A**, **B**, **C** or **D**.

A $400 loss **B** $500 loss **C** $400 profit **D** $500 profit

3 Copy and complete this table.

Cost price	Selling price	Profit or loss
$50	$80	
$125	$105	
	$220	Profit $50
$560		Loss $15
	$380	Profit $82

4 **a** If cost price = $20 and selling price = $45, is there a profit or a loss? How much?

b Find the profit or loss as a percentage of the cost price.

5 A DVD is bought for $28 and sold for $32.

a Find the profit.

b Calculate the profit as a percentage of the *selling price*.

ISBN 9780170350990

EXERCISE 12-07

6 A pair of shoes was bought for $54 and sold for $36.

  a Find the loss.

  b Calculate the loss as a percentage of the cost price.

7 Michael bought a cordless drill worth $280 but he was given a 30% discount. How much did Michael pay for the drill?

8 Find the selling price on each item after a discount of 12.5%.

  a dinner set $240 b vase $56

  c tablet device $320 d shirt $85.60

9 The following items can be bought from an online store, but 10% GST has to be added to find the selling price. Calculate the GST and the selling price for each item.

  a dress $120 b shirt $55 c trousers $86 d shoes $110

  e belt $38 f scarf $24 g necklace $44 h tie $25

10 An online store is having a sale and all goods are reduced by 40%. Find the sale price for each item in Question **9**.

LANGUAGE ACTIVITY

FIND-A-WORD PUZZLE

Make a copy of this grid of letters and then use it to find the words below.

L	A	B	I	T	P	O	C	E	N	V	T
O	P	R	S	I	R	V	O	S	T	I	R
S	O	B	T	S	O	C	F	A	V	N	P
S	T	O	V	E	F	I	B	E	R	C	E
P	E	D	S	U	I	P	T	R	I	R	E
I	S	E	G	A	T	N	E	C	R	E	P
D	O	C	R	U	S	T	Y	E	V	A	R
P	R	I	C	E	R	U	N	D	O	S	T
O	R	M	T	H	U	D	O	R	P	E	R
F	R	A	C	T	I	O	N	S	I	L	T
R	O	L	L	E	R	S	H	I	P	O	R
V	E	N	T	G	N	I	L	L	E	S	A

COST DECIMAL DECREASE LOSS

FRACTION INCREASE PERCENTAGES

PER PRICE PROFIT SELLING

ISBN 9780170350990

PRACTICE TEST 12

Part A General topics

Calculators are not allowed.

1 Evaluate $\frac{4}{9}+\frac{2}{9}$.

2 What is another name for a 90° angle?

3 If $a = 8$, evaluate $10 - 2a$.

4 Find the mode of these scores: 12, 5, 9, 7, 8, 7, 10.

5 If a coin is tossed, what is the probability that it comes up tails?

6 Find the sum of 2.5 and 8.74.

7 Evaluate 5^3.

8 Find the highest common factor of 20 and 12.

9 Simplify $6a - a$.

10 Write the formula for the circumference of a circle with radius r.

Part B Percentages

Calculators are allowed.

12-01 Percentages and fractions

11 Convert 85% to a fraction. Select the correct answer **A**, **B**, **C** or **D**.

A $\frac{1}{85}$ **B** $\frac{43}{50}$ **C** $\frac{100}{85}$ **D** $\frac{17}{20}$

12 Convert 60% to a fraction? Select **A**, **B**, **C** or **D**.

A $\frac{1}{60}$ **B** $\frac{2}{5}$ **C** $\frac{6}{5}$ **D** $\frac{3}{5}$

13 What is $\frac{3}{20}$ as a percentage? Select **A**, **B**, **C** or **D**.

A 20% **B** 15% **C** 30% **D** 60%

12-02 Percentages and decimals

14 Convert each decimal to a percentage.

a 0.02 **b** 0.65 **c** 0.125

15 Write in descending order: 62%, 0.608, 0.06, 64.5%.

12-03 Percentage of a quantity

16 Find each quantity.

a 40% of $420

b 25% of 84 km

12–04 Expressing amounts as fractions and percentages

17 Write 22.5 cm as a percentage of 5 m.

18 What fraction is 24 mL of 6 L?

12–05 Percentage increase and decrease

19 a Decrease 268 kg by 5%.
   b Increase $500 by 40%.

12–06 The unitary method

20 If 28% of my electricity bill is $126, how much is the total bill?

12–07 Profit, loss and discounts

21 Regan bought a suit for $160 and sold it for $220.
   a What was the profit?
   b What was the profit as a percentage of the cost price?

 ISBN 9780170350990

INVESTIGATING DATA

13

WHAT'S IN CHAPTER 13?

IN THIS CHAPTER YOU WILL:

- read, interpret and construct different types of statistical graphs: column graphs, line graphs, divided bar graphs and sector graphs
- find the mean, mode, median and range of a list of scores
- organise data (information) into frequency tables
- read and construct frequency histograms and polygons
- read and construct dot plots
- read and construct stem-and-leaf plots
- find the mean, mode, median and range of scores presented in dot plots and stem-and-leaf plots

* Shutterstock.com/optimarc

ISBN 9780170350990

13–01 Reading and drawing graphs

WORDBANK

data Statistical information, a collection of facts.

column graph A graph that uses columns of different heights to represent data.

line graph A graph that uses lines to represent data.

axes Plural of axis. The two number lines that form the edges of a graph: the horizontal axis and the vertical axis.

divided bar graph A graph in which a rectangle is divided into parts to represent data.

sector graph A graph in which a circle is divided into parts to represent data.

- A graph must have a **title** explaining the information represented by the graph.
- A **column graph** or **line graph** must have **scales** and **labels** on both the horizontal and vertical **axes**.
- A **divided bar graph** or **sector graph** must have **labels** or a **key** describing the categories being represented.

EXAMPLE 1

Caitlin earns \$1200 per week. She spends \$500 on rent, \$120 on food, \$200 on clothes and entertainment, and saves the rest. Show this data on:

a a column graph **b** a sector graph.

SOLUTION

a Savings = \$1200 – \$500 – \$120 – \$200

= \$380

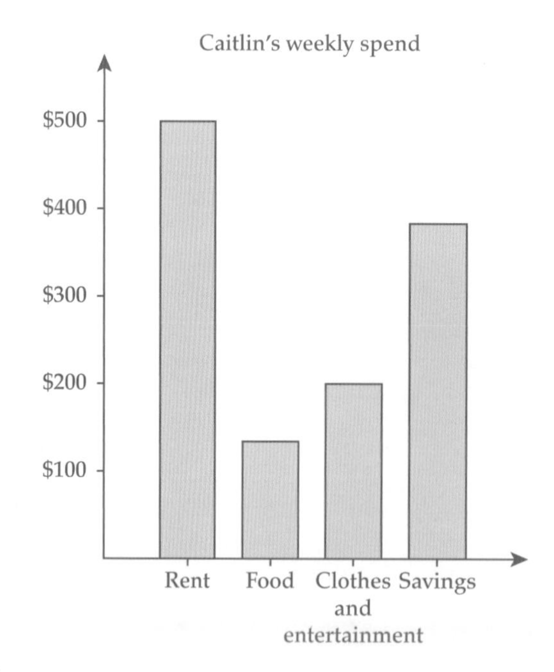

ISBN 9780170350990

b First, calculate the size of each sector.

Rent $= \frac{500}{1200} \times 360° = 150°$

Food $= \frac{120}{1200} \times 360° = 36°$

Clothes/Entertainment $= \frac{200}{1200} \times 360° = 60°$

Savings $= \frac{380}{1200} \times 360° = 114°$

✱ Check that the angles add up to 360°: 150° + 36° + 60° + 114° = 360°.

Then draw and label the sector graph, using the calculated angle sizes.

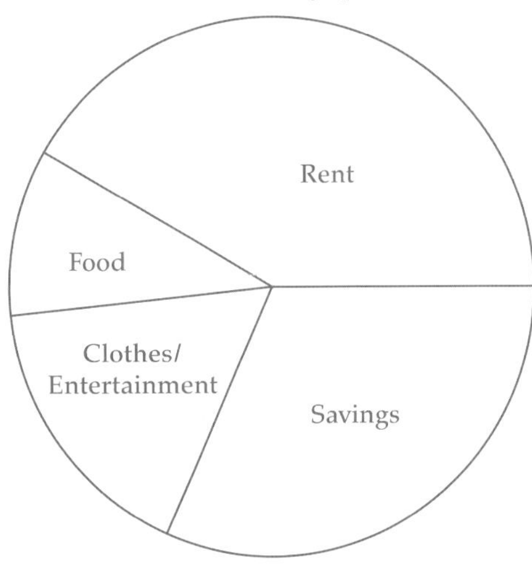

iStockphoto/IPGGutenbergUKLtd

EXERCISE 13-01

1. When drawing a column graph, the columns must be what? Select the correct answer **A, B, C** or **D**.

   A the same width **B** the same height **C** joined **D** different widths

2. What is the angle size of a sector on a sector graph that represents $\frac{1}{4}$ of the data? Select **A, B, C** or **D**.

   A 60° **B** 120° **C** 90° **D** 25°

3. On a divided bar graph, name the shape that is divided. Select **A, B, C** or **D**.

   A square **B** circle **C** triangle **D** rectangle

4. This graph shows the scores of five students in a maths test.

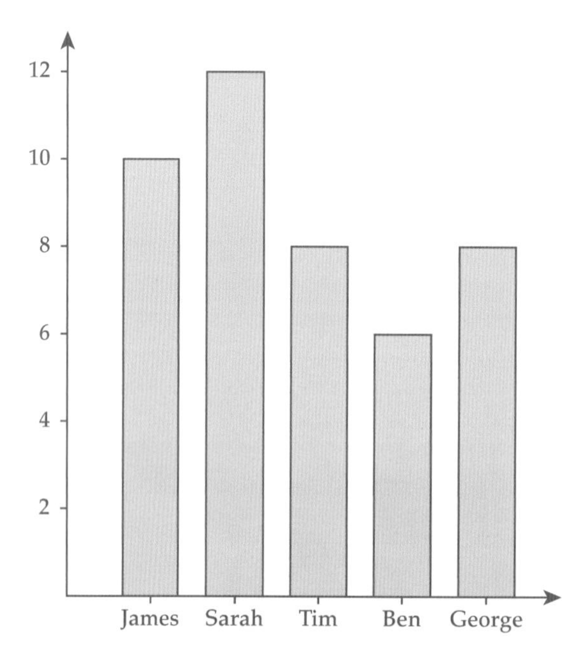

   a What type of graph is drawn?

   b Who scored 8 or more on the test?

   c If 50% was a pass and the test was out of 12, did everyone pass?

5. The following information shows how Lonnie spends a typical day.

   Sleep: 8 hours Work: 6 hours Chores: 2 hours
   Exercise: 1 hour Meals: 2 hours Relaxation: 5 hours

   a Show this information on a sector graph, shading each sector in a different colour.

   b What fraction of Lonnie's day is spent at work and doing chores?

   c What percentage of Lonnie's day is spent relaxing? Answer correct to one decimal place.

   d Represent this information on a divided bar graph 12 cm long.

 ISBN 9780170350990

6 The line graph below shows Liam's journey to visit his friend Uma.

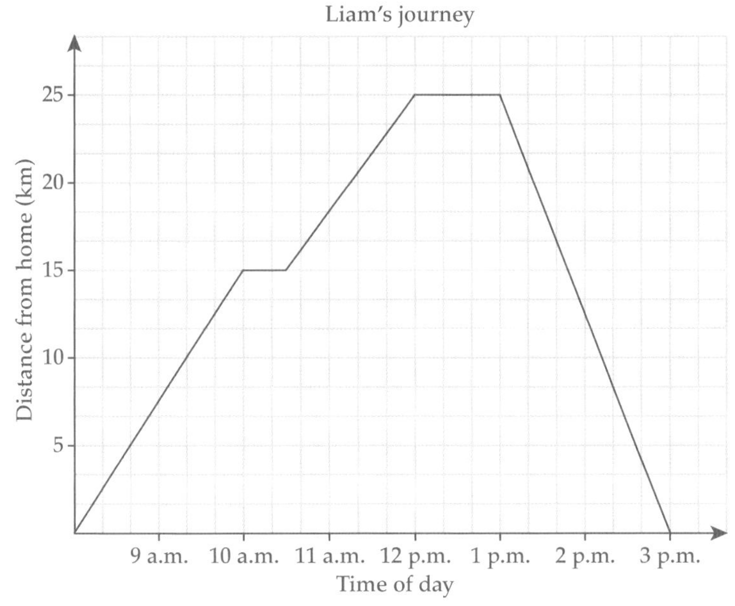

a What time did Liam leave home?

b When did he first stop and rest?

c How far away does Uma live?

d How long did Liam stay at Uma's house?

e How long did it take Liam to travel home?

f What was his average speed on the way home?

7 This line graph shows the number of text messages sent every hour by a group of Year 8 students.

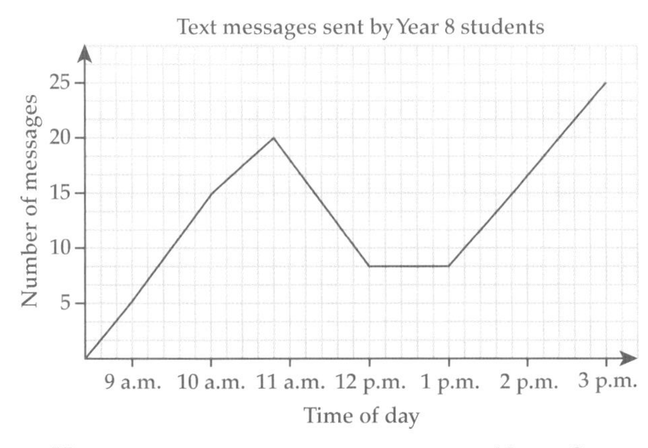

a How many text messages were sent at 10 a.m.?

b What was the greatest number of messages sent and what time were they sent?

c What does the horizontal line from 12 p.m. to 1 p.m. indicate?

d How many messages were sent altogether from 9 a.m. to 3 p.m.?

13–02 The mean and the mode

WORDBANK

mean The average of a set of data scores, with symbol $\bar{x}$, calculated by adding the scores and dividing the sum of the scores by the number of scores.

mode The most popular score(s) in a data set, the score that occurs most often.

In statistics, there are three **measures of location** that measure the central or middle position of a set of data. They are called the **mean**, **mode** and **median**.

$$\text{Mean} = \bar{x} = \frac{\text{sum of scores}}{\text{number of scores}}$$

EXAMPLE 2

Find the mean of these scores: 9, 6, 4, 7, 5, 10, 12, 8.

SOLUTION

$$\begin{aligned}\text{Mean} = \bar{x} &= \frac{\text{sum of scores}}{\text{number of scores}}\\ &= \frac{9+6+4+7+5+10+12+8}{8}\\ &= \frac{61}{8}\\ &= 7.625\end{aligned}$$

✱ Note that the mean is around the centre of the set of scores.

- Mode = the most frequently occurring score(s).
- A set of data can have more than one mode, or no mode at all.

EXAMPLE 3

For each set of data, find the mode.

a 4, 5, 4, 8, 4, 6, 7, 3, 4, 4, 8, 2

b 22, 18, 12, 14, 12, 15, 26, 24, 12, 16, 12, 12

c 3, 7, 7, 6, 3, 7, 9, 5, 3, 7, 4, 10, 3

SOLUTION

a The mode is 4. ← 4 occurs five times and the other scores occur less often.

b The mode is 12. ← 12 occurs more often than the other scores.

c There are two modes: 3 and 7.

ISBN 9780170350990

EXERCISE 13–02

1 Find the mode of this set of scores: 8, 6, 9, 5, 5, 6, 7, 6, 2, 6, 8. Select the correct answer **A**, **B**, **C** or **D**.

A 8 **B** 7 **C** 6 **D** 5

2 Find the mean of the scores in Question **1**. Select **A**, **B**, **C** or **D**.

A 68 **B** 6.2 **C** 618 **D** 6.8

3 Copy and complete this paragraph.

The mean of a set of scores is found by ____________ the scores together and then ____________ by the ____________ of scores. The mode is the score that occurs most ____________. It is the most ____________ score. There can be one or more modes for a set of scores and sometimes there is ____________ mode at all.

4 Find the mean of each set of scores correct to one decimal place.

a 4, 7, 8, 9, 6, 7, 5, 7, 8, 4, 6, 7

b 12, 11, 14, 15, 10, 18, 17, 14

c 28, 32, 25, 26, 33, 28, 22, 34, 27

d 54, 57, 56, 59, 55, 54, 55, 59

e 82, 88, 87, 89, 84, 85, 83

5 **a** Find the mode of each set of scores in Question **4**.

b Which sets of scores in Question **4** have two modes?

6 The following data are the number of tries scored by the Giants rugby league team during 14 matches.

4, 2, 5, 6, 8, 1, 7, 6, 3, 4, 1, 4, 2, 5

Find the mean (correct to two decimal places) and the mode.

Newspix/Peter Wallis

7 Sixteen shopping malls were surveyed on the number of shops in each mall.

22, 16, 18, 24, 12, 18, 16, 24, 15, 21, 19, 25, 18, 16, 22, 16

a Find the mean.

b Find the mode.

ISBN 9780170350990

13–03 The median and the range

WORDBANK

median The middle score when the scores are ordered from lowest to highest.

range The highest score minus the lowest score.

To find the **median:**

- order the scores from lowest to highest
- if there are an odd number of scores, **median = middle score**.
- if there are an even number of scores, **median = average of the two middle scores**.

Range = highest score − lowest score.

The median is another **measure of location**, whereas the range is a **measure of spread**.

EXAMPLE 4

Find the median and range of each set of data.

a 8, 12, 9, 11, 7, 14, 6, 14, 9

b 22, 25, 32, 21, 28, 35, 31, 29, 24, 26

SOLUTION

a Write the scores from lowest to highest:

6, 7, 8, 9, 9, 11, 12, 14, 14

There is an **odd** number of scores (9 scores), so the median will be the one in the middle:

Median = 9 ← There are four scores above 9 and four scores below 9.

Range = 14 − 6 ← highest − lowest

= 8

b 21, 22, 24, 25, 26, 28, 29, 31, 32, 35

There is an **even** number of scores (10 scores), so the median will be the average of the two middle scores.

Median = $\frac{26+28}{2}$ ← The two middle scores are 26 and 28.

= 27 ← The number halfway between 26 and 28 is 27.

Range = 35 − 21

= 14

Notice that the **position** of the middle score is slightly more than half: for nine scores, it is the fifth score and for 10 scores it is the average of the fifth and sixth scores.

ISBN 9780170350990

EXERCISE 13-03

1 Find the range of these scores: 15, 18, 22, 14, 26, 21, 19. Select the correct answer **A**, **B**, **C** or **D**.

A 26 B 14 C 12 D 15

2 What is the median of the scores in Question **1**. Select **A**, **B**, **C** or **D**.

A 18 B 19 C 21 D 22

3 Copy and complete this paragraph.

The range for a set of scores is the ____________ score minus the ____________ score.

The median is found by ordering the scores from ____________ to ____________ and finding the ____________ score for an odd number of ____________.

If there is an ____________ number of scores, the median is the ____________of the two ____________ scores.

4 Find the median and the range for each set of data.

a 7, 9, 12, 2, 5, 3, 8, 7, 6

b 21, 25, 32, 34, 28, 33, 26, 34

c 42, 38, 41, 35, 44, 37, 42, 39, 44, 37, 46

d 9, 6, 7, 8, 12, 5, 9, 4, 11, 6, 4, 5, 7

e 54, 63, 59, 62, 55, 64, 57, 66, 67, 53, 62, 58

f 72, 67, 75, 68, 73, 66, 72, 64, 71, 65, 73

5 These are the daily sales figures for Ali's sports store over 20 days.

345, 256, 390, 420, 385, 456, 390, 560, 486, 320
248, 390, 450, 960, 580, 420, 386, 520, 395, 480

a Find the range of sales figures.

b Find the median sales figure.

c What is the mode?

d Find the mean sales figure.

e If Ali wanted the most likely sales figure for the next day, should he take notice of the median, the mode or the mean?

f Which measure, the mean or the median, shows Ali the middle area of the sales figures?

6 In 12 rounds of golf, Greg scored:

72, 76, 71, 77, 75, 71, 74, 72, 71, 79, 73, 74

Find the mode, range, median and mean for these scores.

13–04 Frequency tables

WORDBANK

frequency The number of times a score occurs in a set of data.

frequency table A table that shows the frequency of each score in a data set.

cumulative frequency A running total of frequencies used for finding the median, combining the frequencies of all scores less than or equal to the given score.

For data in a frequency table:

- a cumulative frequency column can be included to find the median
- an *fx* column can be included to find the mean
- mean $\bar{x} = \frac{\text{sum of } fx}{\text{sum of } f}$

EXAMPLE 5

A class of students was surveyed on the number of pets they own and the numbers are shown below.

2, 3, 5, 3, 0, 3, 2, 1, 4, 5, 6, 4, 2, 5, 2, 1, 0, 5, 2, 1, 3, 2

a Arrange these scores in a frequency table, including columns for cumulative frequency and *fx*.

b For this data set, find:

i the range

ii the mode

iii the median

iv the mean (correct to two decimal places)

SOLUTION

a In the frequency table below, the Tally and Frequency columns show how many times each score appears in the set of data. There are two scores of 0, three scores of 1, six scores of 2, and so on.

Score (x)	Tally	Frequency (f)	Cumulative frequency	fx
0	||	2	2	$2 \times 0 = 0$
1	|||	3	$2 + 3 = 5$	$3 \times 1 = 3$
2	~~||||~~ |	6	11	12
3	||||	4	15	12
4	||	2	17	8
5	||||	4	21	20
6	|	1	22	6
	Totals	22		61

✱ Cumulative frequency is a running total of frequencies, and *fx* means 'frequency (f) × score (x)'.

ISBN 9780170350990

b **i** Range = 6 – 0 = 6 ← highest score – lowest score (from the Score column)

ii Mode = 2 ← The score with the highest frequency is 2.

iii There are 22 scores, so the median is the average of the 11th and 12th scores.

From the cumulative frequency table, it can be seen that the 6th to the 11th scores are all 2s, and the 12th to the 15th scores are all 3s, so the 11th score = 2 and the 12th score = 3.

$$\text{Median} = \frac{2+3}{2}$$
$$= 2.5$$

iv $\text{Mean } \bar{x} = \frac{\text{sum of scores}}{\text{number of scores}}$

$$= \frac{\text{sum of } fx}{\text{sum of } f}$$

$$= \frac{61}{22}$$ ← Use the totals at the bottom of the table.

$$= 2.7727\ldots$$

$$\approx 2.77$$

Shutterstock.com/Eric Isselee

EXERCISE 13-04

1 Which column do you need to find the median in a frequency table? Select the correct answer **A**, **B**, **C** or **D**.

A score **B** frequency **C** *fx* **D** cumulative frequency

2 Which column do you need to find the mean in a frequency table? Select **A**, **B**, **C** or **D**.

A score **B** frequency **C** *fx* **D** cumulative frequency

3 The manager of a shoe store recorded the sizes of shoes sold over a weekend.

5, 6, 8, 7, 7, 6, 4, 8, 4, 3
7, 7, 7, 9, 4, 3, 7, 3, 7, 8
2, 9, 7, 6, 5, 7, 7, 7, 6, 7

a Arrange the results in a frequency table with columns for Score, Tally and Frequency.

b Find the range and the mode for these data.

4 Copy each frequency table below and add the columns for cumulative frequency and *fx*. For each data set, find the range, the mode, the median and the mean (correct to two decimal places).

a

x	f
2	3
3	4
4	8
5	5
6	2

b

x	f
20	5
21	7
22	8
23	6
24	5

5 The number of errors in a spelling test made by 50 students is shown below.

1, 3, 4, 0, 0, 2, 1, 3, 3, 4, 1, 2, 4, 1, 3, 3, 2, 2, 2, 1, 2, 2, 3, 4, 5
3, 2, 2, 2, 3, 2, 3, 3, 2, 2, 1, 0, 1, 0, 2, 3, 3, 2, 4, 2, 2, 2, 4, 5, 1

a Express the data in a frequency table, including an *fx* column.

b Find the mode.

c Calculate the mean number of spelling errors.

6 The number of cars crossing an intersection each minute was counted for 30 minutes and the results are shown below.

2, 2, 0, 2, 5, 0, 4, 4, 0, 1
3, 2, 4, 1, 0, 4, 2, 2, 5, 0
2, 3, 3, 3, 0, 4, 3, 5, 1, 2

Draw a frequency table for this data set, and then find:

a the range

b the mode

c the median

d the mean.

ISBN 9780170350990

13–05 Frequency histograms and polygons

WORDBANK

frequency histogram A column graph that shows the frequency of each score. The columns are joined together.

frequency polygon A line graph that shows the frequency of each score, drawn by joining the top of each column in a histogram.

EXAMPLE 6

This frequency table shows the number of lollies in samples of packets of lollies. Draw a frequency histogram and polygon for the data.

Number of lollies	Frequency
48	3
49	5
50	11
51	4
52	2

SOLUTION

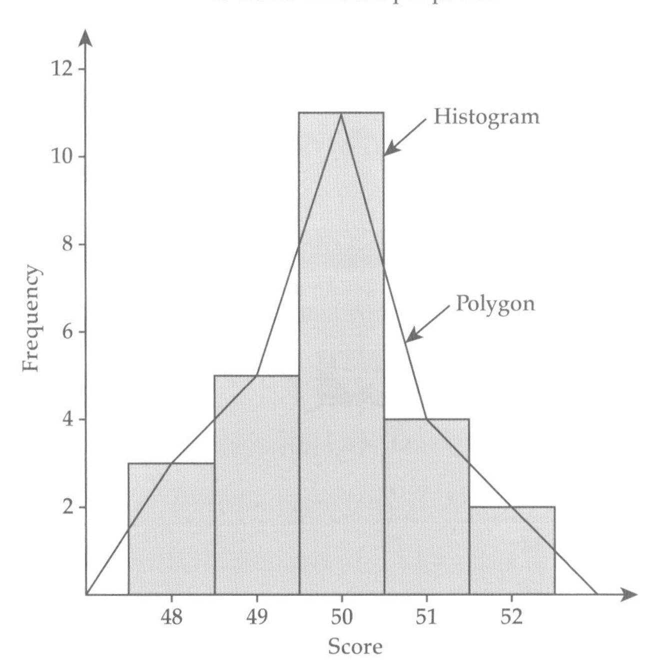

- The **histogram** has columns that are centred on each score, leaving a small gap on the left.
- The **polygon** joins the middle of the top of each column, and starts and ends on the horizontal axis. It is called a polygon because it has the shape of a many-sided figure.

ISBN 9780170350990

EXERCISE 13–05

1 A frequency polygon compares which two things? Select the correct answer **A**, **B**, **C** or **D**.

A tally to frequency
B scores to frequency
C data to scores
D frequency to information

2 What type of graph is a frequency histogram? Select **A**, **B**, **C** or **D**.

A column graph
B divided bar graph
C sector graph
D line graph

3 Is each statement true or false?

a A frequency polygon is a type of line graph.
b A frequency histogram has columns of different widths.
c The columns are all joined together in a frequency histogram.
d A freqency polygon joins the corners of the columns in a histogram.
e A frequency polygon and histogram compare frequency to means.

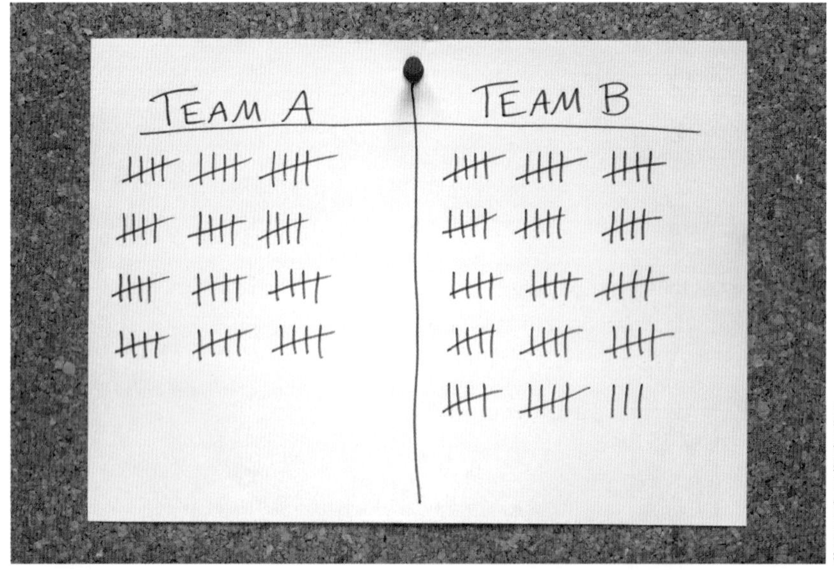

Shutterstock.com/Stephen Rees

4 Draw a frequency histogram and polygon for each frequency table.

a

Score	Frequency
0	2
1	5
2	8
3	6
4	3

b

Score	Frequency
12	2
13	7
14	11
15	8
16	4
17	3

ISBN 9780170350990

EXERCISE 13–05

5 The number of letters delivered to 50 houses in a street is recorded below.

3, 1, 2, 6, 4, 3, 2, 5, 2, 1, 2, 2, 3, 1, 2, 1, 3, 2, 4, 3, 1, 3, 3, 4, 2
2, 1, 3, 6, 2, 2, 1, 3, 4, 3, 2, 1, 2, 5, 2, 5, 3, 1, 1, 2, 3, 2, 4, 2, 3

a Organise the data into a frequency table.

b Draw a frequency histogram and polygon of these data.

c What is the range of the scores?

d What was the most common number of letters delivered per household? What is the statistical name for this value?

6 The number of children in a group of families at a picnic are shown in this frequency table.

Number of children	0	1	2	3	4	5
Frequency	2	5	12	2	1	1

a Draw a frequency polygon for these data.

b Find the mode.

c Find the range.

Shutterstock.com/Serg Zastavkin

13–06 Dot plots

WORDBANK

dot plot A diagram showing the frequency of data scores using dots.

outlier An extreme score that is much higher or lower than the other scores in a data set.

cluster A group of scores that are bunched close together.

A **dot plot** is like a simple **column graph** or **frequency histogram**, except that dots are used instead of columns to show the frequency of each score.

EXAMPLE 7

This dot plot shows the amount of money (in dollars) a group of students spent for lunch at the school canteen yesterday.

Amount spent for lunch ($)

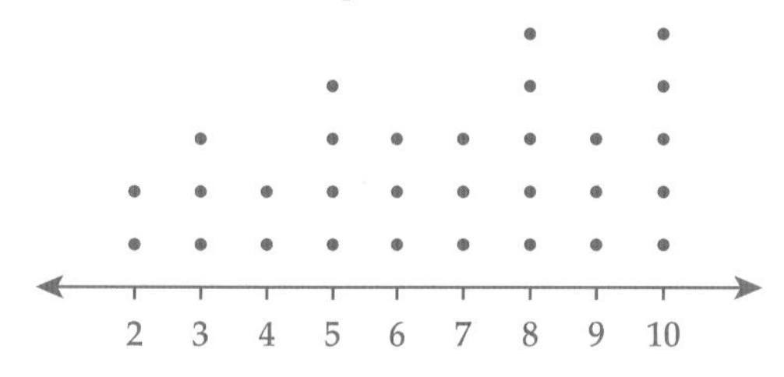

a Find the range.

b What is the mode?

c How many students were in the group?

d Find the median.

e What is the mean (correct to the nearest cent) amount of money spent on lunch?

SOLUTION

a Range = 10 – 2 ← highest score – lowest score

= 8

b Modes = 8 and 10 ← The columns with the most dots.

c Number of students = 30 ← Counting the dots gives 30 dots or students.

d Median $= \dfrac{7+7}{2} = 7$ ← Of the 30 dots, the two middle dots (15th and 16th) are both 7, as shown below.

7

e Mean $= \dfrac{2\times2+3\times3+4\times2+5\times4+6\times3+7\times3+8\times5+9\times3+10\times5}{30}$ ← $\dfrac{\text{sum of scores}}{\text{no. of scores}}$

$= \dfrac{197}{30}$

$= 6.5666\ldots$

$\approx \$6.57$

ISBN 9780170350990

EXERCISE 13–06

1 In a dot plot, the score with the highest column of dots is what? Select the correct answer **A**, **B**, **C** or **D**.

A the range **B** the mean **C** the median **D** the mode

2 In a dot plot, what does the number of dots above each score represent? Select **A**, **B**, **C** or **D**.

A the vertical axis **B** the scale **C** the data **D** the frequency

3 This dot plot shows the weekly pocket money in dollars given to a sample of students.

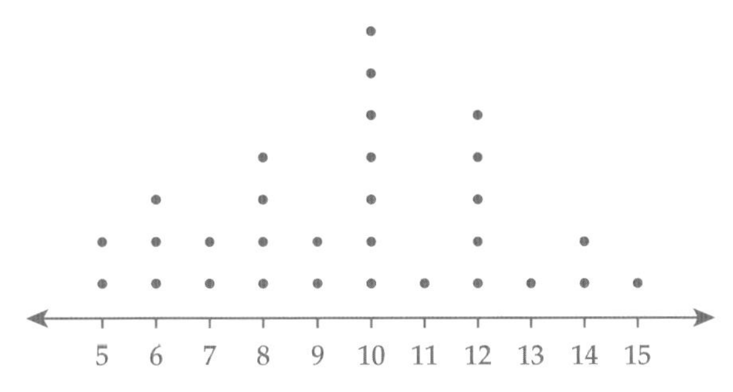

a What was the largest amount of pocket money given?

b What is the mode?

c How many students are in the sample?

d Find the median.

e Find the mean correct to the nearest cent.

4 Besides dots, a dot plot must have what? Select **A**, **B**, **C** or **D**.

A a horizontal axis

B a vertical axis

C no scale

D columns

5 A group of Year 8 students was surveyed on the number of hours spent on the computer each day and the results are shown below.

3, 2, 4, 3, 1, 5, 6, 3, 4
3, 2, 3, 7, 5, 3, 2, 3, 3
1, 6, 7, 2, 3, 2, 3, 4, 6

a Draw a dot plot of these scores.

b What is the modal time spent on the computer?

c Find the range of times.

d Find the median.

e Find correct to one decimal place the mean time spent on the computer.

EXERCISE 13–06

6 This dot plot shows the daily sales in dollars from a newsagent for 2 weeks.

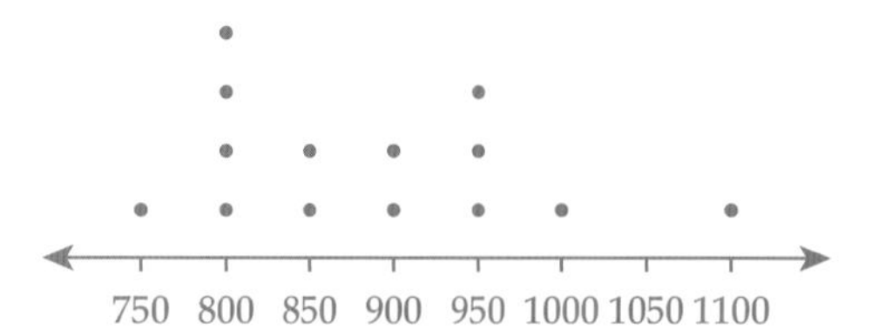

a What is the range?

b Find the mode.

c Calculate the mean correct to the nearest cent.

d Find the median.

e When there is a score that is much lower or higher than all the other scores, it is called an **outlier**. What is the outlier here?

f If scores are grouped close together they are called a **cluster**. Between which two scores is there a cluster of scores here?

Newspix/Lindsay Moller

7 The scores below represent the number of times students go out over the school holidays.

12, 14, 8, 9, 15, 13, 7, 10
14, 11, 9, 12, 13, 12, 8, 12

a Draw a dot plot of these scores.

b Find the range and the mode.

c Find the median.

d Find the mean correct to one decimal place.

e What is the most likely number of times students go out?

ISBN 9780170350990

13–07 Stem-and-leaf plots

WORDBANK

stem-and-leaf plot A table listing data scores where the tens digits are in the stem and the units digits are in the leaf.

This **stem-and-leaf plot** shows the number of dollars spent on food each week by 20 families.

Stem	Leaf
9	5 8
10	8
11	4 5 7 8
12	5 6 6 6 6 8
13	2 3 7
14	3 4 5 6

The values range from $95 to $146, with the stems running from 9 to 14. The score 95 has a stem of 9 (meaning 90) and a leaf of 5.

EXAMPLE 8

For the stem-and-leaf plot above, find:

a the range **b** the mode **c** the mean
d the median **e** any clusters or outliers.

SOLUTION

a Range = $146 – $95 ← highest score – lowest score
= $51

b Mode = $126 ← most common score

c Mean $= \dfrac{95+98+108+...+146}{20}$ ← $\dfrac{\text{sum of scores}}{\text{no. of scores}}$

$= \dfrac{2502}{20}$

= $125.10

d There are 20 scores: the two middle dots (10th and 11th) are both 126.

Median $= \dfrac{126+126}{2}$ ← 12 | 5 6 ⑥ ⑥ 6 8

= $126

e There is a cluster in the 110s and 120s. There are no outliers.

EXERCISE 13-07

1 What does the stem in a stem-and-leaf plot show? Select the correct answer **A**, **B**, **C** or **D**.

A the units digit **B** from 0 to 2 **C** the tens digit **D** none of these

2 What does the leaf in a stem-and-leaf plot show? Select **A**, **B**, **C** or **D**.

A the units digit **B** from 0 to 2 **C** the tens digit **D** none of these

3 This stem-and-leaf plot shows the results of a survey on the number of hours a sample of people shopped per week.

Stem	Leaf
0	5 6 6 8 8 9
1	0 0 1 2 2 3 4 4 4 5 6 7 7 8 8 9
2	1 2 3 4 6 7 8
3	4

a How many people were surveyed?

b What is the mode?

c Find the median.

d Find the mean.

e Find the range.

f Are there any clusters or outliers?

4 The number of cakes sold per day at a cafe is shown.

35, 63, 58, 39, 42, 36, 55, 63
36, 45, 63, 58, 44, 38, 63, 57
36, 47, 52, 63, 65, 54, 62, 37

a Represent these data on a stem-and-leaf plot.

b What is the median number of cakes sold?

c Find the mean number of cakes sold, correct to two decimal places.

d What is the range?

e If I were ordering cakes for the next day, what is the most likely number I would sell?

5 This stem-and-leaf plot shows the distance in kilometres travelled by a group of salespeople last week.

Stem	Leaf
10	0
11	3 5 6 6 7 8 8
12	3 4 4 4 6 8 9
13	2 6 8 9

a Find the range.

b What is the mode?

c Find the median.

d Calculate correct to two decimal places the mean.

e Are there any clusters or outliers?

ISBN 9780170350990

EXERCISE 13–07

6 The daily numbers of hamburgers sold at two local stores are listed below.

Oburgo: 302, 290, 305, 284, 317, 295, 284, 317, 316, 308, 307
Hungry Jill's: 306, 328, 317, 308, 298, 316, 325, 325, 312, 306, 318

- **a** Draw a stem-and-leaf plot for each set of data.
- **b** Which store had the greatest sales for one day? How many?
- **c** Which store sold the least in one day? How many?
- **d** Which store had the greater sales over the 11 days? Show all working.
- **e** Compare the medians of the two stores. Which store's was higher?

Shutterstock.com/Dani Vincek

ISBN 9780170350990

CODE PUZZLE

Use the following table to decode the words and phrases used in this chapter.

1	2	3	4	5	6	7	8	9	10	11	12	13
A	B	C	D	E	F	G	H	I	J	K	L	M

14	15	16	17	18	19	20	21	22	23	24	25	26
N	O	P	Q	R	S	T	U	V	W	X	Y	Z

1 16 – 12 – 15 – 20

2 7 – 18 – 1 – 16 – 8

3 19 – 5 – 3 – 20 – 15 – 18

4 3 – 15 – 12 – 21 – 13 – 14

5 18 – 1 – 14 – 7 – 5

6 13 – 5 – 1 – 14

7 13 – 15 – 4 – 5

8 12 – 5 – 1 – 6

9 19 – 20 – 5 – 13

10 4 – 15 – 20

11 13 – 5 – 4 – 9 – 1 – 14

12 6 – 18 – 5 – 17 – 21 – 5 – 14 – 3 – 25

13 20 – 1 – 2 – 12 – 5

14 1 – 22 – 5 – 18 – 1 – 7 – 5

15 8 – 9 – 19 – 20 – 15 – 7 – 18 – 1 – 13

16 16 – 15 – 12 – 25 – 7 – 15 – 14

ISBN 9780170350990

PRACTICE TEST 13

Part A General topics

Calculators are not allowed.

1 In words, describe a reflex angle.

2 What is the complement of 26°?

3 List the factors of 18.

4 Is 87 a prime or a composite number?

5 Find the range of these scores:
12, 8, 7, 6, 8, 7, 5, 7.

6 Test whether 1258 is divisible by 6. Show working.

7 Evaluate $1\frac{7}{8} \times \frac{1}{5}$.

8 Find the area of this triangle.

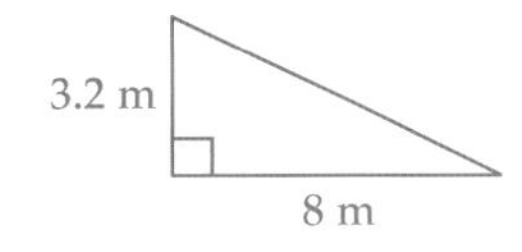

9 Simplify $2ab \times (-6bc)$.

10 Copy and complete: 280 km = ______ m.

Part B Investigating data

Calculators are allowed.

13-01 Reading and drawing graphs

11 Each section of a divided bar graph should be what? Select the correct answer **A**, **B**, **C** or **D**.

A equal **B** increasing **C** different sizes **D** decreasing

12 Ella earns $1080 per week. She spends $324 on rent, $108 on food, $216 on clothes and entertainment and saves the rest.

a Draw a sector graph to illustrate Ella's weekly spending.

b On the graph, what is the angle size of the rent sector?

13-02 The mean and the mode

13 Find the mean of 14, 19, 22, 32, 16, 43 correct to one decimal place.

14 What is the mode of the scores in Question **13**? Select **A**, **B**, **C** or **D**.

A 20.5 **B** 27 **C** 43 **D** no mode

13-03 The median and the range

15 For the scores 3, 4, 6, 7, 8, 10, 10, 10, 12, find:

a the median

b the range.

ISBN 9780170350990

13-04 Frequency tables

16 For the data in this frequency table, find:

a the range **b** the mode **c** the mean correct to two decimal places.

x	f
5	2
6	5
7	8
8	3

13-05 Frequency histograms and polygons

17 This frequency table shows the number of toothpicks in a sample of toothpick boxes. Draw a frequency histogram and polygon for these data.

Number of toothpicks	Frequency
48	2
49	4
50	10
51	7
52	3

13-06 Dot plots

18 For the data shown on this dot plot, find:

a the mean **b** the median **c** the mode **d** the outlier.

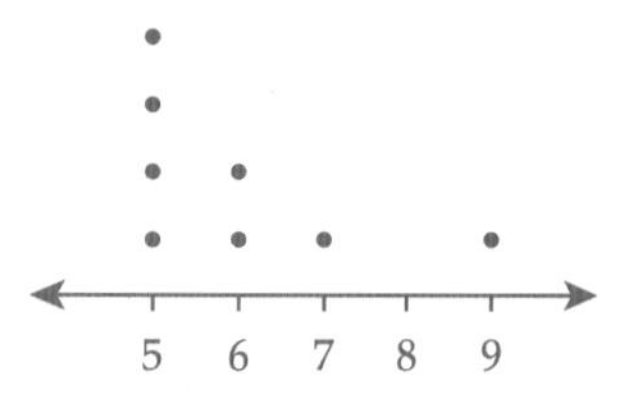

13-07 Stem-and-leaf plots

19 For the data shown on this dot plot, find:

a the range **b** the mode **c** the median **d** where the scores are clustered.

Stem	Leaf
11	8
12	3 5
13	2 4 4 5 7 9

 ISBN 9780170350990

PROBABILITY

14

WHAT'S IN CHAPTER 14?

IN THIS CHAPTER YOU WILL:

- understand probability and words related to chance
- list all of the possible outcomes of a situation (chance experiment)
- calculate the probability of simple events
- calculate the probability of complementary events
- find the relative frequency of an event
- interpret and draw Venn diagrams and use them to solve probability problems
- read two-way tables and use them to solve probability problems

*Shutterstock.com/Sergey Mironov

ISBN 9780170350990

14-01 The language of chance

WORDBANK

probability The chance of an event occurring, written as a fraction between 0 and 1.

outcome A result of a situation involving chance.

event One or more outcomes of a chance experiment.

Here are some words that are used to describe chance:

- An **impossible** event means the event cannot occur; for example, it will snow in Darwin today.
- An **unlikely** event means the event will probably not occur; for example, winning a lotto prize.
- An **even chance** or **50-50 chance** means the event has an equal chance of occurring or not occurring; for example, getting a head when you toss a coin.
- A **likely event** means the event will probably occur. It is more likely to occur than not occur; for example, the next vehicle to pass the school is a car.
- A **certain event** means the event must occur; for example, it will get dark tonight.

The chance of an impossible event occurring is 0, whereas the chance of a certain event occurring is 1. An event that has an even chance is placed halfway between 0 and 1. Likely and unlikely events can be ordered on a number line as follows.

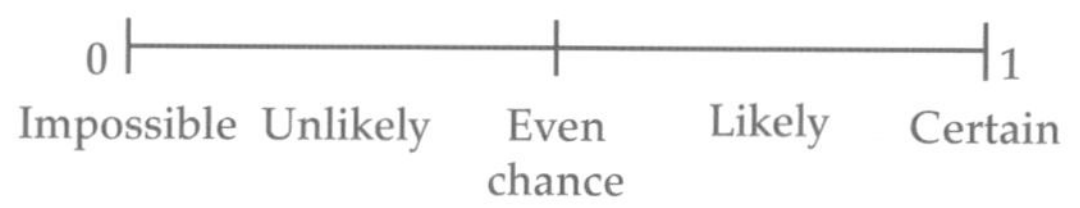

EXAMPLE 1

Using a suitable word, describe the chance of each event below occurring and then mark their position on a number line.

a Joel choosing a black card from 10 black cards.

b Renee winning a raffle if she has no ticket.

c A newborn baby is a boy.

d Rolling a 5 on a die.

SOLUTION

a Certain

b Impossible

c Even chance

d Unlikely

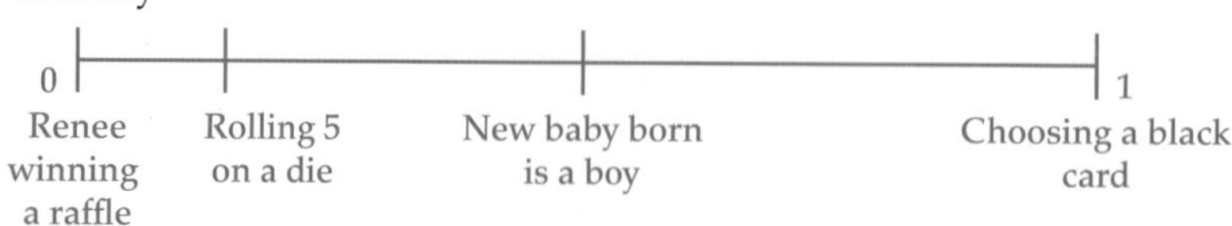

 ISBN 9780170350990

EXERCISE 14-01

1 Which word describes the chance of choosing a blue ball from a bag containing 1 red and 5 blue balls? Select the correct answer **A**, **B**, **C** or **D**.

A Certain **B** Likely **C** Unlikely **D** Impossible

2 Which word describes the chance that it will be hot on a day in winter? Select **A**, **B**, **C** or **D**.

A Certain **B** Likely **C** Unlikely **D** Impossible

3 Describe the events below using a suitable word.

a A summer's day will be hot.
b A newborn baby will be a girl.
c Choosing a white marble from a bag containing red and black marbles.
d A head when a coin is tossed.
e Choosing a red tie from a drawer with 20 blue ties.
f Throwing a 3 when a die is tossed.
g Choosing a red marble from a bag containing 5 black marbles and 1 red marble.
h Throwing an odd number when a die is tossed.
i Throwing a head or a tail when a coin is tossed.
j It will be cold in Canberra on a day in July.

4 Place the events in Question **3a–j** between 0 and 1 on a number line.

5 Is each statement true or false?

a The chance of leaves falling in Autumn is likely.
b It is certain that the Sun will rise tomorrow morning.
c It is impossible to roll a number greater than 4 on a die.
d There is an even chance that the next baby born is a girl.
e It is unlikely that a number less than 6 on a die is rolled.
f It is likely that a day in June in Perth will be cold.

6 Describe an event that matches each probability word.

a even chance **b** impossible **c** likely
d unlikely **e** certain

14–02 Sample spaces

WORDBANK

sample space The set of all possible outcomes in a chance situation. For example, when tossing a coin the sample space is {head, tail}.

equally likely Having exactly the same chance of occurring. For example, head or tail are equally likely when tossing a coin.

random Where each possible outcome is equally likely to occur.

EXAMPLE 2

A die is rolled.

a What is the sample space?

b Are the outcomes equally likely?

c Describe the chance of rolling a number less than 5.

SOLUTION

a The sample space = {1, 2, 3, 4, 5, 6}. ← All the numbers that are on a die.

b Yes, the die is equally likely to land on each number.

c There is a likely chance of rolling a number less than 5, as there are four numbers less than 5 {1, 2, 3, 4} and only two numbers not less than 5 {5, 6}.

EXAMPLE 3

A letter of the alphabet is chosen at random from the vowels only.

a What is the sample space?

b How many outcomes are in the sample space?

SOLUTION

a The sample space = {A, E, I, O, U}.

b Number of outcomes = 5.

Shutterstock.com/Karramba Production

ISBN 9780170350990

EXERCISE 14-02

1 A number from 15 to 27 is selected at random. How many outcomes are possible? Select the correct answer **A**, **B**, **C** or **D**.

A 12 **B** 11 **C** 13 **D** 27

2 A number is selected at random from all the prime numbers between 10 and 20. What is the sample space? Select **A**, **B**, **C** or **D**.

A {11, 12, 13, 15, 17} **B** {11, 13, 15, 17, 19} **C** {11, 13, 15} **D** {11, 13, 17, 19}

3 Write the sample space for each chance situation.

a rolling a die

b choosing a letter of the alphabet

c choosing a number from 5 to 15

d tossing two coins

e choosing a coin in the Australian currency

f choosing a day of the week

g tossing a die and a coin at the same time

h choosing an odd number from 10 to 24

i choosing a note in the Australian currency

j choosing from the factors of 12

4 Write the number of outcomes for each sample space in Question **3**.

5 In a bag of marbles there are 10 red, 8 blue and 4 green marbles. One marble is chosen and its colour is noted.

a List the sample space and count the number of possible outcomes in this sample space.

b Are the outcomes equally likely?

6 **a** If I spin the arrow, what is the sample space?

b Are all outcomes equally likely?

c Is there an even chance of spinning red?

d Is there an even chance of spinning blue or red?

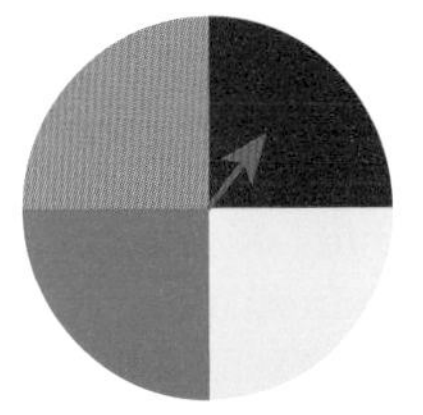

7 **a** If I spin the arrow, what is the sample space?

b Are all outcomes equally likely?

c Is there an even chance of spinning purple?

d Is the chance of spinning green likely or unlikely?

e Is there an even chance of spinning orange?

f Is the chance of spinning purple or green likely or unlikely?

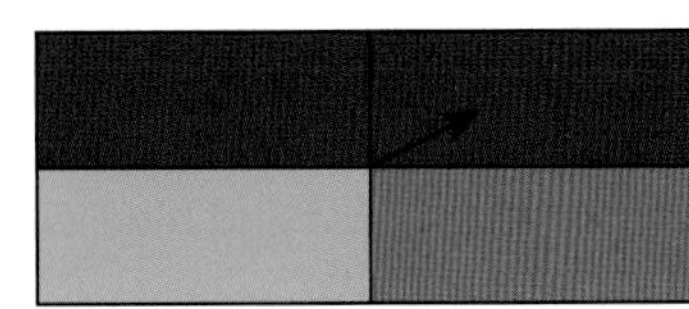

8 **a** Draw a spinner where the outcomes are all equally likely.

b Draw a spinner where the outcomes are not all equally likely.

14–03 Probability

If we roll a die, what is the probability of getting a number less than 3?
Sample space = {1, 2, 3, 4, 5, 6}.
Two of these numbers—1 and 2—are less than 3.
So there are two chances out of six of rolling a number less than 3.

The **probability of an event** has the abbreviation $P(E)$. If all outcomes are equally likely, then:

$$P(E) = \frac{\text{number of outcomes in the event}}{\text{number of outcomes in the sample space}}.$$

EXAMPLE 4

A die is rolled.

a List the sample space. Are all outcomes equally likely?

b Find the probability of rolling:

i 4 **ii** an odd number **iii** 7 **iv** a number less than 7.

SOLUTION

a Sample space = {1, 2, 3, 4, 5, 6}. ← Each number is equally likely to be rolled.

b **i** $P(4) = \frac{1}{6}$ ← One chance out of six.

ii $P(\text{odd}) = \frac{3}{6}$ ← Three odd numbers: 1, 3, 5.

$= \frac{1}{2}$

iii $P(7) = \frac{0}{6}$ ← Cannot roll 7 on a die: impossible.

$= 0$

iv $P(\text{number less than } 7) = \frac{6}{6}$ ← All six numbers on a die are less than 7.

$= 1$ ← Certain: must happen.

EXAMPLE 5

Find the probability of selecting a red or a blue jelly snake at random from a bag containing 3 yellow, 2 white, 6 red and 4 blue jelly snakes.

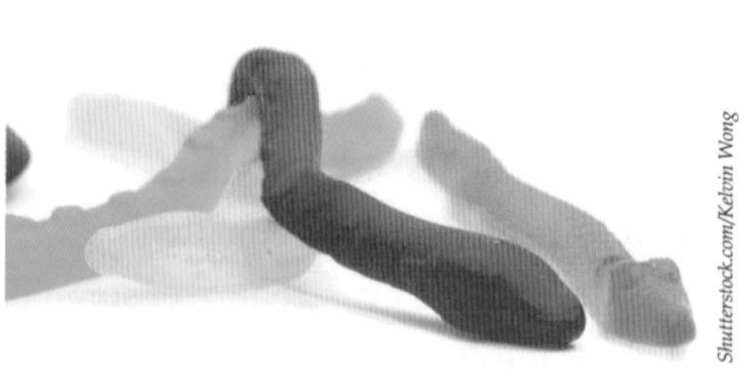
Shutterstock.com/Kelvin Wong

SOLUTION

Total number of jelly snakes = 3 + 2 + 6 + 4 = 15

Total number of red and blue jelly snakes = 6 + 4 = 10

$P(\text{red or blue}) = \frac{10}{15}$

$= \frac{2}{3}$ ← Simplify the fraction if possible.

ISBN 9780170350990

EXERCISE 14–03

1 What is the probability of choosing a red ball from a bag containing 3 red and 5 blue balls? Select the correct answer **A**, **B**, **C** or **D**.

A $\frac{3}{5}$ **B** $\frac{3}{8}$ **C** $\frac{5}{8}$ **D** $\frac{3}{15}$

2 What is the probability of choosing a white sticker from a bag containing 6 red, 8 white and 4 green stickers? Select **A**, **B**, **C** or **D**.

A $\frac{4}{9}$ **B** $\frac{1}{3}$ **C** $\frac{6}{18}$ **D** $\frac{4}{18}$

3 Find the probability of each event.

a Tossing a tail on a coin.

b Rolling a 6 on a die.

c Selecting a blue ribbon from a bag containing 3 red and 8 blue ribbons.

d Choosing a girl in a class of 18 boys and 12 girls.

e Choosing a soft-centred chocolate from a box containing 14 soft- and 16 hard-centred chocolates.

f Choosing a number less than 12 from a set of numbers from 1 to 10.

g Rolling a number greater than 8 on a die.

h Choosing the letter B from the alphabet.

i Rolling a number less than 6 on a die.

j Selecting a red dress from a wardrobe containing 1 black, 6 red and 3 blue dresses.

4 In a bag of lollies, Jenny has 6 red, 8 yellow, 10 green and 4 blue lollies. She selects a lolly from the bag without looking.

a Is each colour equally likely to be selected?

b Find the probability that Jenny selects:

i a red lolly **ii** a green lolly

iii a blue or a yellow lolly **iv** a black lolly

c If Jenny chose a red lolly, ate it and then chooses another lolly, what is the probability that it is another red lolly?

5 A deck of playing cards contains four suits: hearts (♥), diamonds (♦), spades (♠) and clubs (♣).

In each suit there are 13 cards: Ace, 2, 3, 4, 5, 6, 7, 8, 9, 10, Jack, Queen and King.

If a deck of cards is shuffled and Mark selects a card at random, find the probability that he selects:

a a jack **b** a diamond **c** an 8

d a number less than 5 **e** the 6 of spades **f** a club

g the queen of hearts **h** a 3 or a 4 **i** a king or a queen.

14–04 Complementary events

The **complement** of an event is the event **not** taking place. For example, the complement of rolling 3 on a die is not rolling a 3; that is, rolling 1, 2, 4, 5 or 6.

$P(3) = \frac{1}{6}$ and $P(\text{not a } 3) = P(1, 2, 4, 5, 6) = \frac{5}{6}$

So $P(\text{not a } 3) = 1 - P(3) = 1 - \frac{1}{6} = \frac{5}{6}$

Probability is measured on a scale from 0 to 1, where 1 is certain, so if E is an event, then $P(E) + P(\text{not } E) = 1$. The complement of E is written as $\overline{E}$.

- $P(E) + P(\overline{E}) = 1$
- or $P(\overline{E}) = 1 - P(E)$
- or $P(\text{not } E) = 1 - P(E)$

EXAMPLE 6

Jack selects a marble at random from a bag containing 5 red, 3 green and 7 blue marbles.

Find the probability that the marble is:

a green **b** not green

c red or blue **d** not red or blue.

Shutterstock.com/John Brueske

SOLUTION

a Total marbles = 5 + 3 + 7 = 15

$$P(\text{green}) = \frac{3}{15} = \frac{1}{5}$$

b $P(\text{not green}) = 1 - P(\text{green})$

$$= 1 - \frac{1}{5} = \frac{4}{5}$$

* This is correct because out of 15 marbles there are 12 marbles that are not green and $\frac{12}{15} = \frac{4}{5}$.

c $P(\text{red or blue}) = \frac{5+7}{15}$

$$= \frac{12}{15} = \frac{4}{5}$$

d $P(\text{not red or blue}) = 1 - P(\text{red or blue})$

$$= 1 - \frac{4}{5} = \frac{1}{5}$$

ISBN 9780170350990

EXERCISE 14–04

1 What is the probability of rolling a number other than 3 on a die? Select the correct answer **A**, **B**, **C** or **D**.

A $\frac{1}{6}$ **B** $\frac{3}{6}$ **C** $\frac{5}{6}$ **D** $\frac{2}{6}$

2 What would be the complementary event for 'coming first in a race'? Select **A**, **B**, **C** or **D**.

A coming second **B** coming in any place **C** not coming first **D** coming last

3 Describe in words the complementary event for each event.

a tossing a head when a coin is tossed

b rolling a 5 on a die

c selecting 'e' from a set of vowels

d choosing a 5 from the numbers 1 to 9

e selecting a red marble from a bag containing red and blue marbles

f selecting a letter after 't' in the alphabet

g tossing an odd number on a die

4 In a jar of lollies there are 12 Minties, 16 Fantales and 8 Chocolate Eclairs.
Find the probability of selecting:

a a Fantale **b** not a Fantale **c** a Mintie

d not a Mintie **e** a Malteser **f** not a Malteser

g a Fantale or a Chocolate Eclair **h** not a Fantale or a Chocolate Eclair.

5 Jodie has a 25% chance of being elected to student council. What is the probability that she won't be elected to student council? Answer as a percentage. (Remember 1 = 100%.)

6 When Dario drives to work, he notices that there is a 0.6 chance of one set of traffic lights being green and a 0.1 chance of them being amber (yellow). Find the probability (as a decimal) that the traffic light shows:

a red **b** not red **c** not amber **d** not red or amber.

7 The probability of a bus arriving on time is $\frac{1}{3}$ whereas the probability of a train arriving late is $\frac{1}{4}$.

a What is the probability of the bus not arriving on time?

b Would I have a better chance of being on time if I caught the bus or the train?

14–05 Experimental probability

WORDBANK

experimental probability Probability based on the results of an experiment or past statistics.

frequency The number of times something happens.

trial One run or go of a repeated chance experiment; for example, one roll of a die.

IN EXPERIMENTAL PROBABILITY:

$$P(E) = \frac{\text{number of times the event occurred}}{\text{total number of trials}}$$

$$= \frac{\text{frequency of the event}}{\text{total frequency}}$$

EXAMPLE 7

Eloise tosses a coin 60 times and records the results in the table below.

Heads	Tails
38	22

Notice that for 60 tosses of a coin, we would expect 30 heads and 30 tails coming up. In experimental probability, the number expected does not usually happen because we need to run a very large number of trials before we get close to the real probability.

What was the experimental probability of tossing:

a a head **b** a tail?

SOLUTION

a $P(\text{head}) = \frac{38}{60}$

$= \frac{19}{30}$

b $P(\text{tail}) = \frac{22}{60}$

$= \frac{11}{30}$

* Note that $P(\text{head}) + P(\text{tail}) = \frac{19}{30} + \frac{11}{30} = 1$.

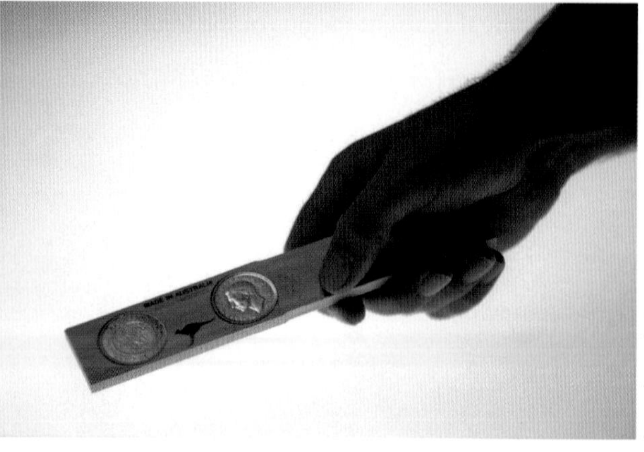

Newspix/Warren Clarke

ISBN 9780170350990

EXERCISE 14–05

1 What is the experimental probability of throwing a head if I toss a coin 8 times and it lands on tails 3 times? Select the correct answer **A**, **B**, **C** or **D**.

A $\frac{3}{8}$ B $\frac{5}{3}$ C $\frac{3}{5}$ D $\frac{5}{8}$

2 For Question **1**, what is the experimental probability of tossing a tail? Select **A**, **B**, **C** or **D**.

A $\frac{3}{8}$ B $\frac{5}{3}$ C $\frac{3}{5}$ D $\frac{5}{8}$

3 **a** If you toss a coin 20 times how many heads would you expect?

b How many tails would you expect?

c Now toss a coin 20 times.

d Record the results in a table.

e Did you get the results you expected? Why?

4 **a** If you roll a die 60 times how many of each number would you expect?

b Roll a die 60 times and record the results in a table.

c Did you get the results you expected?

d What was your experimental probability of rolling:

i 5 **ii** 3 **iii** an even number **iv** 5 or 6

5 Kristy flips a coin 40 times and records the results. She repeats this experiment four times. The results are listed in the table below.

Heads	Tails
28	12
18	22
25	15
19	21
23	17

a Using only the first row of the table, find the experimental probability of flipping a head and a tail.

b Using the results from the entire table, find the experimental probability of flipping a head and a tail.

c Which gives the closer result to the theoretical probability?

d What can you conclude from this?

6 Roll a die 60 times and record the results in a table.

a How many 1s did you expect? Did this happen?

b What is the theoretical probability of rolling a 5?

c What was your experimental probability of rolling a 5?

d Combine the results for the whole class and investigate the experimental probability of each number rolled.

14–06 Venn diagrams

A **Venn diagram** uses circles to group items into categories. Most Venn diagrams involve circles that **overlap**. For example, *A* could represent all singers and *B* could represent all dancers. Therefore, the shaded region would represent ***A* and *B***, which means individuals who are both **singers and dancers**.

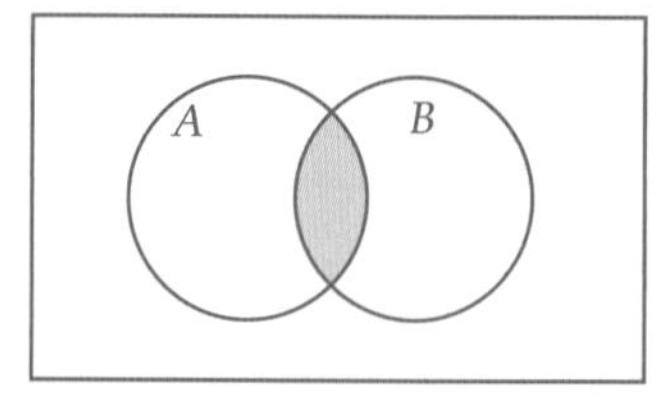

A* or *B means ***A* or *B* or both**

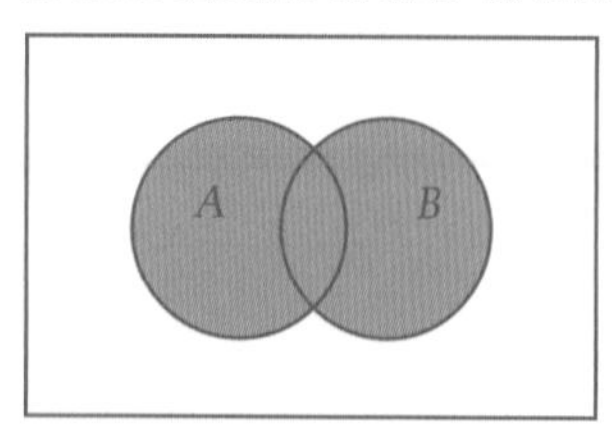

Singers or dancers or both

A* only** means ***A* but not *B

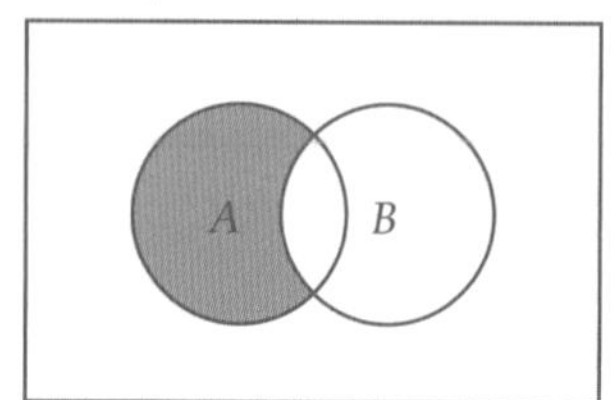

Singers only, not dancers

A Venn diagram may involve two groups that do not overlap.

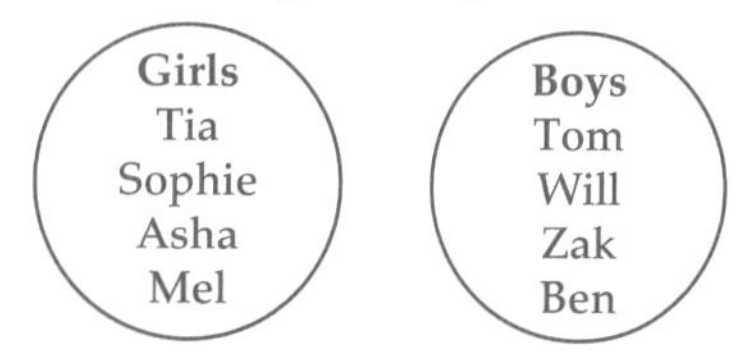

These categories are **mutually exclusive** as it is not possible to be both a boy and a girl.

EXAMPLE 8

A group of 35 students was surveyed on their favourite ice-cream flavour, showing that 25 students like chocolate, 18 like strawberry and 12 like both flavours.

a Show this information on a Venn diagram.

b What is the probability of selecting a student who likes:

i chocolate ice-cream only? **ii** neither chocolate nor strawberry?

SOLUTION

a

- 12 students like both chocolate and strawberry, so write 12 in the overlap.
- This leaves $25 - 12 = 13$ students who like chocolate only.
- This leaves $18 - 12 = 6$ students who like strawberry only.
- $13 + 12 + 6 = 31$, but 35 students were surveyed, so that leaves $35 - 31 = 4$ students who like neither chocolate nor strawberry, so write 4 outside the circles.

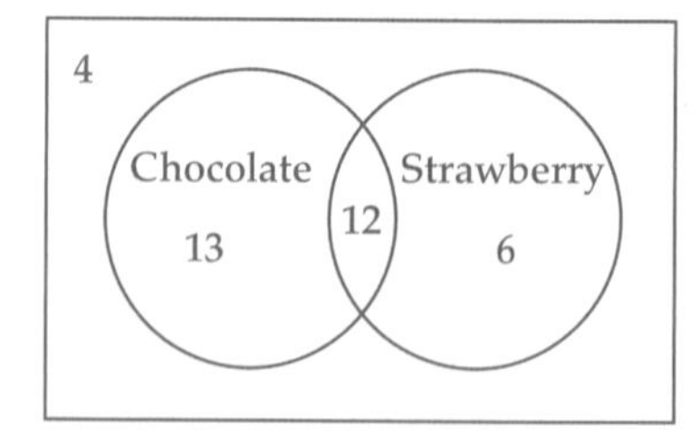

✱ Fill in the centre overlap section first.

b **i** $P(\text{only chocolate}) = \frac{13}{35}$ ← 13 students like chocolate but not strawberry

ii $P(\text{neither chocolate nor strawberry}) = \frac{4}{35}$ ← 4 students like neither

ISBN 9780170350990

EXERCISE 14–06

1 If 40 students are surveyed on whether they prefer soccer or cricket, and 28 like soccer and 18 like cricket, how many like both? Select the correct answer **A**, **B**, **C** or **D**.

A 4 **B** 5 **C** 6 **D** 7

2 In Question **1**, how many students like soccer only? Select **A**, **B**, **C** or **D**.

A 18 **B** 16 **C** 20 **D** 22

3 This Venn diagram shows the number of students who play basketball, hockey and both sports.

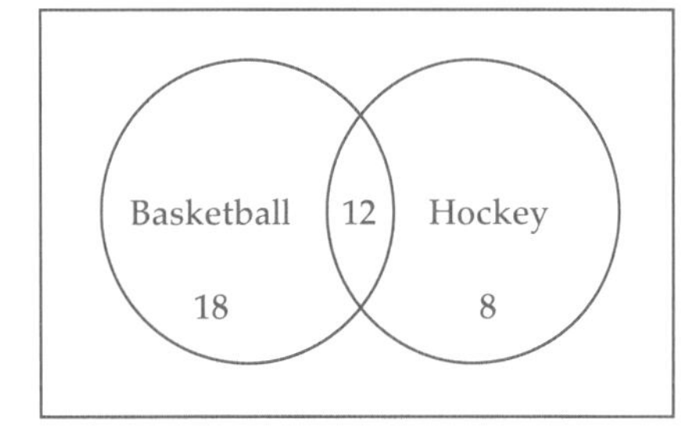

Find the number of students who play:

a basketball

b both sports

c hockey only.

4 This Venn diagram shows whether a group of people prefer fish or pizza on a Friday night.

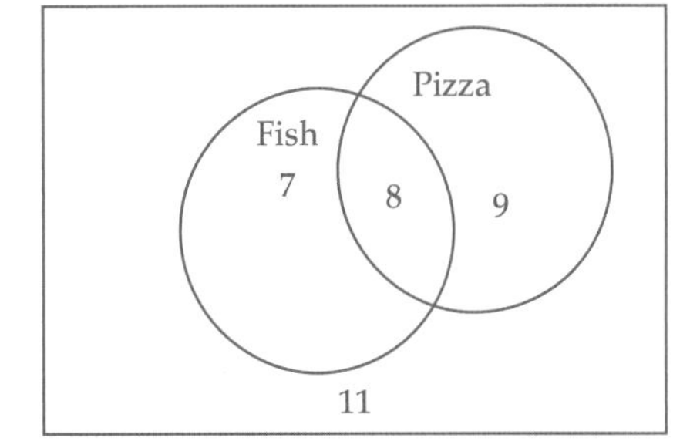

What is the probability that a person chosen at random from this group prefers:

a fish only

b neither pizza nor fish

c pizza or fish but not both

d pizza or fish?

5 Draw a Venn diagram to show that 14 customers buy shampoo, 12 buy conditioner and 8 buy both. What is the probability of randomly selecting a customer who buys:

a shampoo only

b conditioner only?

6 In a survey of 80 travellers, 45 have visited Italy, 52 have visited France and 8 have visited neither country. What is the probability that a traveller has visited:

a both Italy and France

b France only?

7 This Venn diagram shows the number of students playing video games.

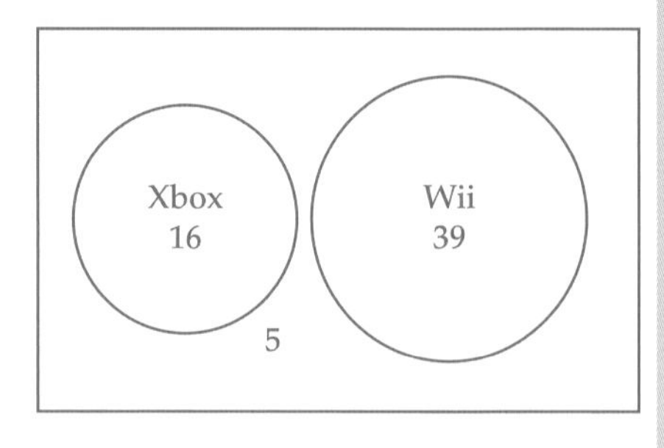

a What is the probability that a student chosen at random:

i plays on the Wii

ii plays on the Xbox

iii plays on both

iv plays on neither the Xbox nor the Wii?

b Is playing on the Xbox and the Wii mutually exclusive for this group of students?

iStockphoto/Mike Cherim

8 120 students in Year 8 were surveyed on their hobbies. 48 liked swimming, 65 liked going to the movies and 36 liked dancing. If 22 liked swimming and the movies, 16 liked swimming and dancing, 12 liked movies and dancing and 8 liked all three activities, draw a Venn diagram to find the probability of a student from this group liking:

a dancing only

b none of these activities

c swimming or dancing.

 ISBN 9780170350990

14-07 Two-way tables

A **two-way table** is useful for grouping items into categories that **overlap**.

EXAMPLE 9

A group of people were surveyed on whether they owned a tablet device. The results are shown in the two-way table below. Copy and complete the table.

	Tablet	No tablet	Total
Female	54	35	89
Male	75	22	
Total	129		

a How many people were surveyed altogether?

b How many females own a tablet device?

c What is the probability that a person selected at random will be a male who does not own a tablet device?

d What is the probability of selecting a person who owns a tablet device?

SOLUTION

	Tablet	No tablet	Total	
Female	54	35	89	
Male	75	22	97	75 + 22
Total	129	57	186	129 + 57
		35 + 22	89 + 97	

a Total surveyed = 186

b No. of female tablet owners = 54

c $P(\text{male, doesn't own tablet}) = \frac{22}{186} = \frac{11}{93}$

d $P(\text{owns tablet}) = \frac{129}{186} = \frac{43}{62}$

Shutterstock.com/Estudi M6

EXERCISE 14–07

1 Copy and complete this two-way table.

	Works full-time	Works part-time	Total
Male	75	15	
Female	48	26	74
Total	123		

a How many females work full-time?

b How many males work part-time?

c Find the probability that a person selected from this group:

i is male

ii works part-time

iii is a female working part-time

iv is male or working full-time?

2 Copy and complete this table showing where in a major city married and single people live.

	Married	Single	Total
Northside	65	27	
Southside	75	19	94
Total	140		

If a person from the survey was randomly selected, what is the probability that the person:

a is married and from Southside

b is single and from Northside

c is single

d is from Northside or married but not both?

3 This two-way table shows whether an audience of adults and children enjoyed a school concert.

	Liked	Disliked	Total
Adult	78	34	112
Child	65	47	
Total		81	

Copy and complete the table.

a Find how many people were:

i children who liked the concert

ii adults who didn't like the concert

b Find the probability of selecting from the audience:

i a child who didn't like the concert

ii a person who didn't like the concert

ISBN 9780170350990

EXERCISE 14–07

4 Copy and complete this table showing whether a group of teenagers liked rock music or hip-hop.

	Hip-hop	Not hip-hop	Total
Rock		9	
Not rock	8		
Total	15		35

What is the probability that a teenager selected at random:

a likes hip-hop but not rock

b likes hip-hop or rock

c likes hip-hop and rock

d likes neither hip-hop nor rock?

iStockphoto/ebstock

14–08 Probability problems

EXAMPLE 10

Two dice are rolled together and the sum of the two numbers is calculated. There are 36 different possible outcomes in the sample space, shown in the table below.

		First die					
	+	1	2	3	4	5	6
Second die	1	2	3	4	5	6	7
	2	3	4	5	6	7	8
	3	4	5	6	7	8	9
	4	5	6	7	8	9	10
	5	6	7	8	9	10	11
	6	7	8	9	10	11	12

Find the probability of rolling a total:

a of 7 **b** that is less than 4

c that is not 9 **d** of 3 or 5.

SOLUTION

a $P(7) = \frac{6}{36} = \frac{1}{6}$ ← There are 6 ways to roll a 7.

b $P(\text{sum} < 4) = \frac{3}{36} = \frac{1}{12}$ ← There are 3 ways to roll a total of 2 or 3.

c $P(\text{not } 9) = 1 - P(9)$

$= 1 - \frac{4}{36}$

$= \frac{32}{36}$

$= \frac{8}{9}$

d $P(3 \text{ or } 5) = \frac{2+4}{36}$

$= \frac{6}{36}$

$= \frac{1}{6}$

EXAMPLE 11

Consider this claim: For the spinner below, the probability of spinning a 5 is $\frac{1}{5}$.

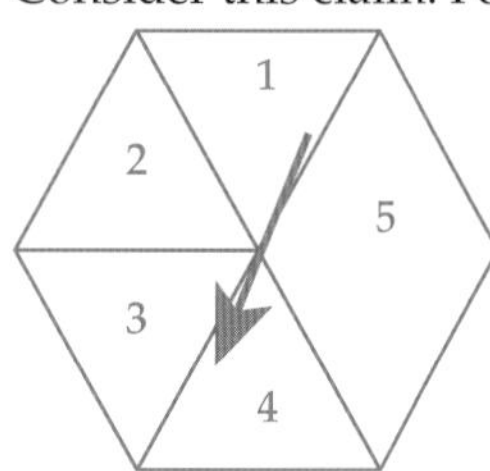

Is this statement correct or incorrect? Justify your answer.

SOLUTION

The statement is incorrect because although there are five numbers, there are actually six equal sections, with the number 5 taking up two of the sections.

 ISBN 9780170350990

EXERCISE 14–08

1 Two dice are rolled together and the sum of the two numbers is calculated. Use the table from Example 10 to find the probability of rolling a total:

a of 4 **b** that is not 5 **c** of 11 or 12 **d** that is greater than 8.

2 In a box of biscuits there are 12 Iced VoVos, 6 Tim Tams and 9 Monte Carlos.

a If I select a biscuit without looking, what is the probability that I will select:

i a Tim Tam?

ii a biscuit that is not a Monte Carlo?

iii an Iced VoVo or a Tim Tam?

b If I choose a Monte Carlo and eat it, what is the probability that I now choose another Monte Carlo?

3 **a** Draw a Venn diagram to represent the following information about a small community.

Total population: 250
Houses with a swimming pool: 120
Houses with a garage: 208
Houses with neither a garage nor a swimming pool: 24

b If I select a house at random, find the probability that I choose:

i a house with both a swimming pool and a garage

ii a house with a garage but no swimming pool

4 **a** Copy and complete this two-way table showing the results of a survey on preferred swimming locations.

	Beach	Pool	Total
Male	46	18	64
Female	28	42	
Total		60	

b What is the probability of selecting:

i a male who prefers the swimming pool?

ii a female who prefers the beach?

c What percentage of males prefer the beach? Answer correct to one decimal place.

5 **a** For the spinner in Example 11, why is the probability of spinning a 4 not $\frac{1}{5}$?

b What is the probability of spinning a 4?

c What other numbers have the same chance of being spun as 4?

d What is the probability of spinning a 5?

6 Justin ordered two Supreme pizzas, one Hawaiian pizza and three Pepperoni pizzas.

a If each pizza is cut into eight equal pieces and Justin chooses one piece of pizza at random, find the probability that it is:

i Supreme **ii** Pepperoni

iii Hawaiian or Pepperoni **iv** not Hawaiian

b If Kamil is hungry and eats two pieces of Supreme, what is the chance that she will next randomly select a piece of Pepperoni?

CROSSWORD PUZZLE

Make a copy of this puzzle, then use the clues to complete the crossword.

Across

3 Must occur, will definitely happen.

4 A chart that describes categories using overlapping circles (two words).

7 A table that shows numbers of items in categories.

8 Probably will not happen.

9 This type of chance has a probability of $\frac{1}{2}$.

Down

1 All of the possible outcomes in a chance situation (two words).

2 Another name for chance.

5 Will probably happen.

6 Cannot happen at all.

ISBN 9780170350990

PRACTICE TEST 14

Part A General topics

Calculators are not allowed.

1 Decrease $280 by 25%.

2 Simplify $3x - 4y - x - 6y$.

3 Find the median of 9, 8, 12, 7, 7 and 6.

4 Copy this diagram and mark two vertically opposite angles.

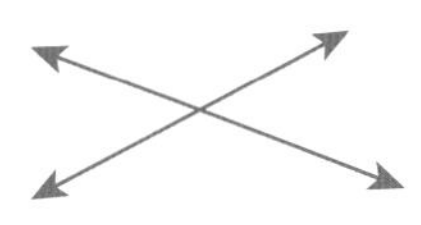

5 Evaluate $\frac{3}{8} \div \frac{15}{12}$.

6 Evaluate $\sqrt{25}$.

7 Convert 2335 to 12-hour time.

8 Factorise $56ab^2 - 8abc$.

9 What order of rotational symmetry has a parallelogram?

10 Evaluate 3×8^0.

Part B Probability

Calculators are allowed.

14–01 The language of chance

11 Which word or phrase describes an event with a probability of $\frac{1}{2}$? Select the correct answer **A**, **B**, **C** or **D**.

A certain **B** impossible **C** likely **D** even chance

12 Which word or phrase describes an event with a probability of 1? Select **A**, **B**, **C** or **D**.

A certain **B** impossible **C** likely **D** even chance

14–02 Sample spaces

13 If Darren selects a letter at random from the consonants in the alphabet, how many outcomes are possible? Select **A**, **B**, **C** or **D**.

A 21 **B** 22 **C** 23 **D** 24

14–03 Probability

14 What is the probability of flipping a head on a coin?

15 What is the probability of rolling a number less than 3 when I toss a die?

14–04 Complementary events

16 In a bag of lollies, there are 12 Snakes, 7 Milk Bottles and 9 Jelly Babies.

What is the probability of not selecting a Snake?

ISBN 9780170350990

14-05 Experimental probability

17 In an experiment, Peter got 24 tails from 40 tosses of a coin.

What is the experimental probability of getting a *head*?

14-06 Venn diagrams

18 This Venn diagram shows the type of food liked by a group of students.

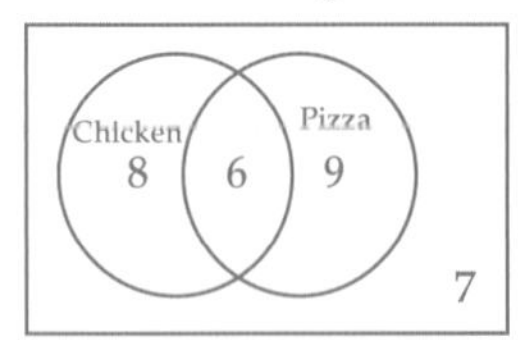

What is the probability that a student chosen at random likes:

a pizza only?

b chicken or pizza?

14-07 Two-way tables

19 Copy and complete this two-way table about the types of chocolates in a box.

	Soft-centred	Hard-centred	Total
Milk chocolate	12	18	
Dark chocolate	16	22	38
Total	28		

a How many chocolates are in the box altogether?

b What is the probability of randomly selecting a soft-centred dark chocolate?

14-08 Probability problems

20 The 10 letters of the alphabet from A to J are written on separate pieces of paper and shuffled. Nina chooses one of these at random. Find the probability that this letter is:

a a vowel

b a consonant

c part of the word PROBABILITY

ISBN 9780170350990

FURTHER ALGEBRA

15

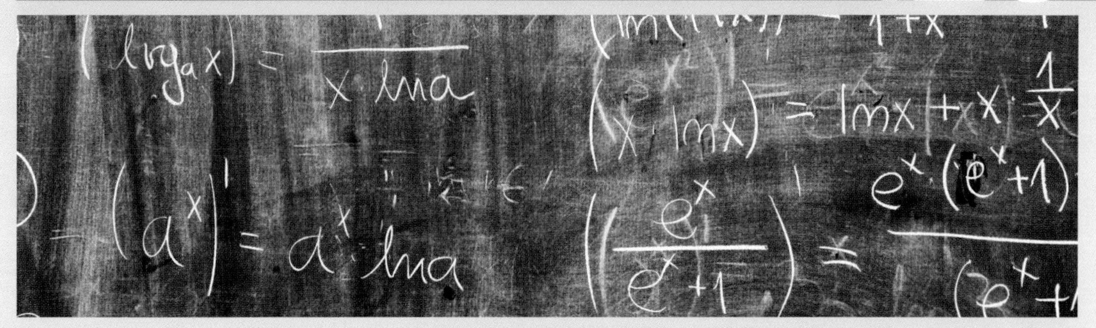

WHAT'S IN CHAPTER 15?

IN THIS CHAPTER YOU WILL:

- expand algebraic expressions
- factorise algebraic expressions
- solve one-step and two-step equations
- solve equations with variables on both sides
- expand and solve equations with brackets

*Shutterstock.com/Robsonphoto

15–01 Expanding expressions

WORDBANK

expand To remove the brackets or grouping symbols in an algebraic expression.

When multiplying 15 by 11 mentally, we know that:

$$15 \times 11 = 15 \times (10 + 1)$$
$$= 15 \times 10 + 15 \times 1$$
$$= 150 + 15$$
$$= 165$$

This idea can be used to expand algebraic expressions.
For example:

$$5 \times (a + 4) = a + 4 + a + 4 + a + 4 + a + 4 + a + 4$$
$$= 5 \times a + 5 \times 4$$
$$= 5a + 20$$

To expand an algebraic expression with brackets, the term outside the brackets must be multiplied by every term inside the brackets.
$a(b + c) = ab + ac$

EXAMPLE 1

Expand each algebraic expression.

a $4(b + 3)$ b $8(m - 5)$ c $5(2w + 3)$

SOLUTION

a $4(b + 3) = 4 \times b + 4 \times 3$
$= 4b + 12$

b $8(m - 5) = 8 \times m - 8 \times 5$
$= 8m - 40$

c $5(2w + 3) = 5 \times 2w + 5 \times 3$
$= 10w + 15$

EXAMPLE 2

Expand each algebraic expression.

a $-4(n + 6)$ b $-7(3a - 5)$ c $-2c(3c - 8)$

SOLUTION

When a **negative number** is outside the brackets, all terms inside the brackets must be multiplied by the negative number.

a $-4(n + 6) = -4 \times n + (-4) \times 6$
$= -4n + (-24)$
$= -4n - 24$

b $-7(3a - 5) = -7 \times 3a + (-7) \times (-5)$
$= -21a + 35$

c $-2c(3c - 8) = -2c \times 3c + (-2c) \times (-8)$
$= -6c^2 + 16c$

ISBN 9780170350990

EXERCISE 15–01

1 Expand $4(2a - 3)$. Select the correct answer **A**, **B**, **C** or **D**.

A $6a - 12$ **B** $8a - 12$ **C** $42a - 7$ **D** $8a - 3$

2 Expand $2m(4m - 1)$. Select **A**, **B**, **C** or **D**.

A $6m - 1$ **B** $6m^2 - 2$ **C** $8m^2 - 4m$ **D** $8m^2 - 2m$

3 Copy and complete each expansion.

a $6(w + 3) = 6 \times$ ___ $+ 6 \times$ ___
$=$ ___ $+ 18$

b $5(a - 4) =$ ___ $\times a - 5 \times$ ___
$= 5a -$ ___

c $3(m + 9) =$ ___ $\times m +$ ___ $\times 9$
$= 3m +$ ___

d $8(2a - 6) =$ ___ $\times 2a -$ ___ $\times$ ___
$=$ ___ $- 48$

e $-4(w + 6) =$ ___ $\times$ ___ $+ (-4) \times$ ___
$=$ ___ $- 24$

f $-7(r - 6) = -7 \times$ ___ $+$ ___ $\times (-6)$
$= -7r +$ ___

4 Expand each expression.

a $5(a + 6)$ b $4(w - 6)$ c $2(3a + 7)$
d $-4(c + 5)$ e $-6(m - 3)$ f $-3(5a - 1)$
g $7(3n + 2)$ h $4(3a - 4)$ i $-7(m + 6)$
j $2(6a - 8)$ k $-5(3n + 6)$ l $-8(2w + 6)$

5 Expand each expression.

a $a(a + 4)$ b $v(2v - 3)$ c $3w(2w + 7)$
d $2m(m - 4)$ e $-a(3a + 5)$ f $-2b(4b - 3)$
g $b(b + 5)$ h $n(n - 4)$ i $2m(m - 6)$
j $-a(2a + 3)$ k $3v(v - 7)$ l $-4r(r + 5)$
m $6w(2w - 4)$ n $-9n(2n + 7)$ o $4m(3m - 6)$

6 Is each equation true or false?

a $3v(v - 8) = 3v^2 - 24$
b $-2n(n + 9) = -2n^2 - 18n$
c $-5m(2m - 1) = -10m^2 - 5m$
d $4r(8 - 2r) = 32r + 8r^2$

7 a Check that $5(a + 4) = 5a + 20$ by substituting $x = 3$ into both sides of the equation and testing whether the values are equal.

b Substitute another value of x into both sides and check whether the values are still equal.

ISBN 9780170350990

15–02 Factorising expressions

WORDBANK

highest common factor (HCF) The largest number or algebraic term that evenly divides into two or more numbers or algebraic terms.

factorise To insert brackets or grouping symbols in an algebraic expression by taking out the highest common factor (HCF); factorising is the opposite of expanding.

The HCF of 12 and 16 is 4 because 4 is the largest factor of both 12 and 16.
The HCF of $8ab$ and $12b^2$ is $4b$ because $4b$ is the largest factor of both $8ab$ and $12b^2$.

To find the highest common factor (HCF) of algebraic terms:

- find the HCF of the numbers
- find the HCF of the variables
- multiply the HCFs together.

EXAMPLE 3

Find the highest common factor of $8ab$ and $12b^2$.

SOLUTION

The HCF of 8 and 12 is 4.

The HCF of ab and b^2 is b.

 Find the HCFs of the numbers and variables separately.

So the HCF of $8ab$ and $12b^2$ is $4 \times b = 4b$.

To factorise an algebraic expression:

- find the HCF of all the terms and write it in front of the brackets
- divide each term by the HCF and write the answers inside the brackets
- $ab + ac = a(b + c)$

To check the answer is correct, expand it.

ISBN 9780170350990

15-02 Factorising expressions

Factorising is the opposite of expanding.

$$\xrightarrow{} \quad \text{Expanding}$$
$$2(a - 5) = 2a - 10$$
$$\xleftarrow{} \quad \text{Factorising}$$

EXAMPLE 4

Factorise each expression.

a $15a + 10$ b $6m - 8mn$ c $12ab - 4b$

d $5xy - 35xz$ e $14n^2 - 6n$ f $30de^2 + 15d^2e$

SOLUTION

a $15a + 10 = 5(\_\_\_ + \_\_\_)$

HCF of $15a$ and $10 = 5$.

$= 5(3a + 2)$

Check each answer by expanding:

$5(3a + 2) = 15a + 10$

b $6m - 8mn = 2m(\_\_\_ - \_\_\_)$

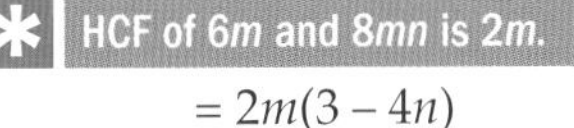

$= 2m(3 - 4n)$

Check each answer by expanding:

$2m(3 - 4n) = 6m - 8mn$

c $12ab - 4b = 4b(\_\_\_ - \_\_\_)$

$= 4b(3a - 1)$

Check by expanding:

$4b(3a - 1) = 12ab - 4b$

d $5xy - 35xz = 5x(\_\_\_ - \_\_\_)$

$= 5x(y - 7z)$

Check by expanding:

$5x(y - 7z) = 5xy - 35xz$

e $14n^2 - 6n = 2n(7n - 3)$

Check by expanding:

$2n(7n - 3) = 14n^2 - 6n$

f $30de^2 + 15d^2e = 15de(2e + d)$

Check by expanding:

$15de(2e + d) = 30de^2 + 15d^2e$

Shutterstock.com/antoniodiaz

ISBN 9780170350990

EXERCISE 15-02

1 Factorise $12x - 24$. Select the correct answer **A**, **B**, **C** or **D**.

A $12(x - 24)$ **B** $4(3x - 6)$ **C** $6(2x - 4)$ **D** $12(x - 2)$

2 Factorise $7mn + 28m$. Select **A**, **B**, **C** or **D**.

A $7(mn + 4m)$ **B** $7m(n + 4)$ **C** $7n(m + 4)$ **D** $7m(n + 4m)$

3 Copy and complete the solution to find the highest common factor of $12xy^2$ and $32xy$.

The HCF of 12 and 32 is: ___.

The HCF of xy^2 and xy is ___.

The HCF of $12xy^2$ and $32xy$ is ___ × ___ = ___.

4 Find the highest common factor of each pair of terms.

a $4bc$ and $12ab$ **b** $12ab$ and $18ac$ **c** $5mn$ and $6m^2$

d $5mn$ and $15b^2$ **e** $4mn^2$ and $6m^2$ **f** $12ab$ and $16c^2$

g $2ab$ and $8bc$ **h** $4mn$ and $12np$ **i** $6a^2$ and $8ab$

5 Copy and complete each factorisation.

a $5a - 10 = 5($___ – ___$)$ **b** $4m + 12 = 4\ ($___ + ___$)$

c $6w + 24 = 6($___$)$ **d** $8n - 16 = 8($___$)$

e $15r - 25rs = 5r($___$)$ **f** $12uv + 16vw = 4v($___$)$

g $6a^2 - 18a = 6a($___$)$ **h** $15bc + 20b^2 = 5b($___$)$

i $12ab + 16bc =$ ___$(3a + 4c)$ **j** $24a^2 - 16ac =$ ___$(3a - 2c)$

6 Factorise each expression, and check your answer by expanding.

a $12a - 6b$ **b** $3m + 9n$ **c** $8ab + 12bc$

d $2a^2 - 6a$ **e** $24ab + 16bc$ **f** $8w - 24wv$

g $16mn + 20np$ **h** $3m^2 - 15mn$ **i** $18bc + 24cd^2$

j $12w^2 - 16vw$ **k** $18bc^2 - 8bc$ **l** $28uv - 7v^2u$

7 **a** Why isn't $30cd^2 - 18c^2d = 3cd(10d - 6c)$ correctly factorised?

b Factorise $30cd^2 - 18c^2d$.

8 Copy and complete each factorisation, and check your answer by expanding.

a $3ab - 6bc + 9abc = 3b($___$)$

b $4w - 12aw + 16vw = 4w($___$)$

c $6a^2 - 12ab + 15ac =$ ___$(2a -$ ___ $+5c)$

d $8mn - 4n^2 + 12mn^2 = 4n($___$)$

ISBN 9780170350990

15–03 Factorising with negative terms

To factorise an algebraic expression with a negative first term:

- include the negative sign when finding the HCF and write it in front of the brackets
- divide each term by the HCF and write the answers inside the brackets.

To check the answer is correct, expand it.

For example, to factorise $-6a + 18$, the HCF will be -6.
To factorise $-5x - 20$, the HCF will be -5.

EXAMPLE 5

Factorise each expression.

a $-6a + 18$

b $-5x - 20$

c $-9y^2 - 12xy$

d $-16mn + 12m^2$

e $-abc - b^2cd$

SOLUTION

a $-6a + 18 = -6(\_\_\_\_ - \_\_\_\_)$
$= -6(a - 3)$
Check by expanding:
$-6(a - 3) = -6a + 18$

 Note that the sign inside each bracket is different from the sign in the question as you are dividing each term by a negative number.

b $-5x - 20 = -5(\_\_\_\_ + \_\_\_\_)$
$= -5(x + 4)$
Check by expanding:
$-5(x + 4) = -5x - 20$

c $-9y^2 - 12xy = -3y(\_\_\_\_ + \_\_\_\_)$
$= -3y(3y + 4x)$
Check by expanding:
$-3y(3y + 4x) = -9y^2 - 12xy$

d $-16mn + 12m^2 = -4m(\_\_\_\_ - \_\_\_\_)$
$= -4m(4n - 3m)$
Check by expanding:
$-4m(4n - 3m) = -16mn + 12m^2$

e $-abc - b^2cd = -bc(\_\_\_\_ + \_\_\_\_)$
$= -bc(a + bd)$
Check by expanding:
$-bc(a + bd) = -abc - b^2cd$

EXERCISE 15-03

1 Factorise $-8x - 32$. Select the correct answer **A**, **B**, **C** or **D**.

A $8(x - 4)$ **B** $-8(x - 4)$ **C** $-8(x + 4)$ **D** $-4(2x + 8)$

2 Factorise $-9mn + 18m$. Select **A**, **B**, **C** or **D**.

A $-9m(n - 2)$ **B** $-9m(n - 2m)$ **C** $9m(n + 2)$ **D** $-9m(n + 2m)$

3 Copy and complete each factorisation.

a $-4m + 8 = -4(\_\_\_)$ **b** $-7w - 42 = -7(\_\_\_)$

c $-12n - 16mn = -4n(\_\_\_)$ **d** $-18uv + 12vw = -6v(\_\_\_)$

e $-5a^2 + 15a = -5a(\_\_\_)$ **f** $-8mn - 12m^2 = -4m(\_\_\_)$

g $-20ab - 12bc = \_\_\_(5a + 3c)$ **h** $-22m^2 + 4mn = \_\_\_(11m - 2n)$

4 Factorise each expression.

a $-8a - 12$ **b** $-6m + 18$ **c** $-5n - 30$

d $-12w + 8$ **e** $-15r - 20$ **f** $-24m + 18$

g $-7n - 21mn$ **h** $-9y + 45ay$ **i** $-12ab - 8bc$

j $-24rs + 16st$ **k** $-18fg - 27gh$ **l** $-20uv + 30vw$

5 **a** What is the HCF of $-4a^2 - 12ab$?

b Hence copy and complete: $-4a^2 - 12ab = \_\_\_(a + 3b)$

6 Factorise each expression.

a $-n^2 - 3n$ **b** $-5b^2 - 20b$ **c** $-8c^2 + 32c$

d $-7m^2 + 56mn$ **e** $-8r^2 - 24rs$ **f** $-4w^2 + 12aw$

g $-15mn - 3n^2$ **h** $-18uv + 9v^2$ **i** $-25bd^2 - 20de$

7 **a** What is the HCF of $-6mn - 9np + 12mnp$?

b So copy and complete: $-6mn - 9np + 12mnp = \_\_\_(2m + 3p - 4mp)$

8 Factorise each expression.

a $-4ab + 12bc - 8abc$ **b** $-6rs - 9st - 18rst$

c $-12uv + 15vw - 18wvu$ **d** $-6ab^2 - 9abc + 24bc^2$

e $-20gh + 15h^2f - 25gfh$ **f** $-27rst + 18s^2t - 36rst^2$

ISBN 9780170350990

15–04 One-step equations

WORDBANK

equation A number sentence that contains algebraic terms, numbers and an equals sign; for example, $x - 3 = 8$.

solve an equation To find the value of the variable that makes the equation true.

solution The answer to an equation, the correct value of the variable.

guess and check A simple method for solving an equation by making an educated guess at the possible value for the variable, then substituting this value into the equation and checking if it is true.

LHS Left-hand side of an equation; for example, in $m + 3 = 15$, the LHS is $m + 3$.

RHS Right-hand side of an equation; for example, in $m + 3 = 15$, the RHS is 15.

inverse operation The 'opposite' process. For example, the inverse operation to adding (+) is subtracting (–).

EXAMPLE 6

Solve each equation using the guess-and-check method.

a $w + 4 = 12$ **b** $m - 2 = 18$ **c** $4n = 20$

SOLUTION

a Try $w = 8$:
Does $8 + 4 = 12$?
YES
So the solution is $w = 8$.

b Try $m = 10$:
Does $10 - 2 = 18$?
No, $10 - 2 = 8$
So it is too low.
Try $m = 20$:
Does $20 - 2 = 18$?
YES
So the solution is $m = 20$.

c Try $n = 7$:
Does $4 \times 7 = 20$?
No, $4 \times 7 = 28$
So it is too high.
Try $n = 5$:
Does $4 \times 5 = 20$?
YES
So the solution is $n = 5$.

Shutterstock.com/legenda

ISBN 9780170350990

15–04 One-step equations

To solve an equation using a more formal algebraic method:

- do the same to both sides of the equation: this will keep it balanced
- use inverse (opposite) operations to simplify the equation (+ and – are inverse operations, × and ÷ are inverse operations)
- write the **solution** as: x = a number.

The solution can be checked by substituting the answer into the original equation.
The two sides of an equation are equal, and when we solve an equation we must keep LHS = RHS at all times. The equation must stay balanced.

EXAMPLE 7

Use inverse operations to solve each equation.

a $m - 5 = 12$

b $b + 6 = 8$

c $4a = 12$

d $\dfrac{x}{3} = -6$

SOLUTION

✱ Do the same inverse operation to both sides of the equation.

a

$$m - 5 = 12$$ ← Opposite of – 5 is + 5.

$$m - 5 + 5 = 12 + 5$$ ← Add 5 to both sides.

$$m = 17$$

Check: $17 - 5 = 12$

b

$$b + 6 = 8$$ ← Opposite of + 6 is – 6.

$$b + 6 - 6 = 8 - 6$$ ← Subtract 6 from both sides.

$$b = 2$$

Check: $2 + 6 = 8$

c

$$4a = 12$$ ← Opposite of × 4 is ÷ 4.

$$\frac{4a}{4} = \frac{12}{4}$$ ← Divide both sides by 4.

$$a = 3$$

Check: $4 \times 3 = 12$

d

$$\frac{x}{3} = -6$$ ← Opposite of ÷ 3 is × 3.

$$\frac{x}{3} \times 3 = -6 \times 3$$ ← Multiply both sides by 3.

$$x = -18$$

Check: $\dfrac{-18}{3} = -6$

These are called **one-step equations** because they require only one step to solve.
When setting out, align the equal signs underneath each other.

ISBN 9780170350990

EXERCISE 15–04

1 What is the LHS of the equation $x + 7 = 12$? Select the correct answer **A**, **B**, **C** or **D**.

A 12 **B** 7 **C** x **D** $x + 7$

2 Solve the equation $x + 7 = 12$. Select **A**, **B**, **C** or **D**.

A $x = 7$ **B** $x = 5$ **C** $x = -5$ **D** $x = 12$

3 Copy and complete this paragraph.

To solve an equation we use ___ operations. Whatever we do to one side of the equation we must do the ___ to the other side. This will keep the equation ___. When solving an equation we must keep the equal signs ___ each other.

4 Write the operation that is the inverse of:

a addition **b** multiplication **c** division **d** subtraction.

5 Solve each equation by using the *guess-and-check* method.

a $x - 5 = 16$ **b** $m + 4 = 8$ **c** $2r = 12$

d $n + 7 = 9$ **e** $\frac{w}{4} = 6$ **f** $m - 8 = 24$

g $5a = -15$ **h** $b + 3 = -12$ **i** $\frac{x}{5} = -6$

6 What is the inverse of:

a adding 6? **b** multiplying by 8? **c** subtracting 7? **d** dividing by 4?

7 Copy and complete the solution to each equation.

a
$$x + 7 = 8$$
$$x + 7 - 7 = 8 - \_\_\_$$
$$x = \_\_\_$$

b
$$m - 2 = 11$$
$$m - 2 + 2 = \_\_\_ + 2$$
$$m = \_\_\_$$

c
$$9v = 36$$
$$\frac{9v}{9} = \frac{36}{\square}$$
$$v = \_\_\_$$

d
$$\frac{x}{4} = 4$$
$$\frac{x}{\square} \times 4 = 4 \times \_\_\_$$
$$x = \_\_\_$$

8 Solve each equation algebraically using inverse operations.

a $n + 6 = 17$ **b** $8b = 56$ **c** $m - 7 = 5$ **d** $\frac{x}{4} = 12$

e $a - 7 = -14$ **f** $\frac{n}{9} = -4$ **g** $5v = 20$ **h** $m - 8 = -14$

i $y + 6 = -12$ **j** $-7r = 49$ **k** $\frac{m}{7} = -8$ **l** $b - 5 = -10$

9 Solve each problem using an equation.

a Jesse thinks of a number, adds 12 and ends up with 54. What was the number he first thought of?

b Zoe thinks of a number, multiplies it by 5 and ends up with 40. What was the number she first thought of?

10 Solve each equation. The solutions are negative or fractions.

a $9 + m = -20$ **b** $n - 12 = -16$ **c** $-4b = -7$ **d** $3m = -17$

e $v - 2 = -8$ **f** $8 + a = -9$ **g** $5t = 19$ **h** $-4m = 6$

15-05 Two-step equations

EXAMPLE 8

Solve each equation.

a $3a - 1 = 14$ **b** $\frac{x}{2} + 6 = 10$ **c** $26 - 2w = 36$

✱ These are called **two-step equations** because they require two steps to solve.

SOLUTION

a

$$3a - 1 = 14$$

$$3a - 1 + 1 = 14 + 1 \quad \longleftarrow \text{Add 1 to both sides.}$$

$$3a = 15$$

$$\frac{3a}{3} = \frac{15}{3} \quad \longleftarrow \text{Divide both sides by 3.}$$

$$a = 5 \quad \longleftarrow \text{This is the solution.}$$

Check: $3 \times 5 - 1 = 14$.

b

$$\frac{x}{2} + 6 = 10$$

$$\frac{x}{2} + 6 - 6 = 10 - 6 \quad \longleftarrow \text{Subtract 6 from both sides.}$$

$$\frac{x}{2} = 4$$

$$\frac{x}{2} \times 2 = 4 \times 2 \quad \longleftarrow \text{Multiply both sides by 2.}$$

$$x = 8$$

Check: $\frac{8}{2} + 6 = 10$

c

$$26 - 2w = 36$$

$$26 - 2w - 26 = 36 - 26 \quad \longleftarrow \text{Subtract 26 from both sides.}$$

$$-2w = 10$$

$$\frac{-2w}{-2} = \frac{10}{-2} \quad \longleftarrow \text{Divide both sides by -2.}$$

$$w = -5$$

Check: $26 - 2 \times (-5) = 36$

iStock.com/annyoshi

ISBN 9780170350990

EXERCISE 15–05

1 Which operation would you do first to solve $2n + 5 = 17$? Select the correct answer **A**, **B**, **C** or **D**.

A -5 **B** $\div 2$ **C** $+5$ **D** $\times 2$

2 Solve $2n + 5 = 17$. Select **A**, **B**, **C** or **D**.

A $n = 11$ **B** $n = 6$ **C** $n = 7$ **D** $n = 5$

3 Copy and complete the solution to each equation.

a
$$2a - 6 = 14$$
$$2a - 6 + \_\_\_ = 14 + \_\_\_$$
$$2a = \_\_\_$$
$$a = \_\_\_$$

b
$$5m + 4 = 19$$
$$5m + 4 - \_\_\_ = 19 - \_\_\_$$
$$5m = \_\_\_$$
$$m = \_\_\_$$

c
$$\frac{x}{4} + 3 = 9$$
$$\frac{x}{4} + 3 - \_\_\_ = 9 - \_\_\_$$
$$\frac{x}{4} = \_\_\_$$
$$x = \_\_\_$$

4 Which operation would you do first to solve each equation?

a $3a + 4 = 10$ **b** $2g - 3 = 15$ **c** $2m + 7 = 11$

d $4v - 3 = 25$ **e** $7x + 5 = 26$ **f** $3c - 4 = -10$

g $8a + 6 = 38$ **h** $9e - 6 = -24$ **i** $5v + 8 = 53$

j $20 - 3y = 11$ **k** $12 - 5m = 47$ **l** $19 - 2n = 5$

5 Solve each equation in Question **4**.

6 Write the operation you would do first when solving each of the following equations and then solve each equation.

a $\frac{x}{2} + 3 = 5$ **b** $\frac{m}{3} - 5 = 3$ **c** $\frac{n}{4} + 5 = 8$

d $\frac{s}{8} - 4 = 5$ **e** $\frac{w}{4} + 7 = 19$ **f** $\frac{x}{5} + 5 = -7$

g $\frac{x}{7} - 6 = -3$ **h** $\frac{s}{3} - 9 = 12$ **i** $\frac{m}{6} + 6 = -4$

7 Solve each problem using an equation.

a Ben thinks of a number. He doubles the number and then subtracts 8 from the number. The result is 28. What is the number?

b Josie thinks of a number. She triples the number and then adds 7 to the number. The result is 52. What is the number?

8 Is each statement true or false?

a $x = 4$ is the solution to $3x - 8 = 4$

b $n = -2$ is the solution to $\frac{n-4}{6} = 1$

c $c = -3$ is the solution to $18 - 2c = 24$

d $m = 6$ is the solution to $3m + 9 = 25$

ISBN 9780170350990

15–06 Equations with variables on both sides

An equation with variables on both sides looks like this: $2a - 6 = a + 8$.
To solve such an equation, we need to rearrange the equation to find the value of the variable; for example, $a = 14$. Such equations usually require three steps to solve and use **inverse operations** to keep the equation **balanced**.

To solve an equation with variables on both sides:

- use inverse operations to move all the variables to the left-hand side (LHS) of the equation
- use inverse operations to move all the numbers to the right-hand side (RHS) of the equation
- then solve the equation.

EXAMPLE 9

Solve each equation.

a $2a - 6 = a + 8$

b $6w - 18 = 4w - 2$

SOLUTION

✱ Move all variables to the LHS of the equation.

a

$2a - 6 = a + 8$

$2a - 6 - a = a + 8 - a$ ← $-a$ from both sides

$a - 6 = 8$ ← simplifying $2a - a$

$a - 6 + 6 = 8 + 6$ ← $+ 6$ to both sides

$a = 14$

These solutions can be checked:
Check LHS = RHS
Substitute $a = 14$:

$$\begin{aligned} \text{LHS} &= 2a - 6 \\ &= 2 \times 14 - 6 \\ &= 22 \\ \text{RHS} &= a + 8 \\ &= 14 + 8 \\ &= 22 \end{aligned}$$

LHS = RHS, so the solution $a = 14$ is correct.

b

$6w - 18 = 4w - 2$

$6w - 18 - 4w = 4w - 2 - 4w$ ← $- 4w$ from both sides

$2w - 18 = -2$ ← simplifying $6w - 4w$

$2w - 18 + 18 = -2 + 18$ ← $+ 18$ to both sides

$2w = 16$

$\frac{2w}{2} = \frac{16}{2}$ ← $\div$ both sides by 2

$w = 8$

Substitute $w = 8$:

$$\begin{aligned} \text{LHS} &= 6w - 18 \\ &= 6 \times 8 - 18 \\ &= 30 \\ \text{RHS} &= 4w - 2 \\ &= 4 \times 8 - 2 \\ &= 30 \end{aligned}$$

LHS = RHS, so the solution $w = 8$ is correct.

ISBN 9780170350990

EXERCISE 15–06

1 To solve $2x - 6 = x + 12$, which operation would you do first? Select the correct answer **A, B, C** or **D**.

A $+x$ B $+6$ C $-x$ D $+12$

2 To solve $3x + 8 = 2x - 5$, which operation would you do first? Select **A, B, C** or **D**.

A $-2x$ B -8 C $+2x$ D -5

3 Write the LHS of each equation.

a $3x - 5 = 2x + 8$ b $4a + 2 = 3a - 4$ c $8 - 3x = x + 4$

4 Write the RHS of each equation in Question **3**.

5 Copy and complete the solution to each equation.

a
$$2m - 4 = m + 8$$
$$2m - \_\_ = 8 + \_\_$$
$$m = \_\_$$

b
$$4n + 2 = n + 17$$
$$4n - \_\_ = 17 - \_\_$$
$$3n = \_\_$$
$$n = \_\_$$

6 Solve each equation.

a $2w - 8 = w + 6$ b $2m + 5 = m - 4$ c $2n - 7 = n - 4$

d $3a + 4 = 2a + 8$ e $3m - 6 = m + 8$ f $4b + 6 = 2b - 8$

g $5m + 6 = 3m - 8$ h $4v - 6 = v + 9$ i $3w - 12 = w + 14$

j $6a - 8 = 3a + 25$ k $7b + 3 = 5b - 9$ l $8n - 16 = 4n + 28$

7 Copy and complete the solution to each equation.

a
$$4m - 6 = 6m - 18$$
$$4m - \_\_ = -18 + \_\_$$
$$-2m = \_\_$$
$$m = \_\_$$

b
$$3v + 6 = 6v - 12$$
$$3v - \_\_ = -12 - \_\_$$
$$-3v = \_\_$$
$$v = \_\_$$

8 Solve each equation.

a $2w + 6 = 5w - 18$ b $3m - 6 = 5m + 24$

c $4n + 2 = 6n - 34$ d $5m - 12 = 9m + 44$

ISBN 9780170350990

15–07 Equations with brackets

An equation with brackets looks like this: $3(m + 5) = 27$.
To solve such an equation, we need to rearrange the equation to find the value of the variable; for example, $m = 4$. Such equations usually require three steps to solve and involve expanding the LHS as the first step.

EXAMPLE 10

Solve each equation.

a $3(m + 5) = 27$

b $5(2a - 4) = 30$

SOLUTION

a

$$3(m + 5) = 27$$
$$3 \times m + 3 \times 5 = 27 \quad \leftarrow \text{expand}$$
$$3m + 15 = 27 \quad \leftarrow \text{simplify}$$
$$3m + 15 - 15 = 27 - 15 \quad \leftarrow -15 \text{ from both sides}$$
$$3m = 12$$
$$\frac{3m}{3} = \frac{12}{3} \quad \leftarrow \div \text{both sides by } 3$$
$$m = 4$$

b

$$5(2a - 4) = 30$$
$$5 \times 2a - 5 \times 4 = 30 \quad \leftarrow \text{expand}$$
$$10a - 20 = 30 \quad \leftarrow \text{simplify}$$
$$10a - 20 + 20 = 30 + 20 \quad \leftarrow +20 \text{ to both sides}$$
$$10a = 50$$
$$\frac{10a}{10} = \frac{50}{10} \quad \leftarrow \div \text{both sides by } 10$$
$$a = 5$$

EXAMPLE 11

Solve $-4(2x + 1) = 20$ and check that your solution is correct.

SOLUTION

$$-4(2x + 1) = 20$$
$$-4 \times 2x + (-4) \times 1 = 20 \quad \leftarrow \text{expand}$$
$$-8x - 4 = 20 \quad \leftarrow \text{simplify}$$
$$-8x - 4 + 4 = 20 + 4 \quad \leftarrow +4 \text{ to both sides}$$
$$-8x = 24$$
$$\frac{-8x}{-8} = \frac{24}{-8} \quad \leftarrow \text{divide both sides by } (-8)$$
$$x = -3$$

To check the solution, substitute $x = -3$:

LHS $= -4(2x + 1)$ RHS $= 20$

$= -4 \times [2 \times (-3) + 1]$

$= 20$

LHS = RHS, so the solution $x = -3$ is correct.

ISBN 9780170350990

EXERCISE 15–07

1 Solve the equation $3(a - 5) = 6$. Select the correct answer **A**, **B**, **C** or **D**.

A $a = 9$ **B** $a = 8$ **C** $a = 7$ **D** $a = -7$

2 Solve $-2(x + 4) = 8$. Select **A**, **B**, **C** or **D**?

A $x = 8$ **B** $x = 4$ **C** $x = -4$ **D** $x = -8$

3 Is each expansion true or false?

a $3(a + 2) = 3a + 2$

b $-3(w - 4) = -3w - 12$

c $4(2x + 3) = 8x + 12$

d $-2(3n - 5) = -6n + 10$

4 Copy and complete the solution to each equation.

a

$$2(n + 5) = 18$$
$$2n + \_\_\_ = 18$$
$$2n + \_\_\_ - \_\_\_ = 18 - \_\_\_$$
$$2n = \_\_\_$$
$$n = \_\_\_$$

b

$$4(2m - 3) = 12$$
$$\_\_\_ - 12 = 12$$
$$8m - \_\_\_ + \_\_\_ = 12 + \_\_\_$$
$$8m = \_\_\_$$
$$m = \_\_\_$$

5 Solve each equation.

a $3(a - 4) = 21$

b $2(n + 3) = 14$

c $5(b + 4) = 25$

d $-6(a + 5) = 12$

e $-4(m + 2) = 28$

f $-8(v - 5) = 24$

g $7(v - 6) = 35$

h $3(2a - 1) = 15$

i $4(3a + 2) = 32$

j $8(3x - 2) = 80$

k $-2(4m - 3) = 66$

l $-5(2c + 4) = 50$

6 For each equation, check whether the given solution is correct by substituting it into the equation.

a $3(b - 4) = 15$; solution: $b = 9$

b $6(w + 5) = 46$; solution: $w = 1$

c $5(2a - 4) = 12$; solution: $a = 3$

d $7(3m + 2) = 77$; solution: $m = 3$

iStock.com/Xavier Marchant

CODE PUZZLE

Use the following table to decode the words used in this chapter.

1	2	3	4	5	6	7	8	9	10	11	12	13
A	B	C	D	E	F	G	H	I	J	K	L	M

14	15	16	17	18	19	20	21	22	23	24	25	26
N	O	P	Q	R	S	T	U	V	W	X	Y	Z

1 1 – 12 – 7 – 5 – 2 – 18 – 1

2 19 – 15 – 12 – 22 – 5

3 16 – 1 – 20 – 20 – 5 – 18 – 14

4 5 – 17 – 21 – 1 – 20 – 9 – 15 – 14

5 22 – 1 – 18 – 9 – 1 – 2 – 12 – 5

6 19 – 15 – 12 – 21 – 20 – 9 – 15 – 14

7 16 – 18 – 15 – 14 – 21 – 13 – 5 – 18 – 1 – 12

8 22 – 1 – 12 – 21 – 5 – 19

9 5 – 24 – 16 – 18 – 5 – 19 – 19 – 9 – 15 – 14

10 5 – 24 – 16 – 1 – 14 – 4

11 6 – 1 – 3 – 20 – 15 – 18 – 9 – 19 – 5

12 2 – 18 – 1 – 3 – 11 – 5 – 20 – 19

ISBN 9780170350990

PRACTICE TEST 15

Part A General topics

Calculators are not allowed.

1 Evaluate 6×0.009.

2 Complete 5 days = _____ hours.

3 Simplify $4a + 4a + 4a$.

4 Simplify $4a \times 4a \times 4a$.

5 Find the range of these scores: 15, 8, 5, 6, 11, 4, 5, 7.

6 Given that $52 \times 6 = 312$, evaluate 5.2×6.

7 Increase $9600 by 20%.

8 Find the value of b in this triangle.

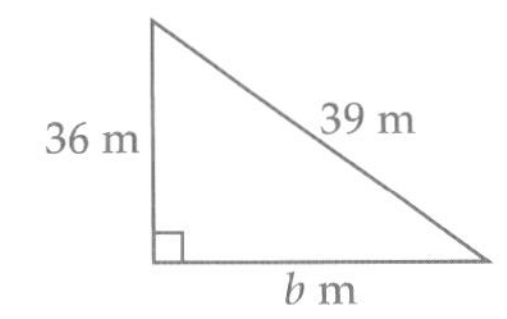

9 Evaluate $-8 + 5 \times (-8)$.

10 What is the probability of selecting the letter I from the word BRILLIANT?

Part B Further algebra

Calculators are allowed.

15-01 Expanding expressions

11 Expand $7(3a + 4)$. Select the correct answer **A, B, C** or **D**.

A $21a + 4$ **B** $28a + 21$ **C** $21a + 28$ **D** $28a + 4$

12 Expand $3m(4m - 6)$. Select **A, B, C** or **D**.

A $12m - 6$ **B** $12m^2 - 18m$ **C** $7m^2 - 6m$ **D** $12m^2 - 18$

15-02 Factorising expressions

13 Factorise $-5a^2 + 20ab$. Select **A, B, C** or **D**.

A $5a(a + 4b)$ **B** $-5(a^2 + 4ab)$ **C** $-5a(a - 4b)$ **D** $-5a(a + 4b)$

14 Factorise $8ab - 24bc^2$.

15-03 Factorising with negative terms

15 Factorise each expression.

a $-5xy - 20y$ **b** $-6ab^2 - 18a^2b$

15-04 One-step equations

16 Solve each equation.

a $w + 5 = 12$ **b** $d - 8 = 4$ **c** $5a = 35$

15-05 Two-step equations

17 Solve each equation.

a $2m - 5 = 13$

b $3x - 6 = 12$

ISBN 9780170350990

15–06 Equations with variables on both sides

18 Solve each equation.

a $3a - 7 = 2a + 8$

b $5x + 4 = 3x - 8$

15–07 Equations with brackets

19 Solve each equation.

a $4(3m - 5) = 16$

b $-2(4x - 1) = 18$

ISBN 9780170350990

RATIOS AND RATES

16

WHAT'S IN CHAPTER 16?

IN THIS CHAPTER YOU WILL:

- find equivalent ratios and simplify ratios
- solve ratio problems
- understand how to read a scale diagram
- interpret scale maps and diagrams
- divide a quantity in a given ratio
- write and simplify rates
- solve rate problems, including those related to speed

Shutterstock.com/Sailorr

ISBN 9780170350990

16–01 Ratios

A **ratio** compares quantities of the same kind, consisting of two or more numbers that represent parts or shares. For example, when mixing ingredients for a cake, we could have a ratio of flour to milk of 3 : 2. This means there are 3 parts of flour to 2 parts of milk. The **order** is important. 3 : 2 is not the same as 2 : 3.
Operations with ratios are similar to operations with fractions.

To find an equivalent ratio, multiply or divide each term by the same number.

EXAMPLE 1

Complete each pair of equivalent ratios.

a $2:5 = 20:\_\_\_$ b $18:12 = \_\_\_:2$

SOLUTION

Examine the term that has been multiplied or divided from LHS to RHS.

a

$2:5 = 20:\_\_\_$ ⟵ $2 \times 10 = 20$

$2:5 = 20:50$ ⟵ Do the same to 5 to complete the ratio: $5 \times 10 = 50$.

b

$18:12 = \_\_\_:2$ ⟵ $12 \div 6 = 2$

$18:12 = 3:2$ ⟵ Do the same to 18 to complete the ratio: $18 \div 6 = 3$.

To simplify a ratio, divide each number in the ratio by the highest common factor (HCF).

EXAMPLE 2

Simplify each ratio.

a $12:20$ b $56:48$ c $0.4:1.6$ d $\frac{1}{3}:\frac{5}{6}$

SOLUTION

a $12:20 = \frac{12}{4}:\frac{20}{4}$ ⟵ Dividing both terms by the HCF 4.

$= 3:5$

✱ Simplifying ratios is similar to simplifying fractions.

b $56:48 = \frac{56}{8}:\frac{48}{8}$ ⟵ Dividing both terms by the HCF 8.

$= 7:6$

c $0.4:1.6 = 0.4 \times 10:1.6 \times 10$ ⟵ Multiplying both terms by 10 to make them whole.

$= 4:16$

$= 1:4$

d $\frac{1}{3}:\frac{5}{6} = \frac{1}{3}\times 6:\frac{5}{6}\times 6$ ⟵ Multiplying both terms by the LCD 6 to make them whole.

$= 2:5$

The LCD (lowest common denominator) of 3 and 6 is 6.

ISBN 9780170350990

EXAMPLE 3

Simplify each ratio.

a 80 cm : 1 m

b 3 kg : 180 g

SOLUTION

a Converting to the same units first:

$$80 \text{ cm} : 1 \text{ m} = 80 \text{ cm} : 100 \text{ cm}$$
$$= 80 : 100$$
$$= \frac{80}{20} : \frac{100}{20}$$
$$= 4 : 5$$

b Converting to the same units first:

$$3 \text{ kg} : 180 \text{ g} = 3000 \text{ g} : 180 \text{ g}$$
$$= 3000 : 180$$
$$= \frac{3000}{60} : \frac{180}{60}$$
$$= 50 : 3$$

EXERCISE 16–01

1 Which ratio is equivalent to 9 : 15? Select the correct answer **A**, **B**, **C** or **D**.

A 5 : 3 **B** 9 : 5 **C** 3 : 5 **D** 5 : 9

2 Simplify 2.5 : 1.5. Select **A**, **B**, **C** or **D**.

A 25 : 15 **B** 5 : 3 **C** 20 : 5 **D** 3 : 5

3 Copy and complete each equivalent ratio.

a 1 : 5 = 9 : ___ **b** 8 : 3 = 24 : ___ **c** 4 : 7 = ___ : 35

d 11 : 9 = ___ : 18 **e** 21 : 36 = 7 : ___ **f** 40 : 70 = ___ : 7

g 12 : 30 = ___ : 15 **h** 50 : 25 = 10 : ___ **i** 16 : 10 = ___ : 5

j 28 : 21 = ___ : 3 **k** 2 : 3 : 7 = 6 : ___ : ___ **l** 32 : 44 : 48 = 8 : ___ : ___

4 For this diagram, write each ratio in simplest form.

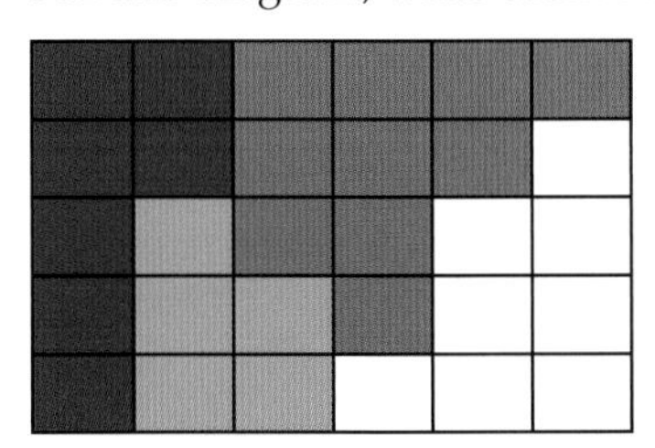

a pink : blue **b** blue : green **c** green : pink **d** blue : white

e white : pink **f** pink : green **g** blue : whole shape **h** whole shape : green

5 Simplify each ratio.

a 14:16 **b** 20 : 28 **c** 15 : 25 **d** 8 : 12

e 9 : 27 **f** 30 : 40 **g** 32 : 48 **h** 120 : 200

i 6 : 9 **j** 18 : 24 **k** 20 : 50 **l** 15 : 45

m 25 : 75 **n** 49 : 63 **o** 80 : 60 **p** 120 : 90

q 12 : 15 : 18 **r** 16 : 24 : 30 **s** 150 : 200 : 450

EXERCISE 16–01

6 Simplify 0.05 cm : 3 m. Select **A**, **B**, **C** or **D**.

A 1 : 6 **B** 1 : 60 **C** 1 : 600 **D** 1 : 6000

7 Simplify each ratio.

a $\frac{1}{2}:1$ **b** 0.2 : 0.5 **c** $2:\frac{1}{4}$ **d** 0.03 : 0.6

e $\frac{3}{4}:\frac{1}{2}$ **f** 0.8 : 2.4 **g** $\frac{1}{5}:\frac{2}{3}$ **h** 4.5 : 0.05

i 2.6 : 0.52 **j** $\frac{2}{5}:\frac{1}{3}$ **k** 12.6 : 6.2 **l** $\frac{7}{8}:\frac{5}{4}$

8 Simplify each ratio by first converting to the same units.

a 10 mm : 4 cm **b** 25 s : 2 min **c** 50 cm : 3 m **d** 150 min : 2 h

e 250 mg : 50 g **f** 3 days : 48 h **g** 1800 mm: 6 m **h** 3500 g : 4 kg

i \$2.45 : \$3.00 **j** $2\frac{1}{2}$ h : 100 min **k** 25 years : 1 century **l** 5 months : 5 years

Shutterstock.com/ankiro

ISBN 9780170350990

16–02 Ratio problems

EXAMPLE 4

The ratio of hair colour for the Year 8 students at Westgate College is blonde : brown = 2 : 5. If 75 Year 8 students have brown hair, how many students have blonde hair?

SOLUTION

Blonde : brown = 2 : 5

5 parts (brown) = 75 students

1 part = 75 ÷ 5 ← Using the unitary method to find one part.

= 15 students

2 parts (blonde) = 2 × 15

= 30 students

30 students have blonde hair.

OR: blonde : brown = 2 : 5 = ____ : 75 ← Using equivalent ratios.

2 : 5 = 30 : 75

30 students have blonde hair.

iStock.com/Mark Bowden

EXAMPLE 5

Amanda is making cupcakes by mixing flour and sugar in the ratio 5 : 3. How much sugar should be mixed with 15 cups of flour?

SOLUTION

Flour : sugar = 5 : 3

5 parts (sugar) = 15 cups

1 part = 15 ÷ 5

= 3 cups

3 parts (sugar) = 3 × 3

= 9 cups

9 cups of sugar are required.

OR

Flour : sugar = 5 : 3 = 15 : ____

5 : 3 = 15 : 9

9 cups of sugar are required.

EXERCISE 16-02

1 In a small mining town, the ratio of women to men is 2 : 5. If there are 40 women in the town, how many men are there? Select the correct answer **A**, **B**, **C** or **D**.

A 100 **B** 16 **C** 116 **D** 140

2 A bushwalking rope is cut in the ratio 3 : 4. The longer piece is 116 m.

a What is the length of the shorter piece?

b What was the original length of the rope?

3 The ratio of teachers to students at a school is 1 : 18. If the school has 64 teachers, how many students are there?

4 In a triangle, the lengths of the sides are in the ratio 3 : 4 : 5. If the longest side is 30 cm long, find the lengths of the other two sides and the perimeter of the triangle.

5 In a Year 8 class, the ratio of boys to girls is 6 : 5.

a If there are 15 girls, how many boys are there?

b If there are 24 boys, how many girls are there?

6 The speed of two trucks is in the ratio 7 : 5. The speed of the slower truck is 60 km/h. Find the speed of the faster truck.

7 The ratio of the Tigers team's wins to losses was 5 : 3. If the team lost 21 games, how many games did it win?

8 Ali and Fahim share the weekly rent of an apartment in the ratio 9 : 7. If Ali pays $270, how much does Fahim pay?

9 The masses of two packets of detergent are in the ratio 3 : 8. If the lighter packet has a mass of 1.5 kg, what is the mass of the heavier one?

10 The heights of two buildings are in the ratio 5 : 4. If the shorter building is 160 m tall, how high is the taller building?

11 A recipe for scones uses sugar, flour and water in the ratio 2 : 7 : 4.

a If 4 cups of sugar are used, how many cups of flour are needed?

b If 10 cups of water are used, how many cups of sugar are needed?

12 Sand and cement are mixed in the ratio 4 : 1 to make concrete. What mass of cement is needed to mix with 90 kg of cement?

13 Jules and Jenny share a cash prize in the ratio of 2 : 3.
If Jenny's share is $120, what is Jules' share?

14 To make a paint colour called Stormy Seas, Norah mixes 4 parts blue, 3 parts purple and 1 part pink. How much pink and purple is needed if Norah uses 600 mL of blue?

15 The ratio of sushi to salad rolls sold at a school canteen was 6 : 5. If 90 sushis were sold, how many salad rolls were sold?

ISBN 9780170350990

16–03 Scale maps and diagrams

WORDBANK

scale diagram A miniature or enlarged drawing of an actual object or building in which lengths and distances are in the same ratio as the actual lengths and distances.

scale The ratio on a scale diagram that compares lengths on the diagram to actual lengths.

scaled length A length on a scale diagram that represents an actual length.

Scale maps and diagrams are an important use of ratios.
A scale of 1: 20 means the actual lengths of the objects are 20 times larger than on the scale diagram.
A scale of 20 : 1 means the actual lengths of the objects are 20 times smaller than on the scale diagram.

The scale on a scale diagram is written as the ratio **scaled length : actual length**.

- The first term of the ratio is usually smaller, meaning that the diagram is a miniature version.
- If the first term of the ratio is larger, then the diagram is an enlarged version.

EXAMPLE 6

If a scale of 1 : 50 has been used, measure and find the actual length in metres represented by each interval.

a   **b**  **c** 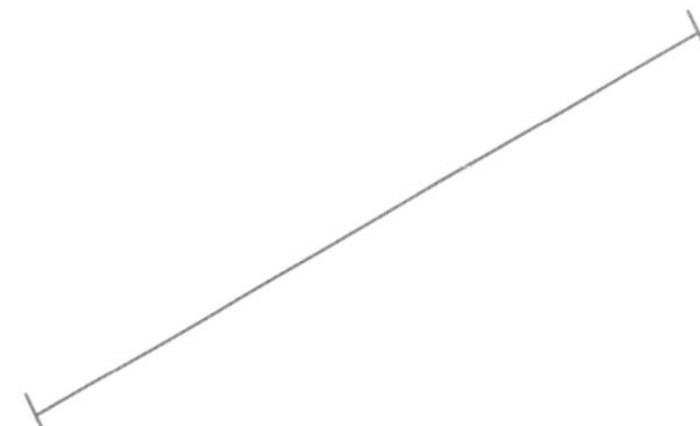

SOLUTION

Scale is 1 : 50

a Scaled length = 3 cm

Actual length = 3 cm × 50 ← Actual length is 50 times larger than the scaled length.

= 150 cm

= 1.5 m

b Scaled length = 3.6 cm

Actual length = 3.6 cm × 50

= 180 cm

= 1.8 m

c Scaled length = 5.8 cm

Actual length = 5.8 cm × 50

= 290 cm

= 2.9 m

EXAMPLE 7

Given the scale in each diagram, measure and calculate the actual length of each object.

a

Scale 1.5 : 1

Shutterstock.com/Sofiaworld

b

iStockphoto/pagadesign

Scale 1 : 35

SOLUTION

a Scaled length = 10.1 cm ← Measuring across the top for greatest length.

Butterfly's length = 10.1 cm ÷ 1.5 ← Divide, as the scale diagram is an enlargement.

≈ 6.73 cm

b Scaled length = 9.8 cm

Car's length = 9.8 cm × 35 ← Multiply, as the actual car is 35 times larger.

= 343 cm

= 3.43 m

ISBN 9780170350990

EXERCISE 16–03

1 What is the actual length of an interval if the scaled length is 7 cm and the scale is 1 : 20? Select the correct answer **A**, **B**, **C** or **D**.

A 20 cm **B** 14 cm **C** 140 cm **D** 1.4 cm

2 What is the length of a beetle if its scaled length is 4 cm and the scale is 8 : 1. Select **A**, **B**, **C** or **D**.

A 4 cm **B** 32 cm **C** 0.25 cm **D** 0.5 cm

3 Find the actual length that each interval represents in metres if a scale of 1 : 40 has been used.

a **b** **c**

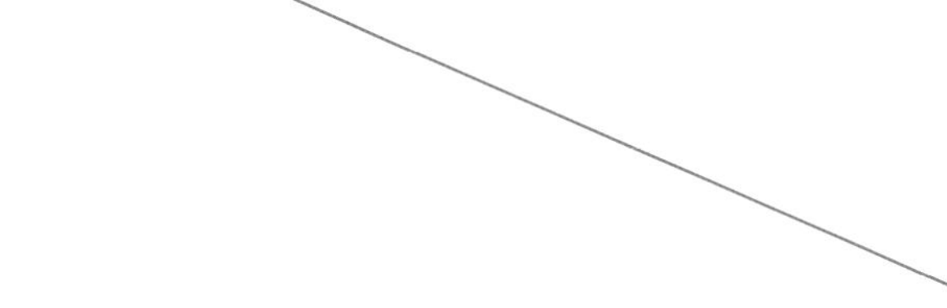

4 Find the actual length of each object.

a

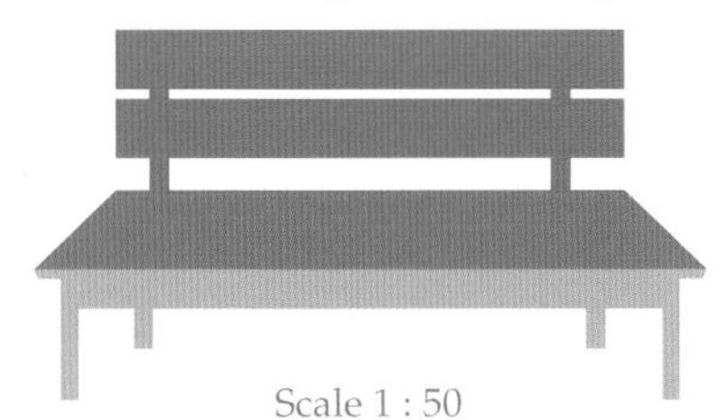

Scale 1 : 50

b

Scale 1 : 750

c

Scale 1 : 30

d

Scale 1 : 5

5 A spider is drawn to a scale of 4 : 1.

a How long are its legs if they are drawn 12 cm long?

b How long is its body if it is drawn 16.4 cm long?

EXERCISE 16–03

6 A scale drawing of a house is shown below. The scale used is 1 : 120.

Bedroom 2
Bathroom
Study
Dining room
Ensuite
Hall way
Lounge room
Main Bedroom
Kitchen

a What is the scaled length and width of the kitchen on the plan?

b What is the actual length and width of the kitchen?

c What are the scaled dimensions of the bathroom?

d What are the actual dimensions of the bathroom?

e If the hallway, bathroom and kitchen are being tiled at a cost of $35.90/m^2, what will be the total cost?

iStock.com/CandyBoxImages

 ISBN 9780170350990

EXERCISE 16–03

7 For this map of Canberra, measure the distance between each pair of locations below and use the map's scale to calculate the actual distance correct to the nearest 0.1 km.

a Parliament House and Questacon

b Mt Pleasant and the Australian War Memorial

c Springbank Island and the Australian National University

d Australian National Botanic Gardens and Vernon Circle

e Commonwealth Park and the tip of Weston Park

f Questacon and Mount Pleasant

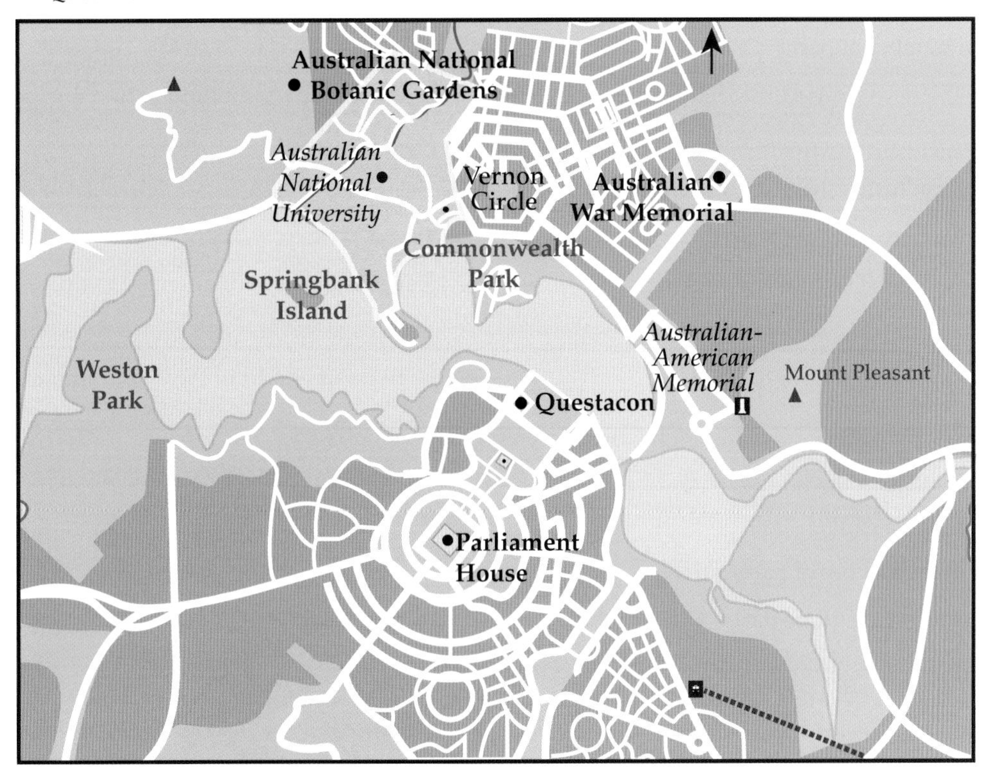

Scale: 1 cm = 0.7 km

ISBN 9780170350990

16–04 Dividing a quantity in a given ratio

To divide a quantity in a given ratio:

- find the total number of parts by adding the terms of the ratio
- find the size of one part by dividing the quantity by the number of parts (unitary method)
- multiply to find the shares required.

EXAMPLE 8

Divide a profit of \$2500 between Santosh and Raj in the ratio 2 : 3.

SOLUTION

Total number of parts = 2 + 3 = 5

One part = \$2500 ÷ 5 ← Using the unitary method to find one part.

= \$500

Santosh's share = 2 × \$500 ← 2 parts

= \$1000

Raj's share = 3 × \$500 ← 4 parts

= \$1500

Check: \$1000 + \$1500 = \$2500 ← The two shares add to the whole amount.

EXAMPLE 9

Divide \$740 in the ratio 3 : 5 : 2.

SOLUTION

Total number of parts = 3 + 5 + 2 = 10

One part = \$740 ÷ 10

= \$74

3 parts = 3 × \$74 = \$222

5 parts = 5 × \$74 = \$370

2 parts = 2 × \$74 = \$148

Check: \$222 + \$370 + \$148 = \$740

The parts add to the whole amount.

Dreamstime/Robyn Mackenzie

ISBN 9780170350990

EXERCISE 16–04

1 Find the amount of one part if \$240 is divided in the ratio 5 : 3. Select the correct answer **A**, **B**, **C** or **D**.

A \$120 **B** \$30 **C** \$80 **D** \$48

2 Copy and complete this table.

Ratio	Total parts	Total amount	One part	New ratio
3 : 5	3 + 5 = 8	\$640	\$640 ÷ 8 = \$80	\$240 : \$_____
4 : 3		\$5600		
2 : 7		\$720		
5 : 2		\$7700		

3 Copy and complete this working to divide \$700 in the ratio 4 : 3.

Total number of parts = 4 + __ = __

One part = \$700 ÷ __

= \$ __

New ratio = 4 × __ : 3 × __

= \$__ : \$__

4 If \$6400 is divided between Nuraan and Hadieya in the ratio 2 : 3, what does Hadieya receive?

5 Over a football season, the Bulldogs team had 3 wins for every 2 losses.

a If they played 45 matches, how many wins did they have?

b If they had to win 70% of their games to progress to the final series, did they play in the finals?

6 Lee and Nathan own a computer business. They share their profits each year in the ratio 4 : 3. How much do they each receive if the profit is \$35 000?

7 Georgia, Lucy and Megan won \$1800 in the town raffle. They wanted to share it in the same ratio as the number of tickets each bought, which was 24 : 18 : 6, respectively.

a Simplify the ratio.

b How much does Lucy receive?

c Georgia donated 15% of her share to charity. How much did the charity receive?

8 Ante and Josh share 28 chocolates in the ratio 4 : 3. How many chocolates does each person receive?

9 Charlie is 12 years old and Lola is 8 years old. They were given \$130 to be shared in the ratio of their ages. How much should Lola get?

10 Sophie invests \$15 000 in a business and Claire invests \$25 000.

a Simplify the ratio 15 000 : 25 000.

b If the profit at the end of the year is \$150 000, how much should each receive if the profits are shared in the same ratio as their investments?

16–05 Rates

A **rate** compares two quantities of different types or different units of measure. For example, heartbeat is measured in beats/minute and the cost of petrol is measured in cents/litre. 90 beats **per** minute means 90 beats in 1 minute and is written as 90 beats/minute.

EXAMPLE 10

Simplify each rate.

a \$4.50 for 3 kg of apples
b 255 words in 5 min of typing
c 210 m in 7 s of cycling
d a heartbeat of 280 beats in 4 min

SOLUTION

a \$4.50 for 3 kg = \$4.50 ÷ 3 kg ← divide by 3 to find one unit
= \$1.50/kg

b 255 words in 5 min = 255 words ÷ 5 min ← divide by 5
= 51 words/min

c 210 m in 7 s = 210 m ÷ 7 s ← divide by 7
= 30 m/s

d 280 beats in 4 min = 280 beats ÷ 4 min ← divide by 4
= 70 beats/min

iStock.com/Razvan

ISBN 9780170350990

EXERCISE 16–05

1 Simplify the rate \$400 for 16 hours of work. Select the correct answer **A**, **B**, **C** or **D**.

A \$25/h **B** \$40/h **C** \$16/h **D** \$20/h

2 Simplify the rate 160 words in 3.2 minutes. Select **A**, **B**, **C** or **D**.

A 20 words/min **B** 60 words/min **C** 30 words/min **D** 50 words/min

3 Write the units used to measure each rate.

a speed of a car
b cost of fruit
c typing speed
d cost of petrol
e speed of a runner
f cost of posting a parcel
g cricket team's run rate
h an employee's wage

4 Copy and complete each rate.

a \$48 for 3 hours work is a rate of \$____/h.
b \$4.50 for 2 kg oranges is a rate of \$____/kg.
c 560 beats in 7 min is a rate of ____ beats/min.
d 440 students with 20 teachers is a rate of ____ studcnts/teacher.
e \$720 for 6 nights is a rate of \$____/night.
f 1260 metres in 9 s is a rate of ____ m/s.
g \$60.50 for 50 km is a rate of \$____/km.
h 432 runs for 8 wickets is a rate of ____ runs/wicket.

5 Simplify each rate.

a 48 goals in 8 matches
b \$360 for 4 days
c \$42 for 6 kg
d 252 runs for 9 wickets
e 480 m in 5 s
f 875 students for 25 teachers
g \$63 000 for 70 hectares
h 12 000 revolutions in 8 min

6 Tyler went on a road trip and used 280 litres which cost \$320. Find the cost of petrol in dollars per litre correct to the nearest cent.

7 Australia's land area is approximately 7 682 300 km^2. Calculate correct to two decimal places Australia's population density in persons/km^2 if the population is 24 100 000.

8 Jacinta works 5 hours at a chemist and earns \$94.50. What is her hourly rate of pay?

9 An electrician took 3 hours to complete a job. If he charged \$171, calculate his hourly rate.

10 Erin's car travelled 440 km in $4\frac{1}{2}$ hours. What was her average speed in km/h correct to one decimal place?

11 Which typing rate is faster: 520 words in 5 min or 729 words in 9 min?

16–06 Rate problems

To solve a rate problem, write the units in the rate as a fraction $\frac{x}{y}$.

- To find x (the numerator amount), **multiply** by the rate.
- To find y (the denominator amount), **divide** by the rate.

EXAMPLE 11

Bella earns $17.50 per hour as a data entry operator.

a How much does she earn for working 9 hours?

b If Bella earned $542.50 last week, how many hours did she work?

SOLUTION

The units of the rate are $\frac{\$}{h}$. ← writing the units as a fraction

a To find $, multiply by the rate:

Pay = 9 × $17.50 ← Bella earns $17.50 for one hour, so multiply by 9.

= $157.50

b To find hours, divide by the rate:

Number of hours = $542.50 ÷ $17.50 ← $17.50 for one hour, so divide by $22.80

= 31

Bella worked 31 hours last week.

EXAMPLE 12

Darian can type at an average rate of 52 words per minute. How long should it take him to type a 5000-word essay? Answer to the nearest minute.

SOLUTION

The units of the rate are $\frac{\text{words}}{\text{min}}$.

To find minutes, divide by the rate:

Number of minutes = 5000 ÷ 52

= 96.1538…

≈ 96 min

Darian should take 96 minutes (or 1 h 36 min) to type a 5000-word essay.

ISBN 9780170350990

EXERCISE 16–06

1 A factory makes toys at a rate of 23 toys per minute.

  a How many toys are produced in one hour?

  b How long to the nearest minute will it take to produce 1600 toys?

2 Sarah earns $24/h working in a boutique.

  a Write the units in the rate as a fraction.

  b How much will Sarah earn if she works 26 hours?

  c How long will it take Sarah to earn $360?

3 Stefan washes 6 cars in 2 hours 42 minutes.

  a Convert 2 hours 42 minutes to minutes.

  b What is Stefan's car-washing rate in min/car?

  c How long will it take him to wash 10 cars? Answer in hours and minutes.

  d How many cars can he wash in 6 hours? Answer to the nearest whole number.

4 Liam earns $22.40 per hour working in a hardware store. If he worked 7 hours a day for 11 days, what was his total pay?

5 Rowena scores goals at an average rate of 6 per game. If she plays 14 games in a season, how many goals will she score altogether?

6 Goran's heart rate is 72 beats per minute. How long will it take to beat 2880 times?

7 James works part time at a pizza shop at a rate of $23.40/h. How much will he earn if he works for 6 hours a night for 6 nights?

8 Nelly can run 70 m in 14 s.

  a What is her speed in m/s?

  b How far can Nelly run in 3 s?

  c How many seconds will it take her to run 175 m?

9 Anna types 92 words per minute.

  a How many words will Anna type in 20 minutes?

  b How long will it take Anna to type 4140 words?

10 Hoa paid tax at the rate of 32c per dollar of income earned.

  a Write the units in the rate as a fraction.

  b How much tax does he pay on his income of $62 400? Answer in dollars.

  c What is his income if he pays $17 440 in tax?

11 Elyse's van travelled 306 km on 36 litres of petrol. What is its fuel consumption in km/L?

12 Hand-made chocolates cost $22.90/kg.

  a How much will 250 *grams* of chocolates cost? Answer to the nearest cent.

  b How much chocolate could you buy for $45? Answer correct to two decimal places.

16–07 Speed

Speed is a special rate that compares distance travelled with time taken. The units are kilometres per hour (km/h) or metres per second (m/s).

THE SPEED FORMULA

$$\text{Speed} = \frac{\text{Distance travelled}}{\text{Time taken}}$$

$$S = \frac{D}{T}$$

You can memorise this triangle to help you remember this rule.
If you want to find speed, S, cover S with your finger. You are left with $\frac{D}{T}$.

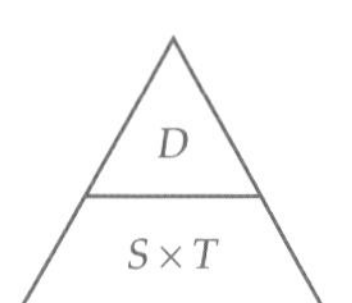

$$\text{So, } S = \frac{D}{T}$$

If you want to find distance, D, cover D and you are left with $S \times T$.

$$\text{So, } D = S \times T$$

Similarly, if you want to find time, T, cover T and you are left with $\frac{D}{S}$.

$$\text{So, } T = \frac{D}{S}$$

EXAMPLE 13

An interstate train travels 564 km in 6 hours.

a What is its average speed for the journey?

b How long would it take to travel 1081 km?

c How far would it travel in $8\frac{1}{2}$ hours?

SOLUTION

a $S = \frac{D}{T}$

$= \frac{564 \text{ km}}{6 \text{ h}}$

$= 94$ km/h

b Rearrange the formula $S = \frac{D}{T}$ to find time, T.

$T = \frac{D}{S}$

$= \frac{1081}{94}$ ← $D = 1081, S = 94$

$= 11.5$ h

c Rearrange the formula $S = \frac{D}{T}$ to find distance, D.

$D = ST$

$= 94 \times 8.5$ ← $S = 94, T = 8.5$

$= 799$ km

 ISBN 9780170350990

EXERCISE 16–07

1 If it takes Melanie 4 hours to travel 340 kilometres, what is her average speed? Select the correct answer **A**, **B**, **C** or **D**.

A 80 km/h **B** 82 km/h **C** 85 km/h **D** 170 km/h

2 If a train is travelling at 120 km/h, how long will it take to travel 900 km? Select **A**, **B**, **C** or **D**.

A 8.5 h **B** 9 h **C** 9.5 h **D** 7.5 h

3 Calculate the average speed for:

a 80 km in 4 hours **b** 350 km in 7 hours **c** 480 km in 12 hours

d 550 km in 5 hours **e** 18 km in 6 hours **f** 1680 km in 8 hours.

4 Match each speed calculated in Question **4** with a type of travel described below.

A riding a bicycle **B** driving on a freeway **C** walking

D driving in a school zone **E** riding on a bullet train **F** driving on a street

5 Find the:

a time if the speed is 90 km/h and the distance is 540 km

b distance if the speed is 85 km/h and the time is 5 hours

c speed if the distance is 420 m and the time is 7 s

d time if the speed is 150 m/s and the distance is 750 m

e distance if the speed is 72 m/s and the time is 28 s.

6 Renee set out on a road trip from Bulladoo to Gerang, a distance of 756 km. She travelled at an average speed of 75 km/h for 3 hours until she stopped for a break. Then she drove for 315 km for $3\frac{1}{2}$ hours.

a How far did she travel for the first leg of her journey before her break?

b What was her average speed for the second leg of her journey?

c How far does she still have to travel?

d How long before she arrives in Gerang if she travels the last part of her trip at 80 km/h? Answer in hours and minutes.

7 **a** If Zak jogged at an average speed of 3 m/s, how long would he take to jog 1 km? Answer in minutes and seconds.

b Zak slowed down to 2 m/s and jogged for half an hour. How far did he go in this time? Answer in kilometres.

8 A racing car driver does one lap of a 5.5 km race track in 2 minutes.

a How far would the driver travel in 60 minutes?

b What is the speed in km/h?

9 Ngaire takes 1 hour to cycle 10 km and then walks for another hour, travelling a further 6 km.

a What is the total distance travelled?

b How much time has she taken for the whole distance?

c Find her average speed.

COMPOUND WORDS

The words below are compound words: they are made up of two smaller words.

Break each compound word into two smaller words and write the ratio used.

For example, QUEENSLAND = QUEENS LAND = 6 : 4 = 3 : 2.

1 BRAINWAVE

2 OVERALL

3 HIGHCHAIR

4 CHAIRMAN

5 TOMBSTONE

6 SWIMWEAR

7 PREMIERSHIP

8 FRYPAN

9 DASHBOARD

10 COBWEB

11 SCHOLARSHIP

12 GRAVEYARD

13 MILESTONE

14 BROADBAND

15 CLOCKWORK

16 CYCLEWAY

17 PINEAPPLE

18 ROLLERSKATE

19 MOCKINGBIRD

20 MASTERPIECE

ISBN 9780170350990

PRACTICE TEST 16

Part A General topics

Calculators are not allowed.

1 Evaluate $28 + 6 \times 3 \div 2$.

2 If $y = -3$, evaluate $16 - 4y$.

3 Find the range of 4, 9, 15, 11 and 6.

4 Find the perimeter of this shape.

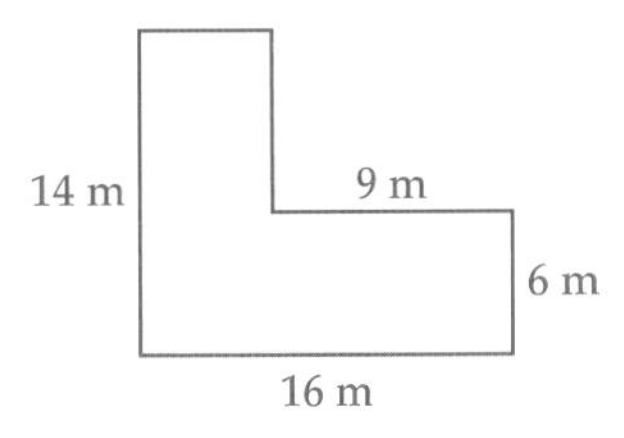

5 Evaluate $\frac{3}{4} + \frac{2}{3}$.

6 What percentage is \$350 of \$1400?

7 Convert 2218 to 12-hour time.

8 Simplify $5x - 3y - 2y + 8x$.

9 Round \$1126.546 to the nearest cent.

10 What is the probability of the next baby born being a girl?

Part B Ratios and rates

Calculators are allowed.

16-01 Ratios

11 Which ratio does not simplify to 4 : 3? Select the correct answer **A**, **B**, **C** or **D**.

A 16 : 12 **B** 28 : 24 **C** 8 : 6 **D** $1 : \frac{3}{4}$

12 Simplify 3 : 18. Select **A**, **B**, **C** or **D**.

A 3 : 9 **B** 3 : 6 **C** 1 : 6 **D** 1 : 15

13 Copy and complete: 3 : 7 : 2 = 15 : __ : 10.

16-02 Ratio problems

14 If the ratio of adults : children is 3 : 5 and there are 75 children, how many adults are there?

16-03 Scale maps and diagrams

15 What is the actual length of an interval if the scaled length is 5 cm and the scale is 1 : 15?

16 What is the actual length of a spider if its scaled length is 9 cm and the scale is 6 : 1?

16-04 Dividing a quantity in a given ratio

17 Divide \$7196 between Tanya and Mikayla in the ratio 3 : 4.

16–05 Rates

18 Simplify each rate.

a 63 goals in 9 matches

b $625 for 5 days

16–06 Rate problems

19 Sophie washes 5 cars in 2 hours.

a What is her car-washing rate in minutes per car?

b How many cars can she wash in 4 days working 6 hours per day?

16–07 Speed

20 If a train travels 120 km in 1 h 30 min, what is its average speed in km/h?

ISBN 9780170350990

GRAPHING LINES

17

WHAT'S IN CHAPTER 17?

IN THIS CHAPTER YOU WILL:

- identify points and quadrants on a number plane
- use a linear equation to complete a table of values
- graph tables of values on the number plane
- graph linear equations on the number plane
- find the equation of horizontal and vertical lines

17-01 The number plane

A **number plane** is a grid for plotting points and drawing graphs.
It has an ***x*-axis** which is horizontal (goes across) and a ***y*-axis** which is vertical (goes up and down).
The **origin** is the centre of the number plane.
The number plane is divided into four quadrants (quarters).

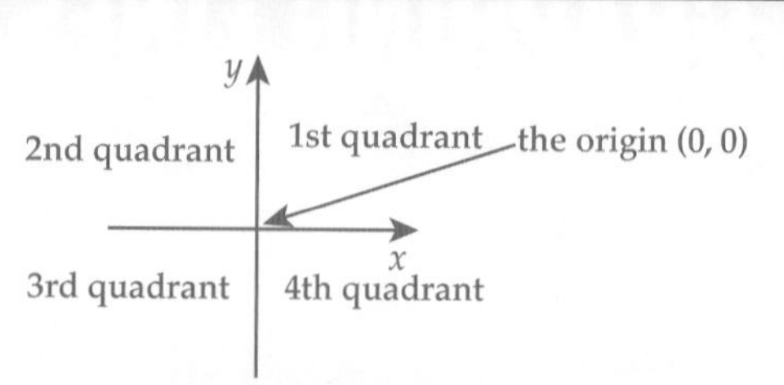

To plot a point (x, y) on the number plane:

- start at the origin (0, 0)
- for the ***x*-coordinate**, move left for a negative number or right for a positive number
- for the ***y*-coordinate**, move down for a negative number or up for a positive number.

EXAMPLE 1

Plot each point on a number plane and state which quadrant it is in or what axis it is on.

$A(-1, 3)$, $B(2, 6)$, $C(-3, -5)$, $D(4, -3)$, $E(0, -1)$, $F(-5, 0)$

SOLUTION

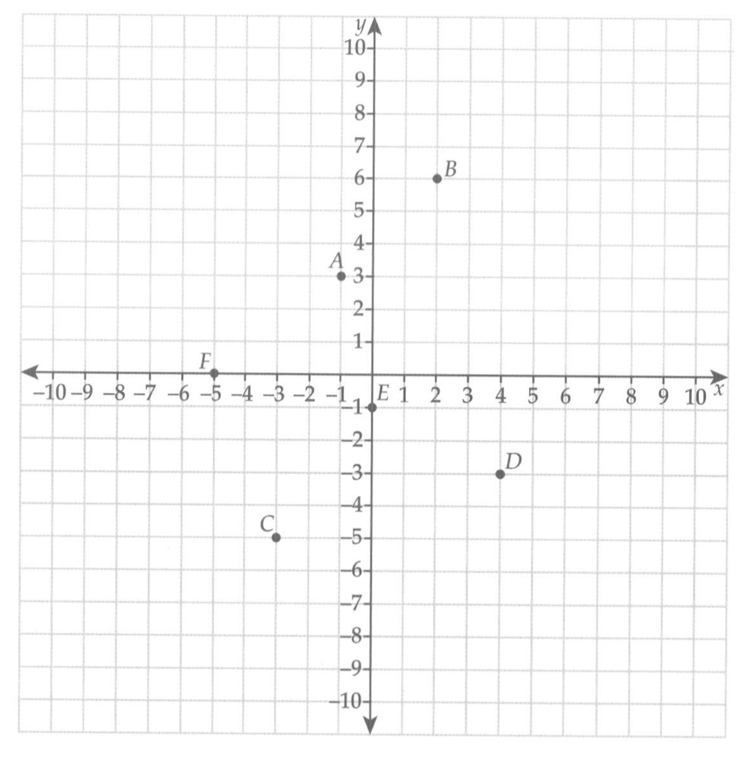

A: 2nd quadrant ← 1 left, 3 up

B: 1st quadrant ← 2 right, 6 up

C: 3rd quadrant ← 3 left, 5 down

D: 4th quadrant ← 4 right, 3 down

E: *y*-axis ← 1 down

F: *x*-axis ← 5 left

ISBN 9780170350990

EXERCISE 17–01

1 To plot the point (3,–5) on the number plane, in which direction should you move from the origin? Select the correct answer **A**, **B**, **C** or **D**.

A 3 left, 5 down **B** 3 right, 5 down **C** 3 left, 5 up **D** 3 right, 5 up

2 Copy and complete this table.

Point	Move left or right?	Move up or down?
(–2, 6)		
(–4, –5)		
(3, –4)		
(1, 6)		

3 In which direction are the quadrants on the number plane numbered: clockwise or anticlockwise?

4 **a** Write the coordinates of each point shown on the following number plane.

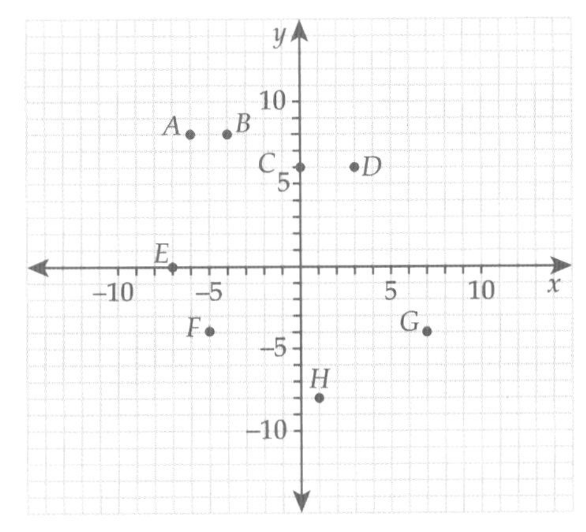

b Name the quadrant in which each point lies.

i *B* **ii** *F* **iii** *D* **iv** *G*

5 Plot each point on a number plane.

A(–2, 6), *B*(5, 9), *C*(–1, –7), *D*(4, –6), *E*(6, –4), *F*(0, –2), *G*(8, 0), *H*(–5, –4)

6 Which of the points in Question **5** are in the fourth quadrant?

7 In which quadrant does the point (–2, 3) lie?

8 Plot the points below on a number plane and join them as you go.

(6, 6), (–4, 6), (–4, 4), (–1, 4), (–1, –3), (6, –3), (6, 6), Stop

(–5, –3), (–5, –1), (–4, 0), (–3, 4), (–1, 4), (–1, –3), (–5, –3), Stop

(–4, –5), (–3, –3), (–2, –5), (–4, –5), Stop, (2, –5), (3, –3), (4, –5), (2, –5), Stop

What have you drawn? (Hint: The triangles are the wheels.)

EXAMPLE 2

Complete each table of values using the equation given.

a $y = x + 2$

x	1	2	3	4
y				

b $y = 3x - 1$

x	–1	0	1	2
y				

c $d = 4 - 2c$

c	–1	0	1	2
d				

SOLUTION

Substitute the x-values from the table into each equation.

a $y = x + 2$
When $x = 1$, $y = 1 + 2 = 3$
When $x = 2$, $y = 2 + 2 = 4$ and so on.

x	1	2	3	4
y	3	4	5	6

b $y = 3x - 1$
When $x = -1$, $y = 3 \times (-1) - 1 = -4$
When $x = 0$, $y = 3 \times 0 - 1 = -1$ and so on.

x	–1	0	1	2
y	–4	–1	2	5

c $d = 4 - 2c$
When $c = -1$, $d = 4 - 2 \times (-1) = 6$
When $c = 0$, $d = 4 - 2 \times 0 = 4$ and so on.

c	–1	0	1	2
d	6	4	2	0

ISBN 9780170350990

EXERCISE 17–02

1 If $y = x + 3$, what is the y-value when $x = -1$? Select the correct answer **A**, **B**, **C** or **D**.

A -2 **B** -1 **C** 2 **D** 1

2 If $y = 6 - x$, what is the y-value when $x = 3$? Select **A**, **B**, **C** or **D**.

A -6 **B** 3 **C** 6 **D** -3

3 If $y = x - 8$, find y when:

a $x = 2$ **b** $x = 4$ **c** $x = 6$ **d** $x = 8$

e $x = -1$ **f** $x = -2$ **g** $x = -4$ **h** $x = 0$

4 If $y = 12 - 2x$, find y when:

a $x = 4$ **b** $x = 6$ **c** $x = 3$ **d** $x = 10$

e $x = -1$ **f** $x = -3$ **g** $x = -6$ **h** $x = -2$

5 Copy and complete each table of values.

a $y = 2x$

x	0	1	2	3
y				

b $y = x - 1$

x	4	3	2	1
y				

c $y = x \div 2$

x	10	8	6	4
y				

d $y = x + 3$

x	0	1	2	3	4
y					

e $p = 3n$

n	1	2	3	4
p				

f $b = 5 - a$

a	−3	−2	−1	0	1	2
b						

g $e = 2 - d$

d	0	1	2	3	4
e					

h $z = \frac{y}{3}$

y	12	9	6	3	0	−3	−6
z							

i $v = 2u + 5$

u	1	2	3	4
v				

j $h = 10 - 3f$

f	−3	−2	−1	0	1	2
h						

ISBN 9780170350990

EXAMPLE 3

Graph this table of values on a number plane.

x	−2	−1	0	1	2
y	3	4	5	6	7

SOLUTION

Reading the table of values in columns, we get the coordinates of the points.

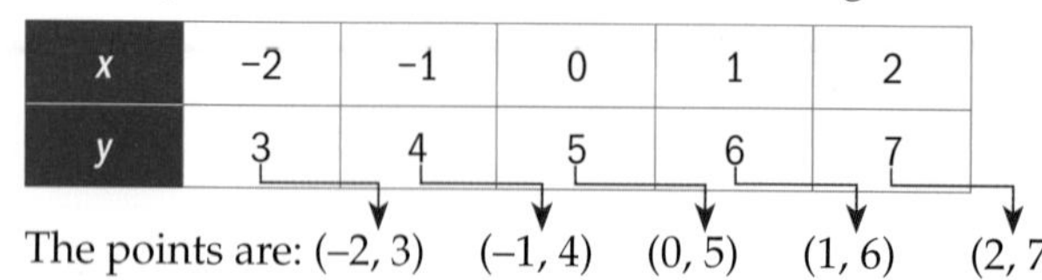

x	−2	−1	0	1	2
y	3	4	5	6	7

The points are: (–2, 3) (–1, 4) (0, 5) (1, 6) (2, 7)

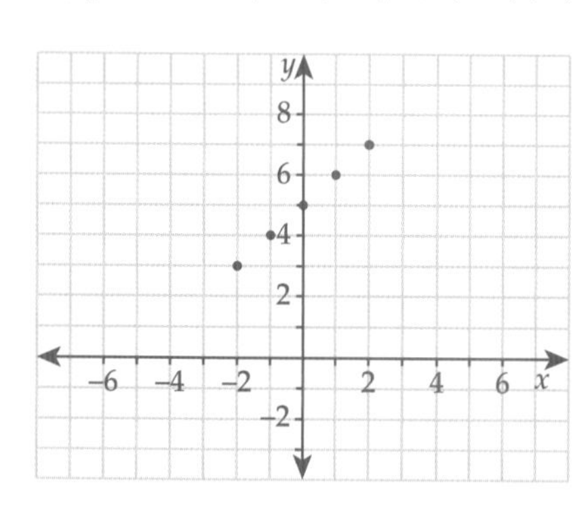

(–2, 3) is 2 left and 3 up

(–1, 4) is 1 left and 4 up

(0, 5) is 0 and 5 up

(1, 6) is 1 right and 6 up

(2, 7) is 2 right and 7 up

✱ Notice that the points form a straight-line pattern.

EXAMPLE 4

Graph this table of values after completing it.

$y = x - 4$

x	−2	0	2	4
y				

SOLUTION

Substitute each x-value into the formula $y = x - 4$.

x	−2	0	2	4
y	−6	−4	−2	0

The points are (–2, –6), (0, –4), (2, –2), (4, 0).

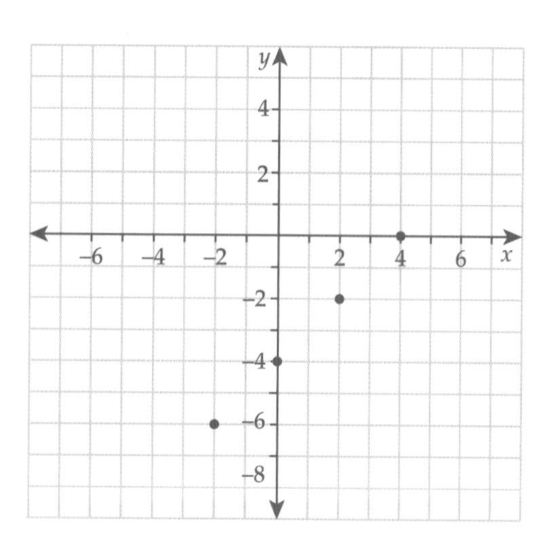

ISBN 9780170350990

EXERCISE 17-03

1 If $y = x + 5$, what is the y-value when $x = -2$? Select the correct answer **A**, **B**, **C** or **D**.

A -2 B -3 C 3 D 2

2 If $y = 4 - 2x$, what is the y-value when $x = -1$? Select **A**, **B**, **C** or **D**.

A 6 B 2 C -6 D -2

3 Graph each table of values on a number plane.

a

x	1	2	3	4
y	3	4	5	6

b

x	-1	0	1	2
y	5	4	3	2

c

x	-1	-1	3	3
y	3	-2	3	-2

d

x	-4	-2	1	1
y	2	2	2	6

4 What pattern or shape would be formed if you joined each set of points graphed in Question **3**?

5 Graph each table of values on a number plane and state which graphs form a straight-line pattern.

a

x	-4	-2	1	3
y	2	5	5	7

b

x	-2	-1	4	6
y	6	4	8	2

c

x	-3	-1	3	4
y	1	3	7	8

d

x	-4	-2	0	3
y	6	6	2	2

6 Copy and complete each table of values and then graph the values on a number plane.

a $y = x + 2$

x	0	1	2
y			

b $y = x - 1$

x	0	1	2
y			

c $y = 2x + 3$

x	0	1	2
y			

d $y = 3x - 1$

x	0	1	2
y			

17–04 Graphing linear equations

WORDBANK

linear equation An equation that connects two variables, usually x and y, whose graph is a straight line.

y-intercept The value where a line crosses the y-axis.

To graph a linear equation:

- complete a table of values using the equation
- plot the points from the table on a number plane
- join the points to form a straight line
- extend the line and place arrows on both ends of the line
- label the line with the equation.

The arrows on the line show that the line extends forever on both ends.
To graph a linear equation, it is best to find three points on the line using a table of values.
We can substitute any x-values into the linear equation but $x = 0$, $x = 1$ or $x = 2$ are usually easiest to use.

EXAMPLE 5

Graph each linear equation and find the y-intercept of each line.

a $y = x + 3$ **b** $y = 2x - 1$

SOLUTION

a Complete a table of values for $y = x + 3$.

x	0	1	2
y	3	4	5

Graph the table of values, rule the line and label it with the equation.

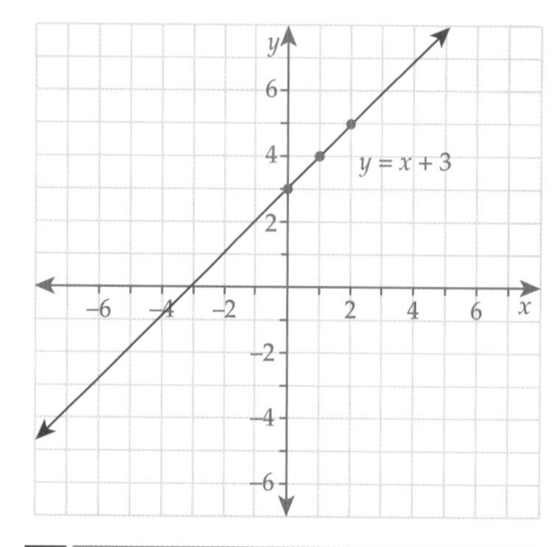

Draw arrows on the ends of the line because a line has an infinite number of points and goes on endlessly in both directions.

The line crosses the y-axis at 3, so its y-intercept is 3.

ISBN 9780170350990

b Complete a table of values for $y = 2x - 1$.

x	1	2	3
y	1	3	5

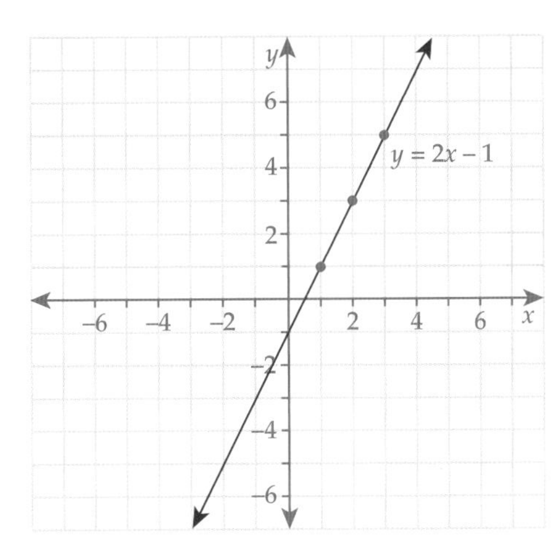

The line crosses the y-axis at –1, so its y-intercept is –1.

EXERCISE 17–04

1 How many points are best for graphing a linear equation? Select the correct answer **A**, **B**, **C** or **D**.

A 1 **B** 2 **C** 3 **D** 4

2 Copy and complete these sentences.

To graph a linear equation, draw a table of ______ and then plot each point from the ______ on a number plane. Join the ______ to form a straight ______.

3 Find the y-intercept of each line.

a

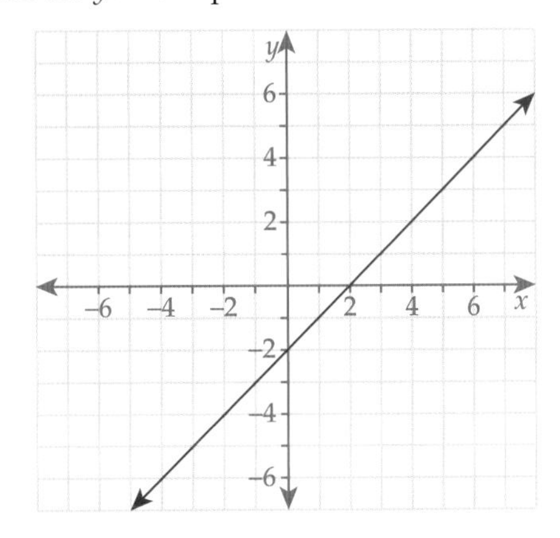

b

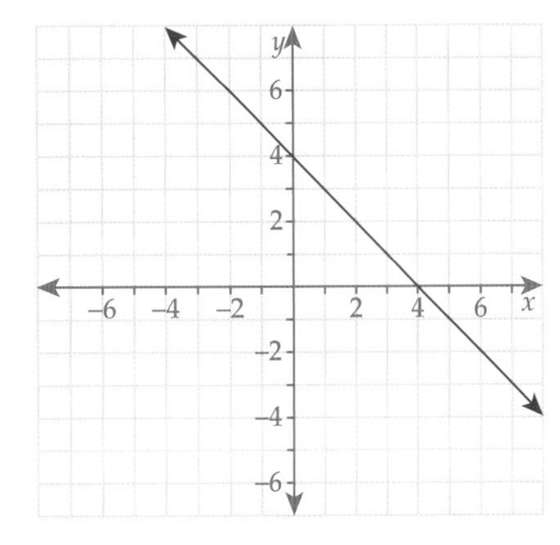

ISBN 9780170350990

EXERCISE 17–04

4 Graph each linear equation on a number plane after copying and completing the table of values, and then find the y-intercept of each line.

a $y = x + 5$

x	0	1	2
y			

b $y = x - 3$

x	0	1	2
y			

c $y = 2x + 1$

x	0	1	2
y			

d $y = 4x - 2$

x	0	1	2
y			

5 Graph each linear equation on a number plane after completing a table of values, and then find the y-intercept of each line.

a $y = 2x + 4$ b $y = 4x - 2$ c $y = 6 - x$

d $y = 5 - 2x$ e $x + y = 5$ f $y = -2x + 1$

Shutterstock.com/Zhukov Oleg

ISBN 9780170350990

17–05 Horizontal and vertical lines

WORDBANK

horizontal A line that is flat, parallel to the horizon.

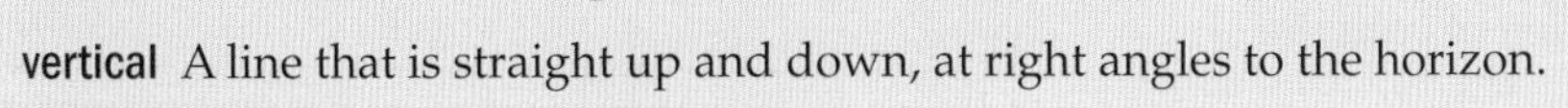

vertical A line that is straight up and down, at right angles to the horizon.

constant A number, not a variable.

***x*-intercept** The value where a line crosses the *x*-axis.

EXAMPLE 6

Find the equation of the line represented by these points:

(2, –4), (2, –1), (2, 0), (2, 2), (2, 5).

SOLUTION

Plot the points on a number plane and join them.

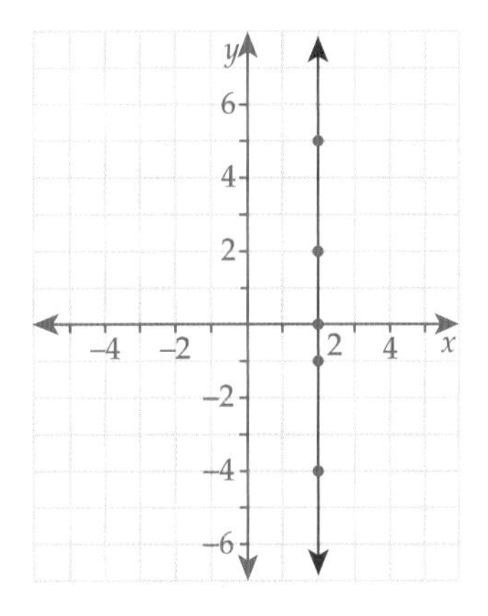

Every point on this vertical line has an x-value of 2, whereas y can take any value, so the equation of this line is $x = 2$.

All **vertical lines** have equation 'x = a number' and this number is called a **constant**. The constant is the x-intercept of the line.

A **vertical line** has equation $x = c$, where c is a constant (number).

EXAMPLE 7

Find the equation of the line represented by these points:

(–1, 3), (2, 3), (5, 3), (0, 3).

SOLUTION

Plot the points on a number plane and join them.

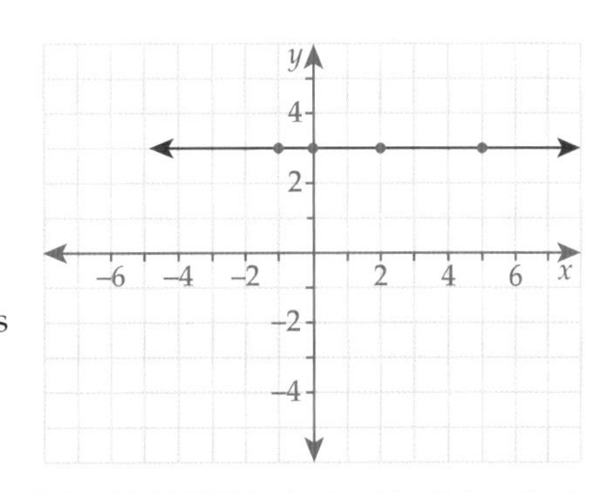

Every point on this horizontal line has a y-value of 3, whereas x can take any value, so the equation of this line is $y = 3$.

All **horizontal lines** have equation 'y = a number' and this number is called a **constant**. The constant is the y-intercept of the line.

A **horizontal line** has equation $y = c$, where c is a constant (number).

ISBN 9780170350990

EXERCISE 17-05

1 What type of line is $x = 4$? Select the correct answer **A**, **B** or **C**.

A horizontal **B** diagonal **C** vertical

2 Write the equation of each line.

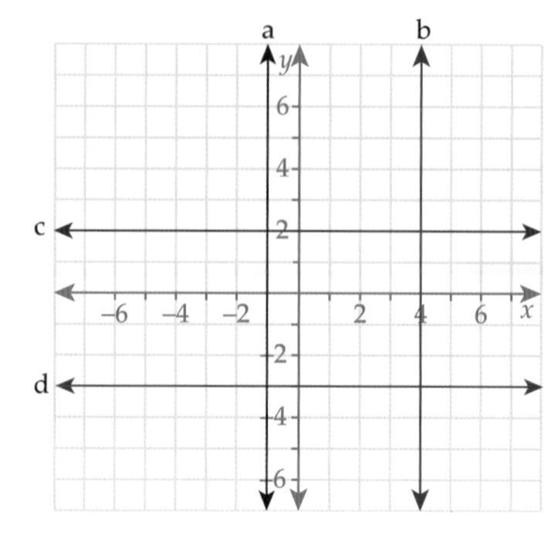

3 Plot each set of points, then write the equation of the line that passes through those points.

a (1, –4), (1, 0), (1, 5), (1, 3), (1, –2)

b (3, 2), (4, 2), (–3, 2), (–1, 2), (6, 2)

c (5, –3), (0, –3), (2, –3), (–1, –3), (1, –3)

d (–2, 4), (–2, 0), (–2, 1), (–2, 5), (–2, –3)

4 Graph each line on the same number plane.

a $x = 5$ **b** $y = -2$ **c** $x = -4$ **d** $y = 4$

5 Which point lies on the line $y = -2$? Select **A**, **B**, **C** or **D**.

A (–2, 4) **B** (3, –2) **C** (–2, 2) **D** (–2, 0)

6 What special name is given to the line with equation:

a $x = 0$ **b** $y = 0$?

7 Write the equation of the line that is:

a horizontal with a y-intercept of 5

b vertical with an x-intercept of –3

c the horizontal line passing through (–2, 3)

d the vertical line passing through (4, –1).

8 Graph the lines $x = 2$ and $y = -4$ on the same number plane and write their point of intersection.

ISBN 9780170350990

FIND-A-WORD PUZZLE

Copy this puzzle, then find the words listed below. Then, use the first 24 remaining letters to spell out three words (5 letters, 8 letters, 11 letters).

S	E	T	A	N	I	D	R	O	O	C	H	P	O	I
E	U	L	A	V	N	T	P	L	L	O	O	T	T	I
N	G	B	P	A	N	D	A	E	M	L	R	I	A	P
O	Q	N	S	I	U	C	M	Q	A	I	I	R	C	M
E	Q	U	A	T	I	O	N	N	I	N	Z	A	Q	U
R	V	U	A	T	I	V	K	C	Q	E	O	R	M	A
B	S	I	R	D	J	T	D	N	N	D	N	A	E	Y
P	R	E	R	N	R	E	U	U	A	A	T	E	K	M
F	V	S	A	I	R	A	M	T	T	C	A	N	G	B
A	L	Z	P	E	G	B	N	J	E	H	L	I	E	L
N	W	O	D	L	E	H	O	T	T	P	N	L	L	E
J	E	R	P	R	A	I	T	W	U	A	N	Y	B	F
E	O	J	S	U	N	N	G	I	P	R	E	I	A	T
V	E	H	T	N	F	F	E	H	F	G	T	Y	T	C
X	J	I	R	D	A	P	L	K	T	Y	X	Z	K	I

COORDINATES
GRAPH
LEFT
NUMBER
PLANE
SUBSTITUTE
VERTICAL
DOWN
HORIZONTAL
LINE
ORDERED
QUADRANT
TABLE
EQUATION
JOIN
LINEAR
PAIR
RIGHT
VALUE

ISBN 9780170350990

PRACTICE TEST 17

Part A General topics

Calculators are not allowed.

1 Evaluate 0.08×0.004.

2 Complete: 5.5 hours = _____ minutes.

3 Simplify $\frac{a+a}{a \times a}$.

4 Simplify $6 \times m \times n - 2 \times m \times n$.

5 Write the factors of 30.

6 What is the perimeter of a square of side length 9 cm?

7 Increase $450 by 20%.

8 Find the value of b if the area of this rectangle is 72 m^2.

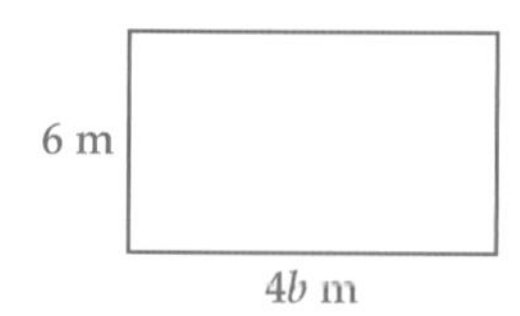

9 Decrease $850 by 10%.

10 What is the probability of selecting a red king from a standard deck of cards?

Part B Graphing lines

Calculators are allowed.

17-01 The number plane

11 The point (–3, –5) is in which quadrant of the number plane? Select the correct answer **A**, **B**, **C** or **D**.

A first **B** second **C** third **D** fourth

17-02 Tables of values

12 Copy and complete the table of values for $y = 4x - 2$.

x	0	1	2
y			

17-03 Graphing tables of values

13 Graph this table of values on a number plane.

x	0	1	2
y	2	4	6

17-04 Graphing linear equations

14 Graph the linear equation $y = 3x - 1$ on a number plane and write its y-intercept.

17-05 Horizontal and vertical lines

15 Graph each line on the same number plane and write the coordinates of their point of intersection.

a $x = -3$ **b** $y = 2$

ISBN 9780170350990

CHAPTER 1

Exercise 1-01

1 B
2 B
3 36, 64
4 a Add 10 and then 1
b Add 10 and subtract 2
c Add 20 and subtract 1
5 a 10 b 10 c 20
d 10 e 1 f 1
6 a 62 b 38 c 275 d 331
e 65 f 499 g 90 h 277
i 751 j 1136 k 378 l 4683
7 a 80 b 280
c 130 d 180 or 190
e 520 or 530 f 3870
g 730 h 8490
i 21 010
8 a 80 b 280 c 130
d 186 e 524 f 3875
g 726 h 8495 i 21 016
9 a 208 b 414 c 243 d 395
e 240 f 670 g 410 h 610
10 a

3	6	6
8	5	2
4	4	7

b

4	7	7
9	6	3
5	5	8

Exercise 1-02

1 D
2 B
3 a 813 b 553 c 103 d 149
e 3214 f 5730 g 422 h 901
i 4319 j 3525 k 939 l 730
4 833
5 351
6 $278
7 a 1062 cm b $3872
8 $30.15
9 26 073
10 a $553 b $553.32

Exercise 1-03

1 C
2 A
3 a Subtract 10 and add 1
b Subtract 10 and then 2
c Subtract 20 and then 1
4 a 10 b 10 c 20
d 10 e 2 f 1
5 a 10, 1, 75 b 40, 1, 317
6 a 58 b 61 c 106
d 68 e 448 f 716
g 60 h 606 i 1007
j 374 k 6481 l 3337
7 a 60 b 80 c 470
d 820 e 1040 f 4510
8 a 55 b 81 c 471
d 818 e 1035 f 4505
9 30, 39
10 a 267 b 248 c 165 d 226
e 419 f 243 g 215 h 265
i 253 j 357 k 443 l 147

Exercise 1-04

1 B
2 D
3 a 44 b 145 c 56 d 25
e 2058 f 7181 g 342 h 567
i 5043 j 6551 k 3358 l 15 454
4 2164
5 $377
6 a 190 km b 184 km
7 11 712
8 62 passengers
9 a $25.40 b $24.60
10 a 570 b 572

Exercise 1-05

1 C
2 A
3 a Double twice
b Add three zeros
c ×10 and subtract the number
d Double three times
4 a 52 b 45 000 c 90 d 176
e 470 f 108 g 36 200 h 171
5 a 20 b 42 c 24 d 30
e 28 f 45 g 48 h 60
i 56 j 40 k 49 l 72
6 a 56 b 260 c 6500 d 112
e 132 000 f 342 g 56 080 h 120
i 198 j 292 k 90 l 135
7 a 5, 10, 180 b 25, 100, 3100
8 a 500 b 390 c 2400 d 1000
e 700 f 180 g 1400 h 11 000
i 400 j 30 k 600 l 2700

Exercise 1-06

1 B
2 D
3 a 332 b 392 c 632 d 1308
e 2624 f 2915 g 5396 h 47 412
i 3192 j 7608 k 18 504 l 14 832
4 a 7744 b 24 075 c 4750 d 8946
e 16 250 f 44 928 g 27 680 h 235 200
5 154 hours
6 1088 seeds
7 a 24 b 432 c 8760
8 a T b F c T
d T e T f F

ISBN 9780170350990

9	a 140	b 506	c 567	d 328
	e 350	f 416	g 150	h 1164
	i 340	j 570	k 2960	l 6120
	m 8100	n 28 800	o 41 000	p 364 000

Exercise **1-07**

1 A
2 D

3	a 5	b 4	c 4	d 5
	e 6	f 7	g 5	h 5
	i 11	j 9	k 7	l 8
	m 5	n 9	o 5	p 9

4 a Halve the number twice
b ÷ 10 and then double
c ÷ 10 and then halve

5	a 284	b 274	c 390	d 170
	e 87	f 81	g 118	h 72
	i 84	j 23	k 52	l 17
6	a 65	b 543	c 68	d 45.8
	e 1.256	f 23.4	g 6.78	h 24.5
	i 405	j 2.18	k 4.9	l 0.321
7	a F	b T	c T	d T
	e F	f F	g T	

Exercise **1-08**

1 A
2 B

3	a 21	b 102	c 702	d 112
	e 63	f 206	g 2095	h 833
	i 305	j 6857	k 703	l 3021

4 \$21
5 \$59
6 14
7 25

8	a $29\frac{2}{3}$	b $120\frac{2}{3}$	c $136\frac{5}{7}$	d $77\frac{1}{4}$
	e $649\frac{1}{5}$	f $806\frac{5}{9}$	g $3061\frac{2}{3}$	h $331\frac{4}{5}$

9 a 408 mm b 4 mm
10 \$8300

Language activity

Across

1 addition
6 of
7 add
9 product
12 equal
13 twice

Down

2 difference
3 tens
5 day
7 ones
8 division
10 digit
11 value

Practice test **1**

Part A

1 4
2 360°
3 1040 mm
4 $\frac{1}{6}$
5 86 m
6 200
7 75%
8 West
9 $x + 1$
10 31 days

Part B

11 C
12 A

13	a 785	b 1765
14	a 437	b 2025
15	a 352	b 2257
16	a 140	b 1080
17	a 288	b 4560
18	a 357	b 27.5
19	a 1557	b $710\frac{1}{2}$

CHAPTER 2

Exercise **2-01**

1 A
2 D
3 Divisible by:
a 2 b 5
c 2, 5 and 10 d 2
e 2, 5 and 10 f 2
4 a The sum of its digits is divisible by 3.
b The sum of its digits is divisible by 9.
5 Divisible by:
a 3 and 9 b Neither
c 3 and 9 d 3
e 3 f 3 and 9
6 Divisible by:
a 4, 6 and 8 b None
c 4 and 6 d 4 and 8
e 6 f None
7 a Yes b Yes c Yes d No
8 a Yes b Yes c No
9 15, 30, 45

Exercise **2-02**

1 D
2 A
3 11, 13, 17, 19, 23, 29
4 32, 33, 34, 35, 36, 38, 39, 40, 42, 44, 45, 46, 48, 49
5 g 2, 3, 5, 7, 11, 13, 17, 19, 23, 29, 31, 37, 41, 43, 47, 53, 59, 61, 67, 71, 73, 79, 83, 89, 97, 101, 103, 107, 109, 113

ISBN 9780170350990

6 3 and 5, 5 and 7, 11 and 13, 17 and 19, 29 and 31, 41 and 43, 59 and 61, 71 and 73, 101 and 103, 107 and 109

7 **a** C **b** C **c** P **d** P
e C **f** C **g** C **h** C
i P **j** P **k** C **l** C

8 **a** P **b** C **c** C **d** P

9 **a** 43 **b** 156 or 174 or 192

Exercise 2-03

1 B

2 D

3 **a** Base 5, Index 8 **b** Base 7, Index 4
c Base 3, Index 9 **d** Base 4, Index 2
e Base 8, Index 5 **f** Base 3, Index 1
g Base 15, Index 7 **h** Base 20, Index 4

4 **a** 390 625 **b** 2401 **c** 19 683
d 16 **e** 32 768 **f** 3
g 15 **h** 160 000

5 **a** 2^5 **b** 5^4 **c** 8^6 **d** 9^5
e 4^4 **f** 11^6 **g** 7^1 **h** 21^4

6 **a** 32 **b** 625 **c** 262 144
d 59 049 **e** 256 **f** 1 771 561
g 7 **h** 194 481

7 **a** 6 **b** 4 **c** 4^2 **d** 2^3

8 **a** 9 **b** 10 **c** 5 **d** 6
e 7 **f** 10 **g** 2 **h** 1
i 12 **j** 8 **k** 14 **l** 9

9 **a** 10.72 **b** 6.16 **c** 4.21
d 4.90 **e** 23.58 **f** 4.75

10 **a** 900, 4900, 90 000 **b** 30, 70, 300

11 **a** 27 000, 125 000, 8 000 000 **b** 30, 50, 200

Exercise 2-04

1 C

2 D

3 **a** 5 **b** 4 **c** $5 \times 5 \times 5 \times 5$

4 **a** $(7 \times 7 \times 7 \times 7 \times 7 \times 7 \times 7 \times 7) \times (7 \times 7 \times 7 \times 7 \times 7)$
b $(9 \times 9 \times 9 \times 9 \times 9 \times 9) \times (9 \times 9 \times 9)$
c $(10 \times 10 \times 10) \times (10 \times 10)$

5 **a** 3^8 **b** 6^7 **c** 5^9 **d** 7^{12}
e 9^8 **f** 10^8 **g** 8^{10} **h** 12^{12}
i 20^8 **j** 5^6 **k** 3^{11} **l** 2^7

6 $(6 \times 6 \times 6 \times 6) \times 6 \times (6 \times 6 \times 6 \times 6 \times 6) = 6^4 \times 6 \times 6^5 = 6^{10}$

7 **a** F **b** T **c** F **d** T

8 No, 6912

9 **a** 1024 **b** 129 600 **c** 1125 **d** –2048

Exercise 2-05

1 B

2 D

3 **a** 8 **b** 3 **c** $8 \times 8 \times 8$

4 **a** $\dfrac{7\times7\times7\times7\times7\times7\times7\times7\times7\times7\times7\times7}{7\times7\times7\times7\times7}$

b $\dfrac{9\times9\times9\times9\times9\times9\times9}{9\times9\times9}$

c $\dfrac{10\times10\times10\times10\times10\times10\times10\times10\times10}{10\times10}$

5 **a** 3^2 **b** 6^5 **c** 5^3 **d** 7^4
e 9^3 **f** 10^4 **g** 8^2 **h** 12
i 20^4 **j** 5^5 **k** 3^{10} **l** 6^4

6 **a** T **b** F **c** F **d** F

7 No, $\frac{27}{64}$

Exercise 2-06

1 C

2 B

3 **a** 2, 20 **b** 4, 8

4 **a** 3^{12} **b** 5^{12} **c** 9^{10} **d** 11^{12}
e 2^{21} **f** 5^{24} **g** 8^{12} **h** 3^{30}
i 4^{18} **j** 7^{24} **k** 10^{15} **l** 6^{21}
m 5^{20} **n** 7^9 **o** 4^{18} **p** 8^{10}

5 **a** F **b** T **c** F

Exercise 2-07

1 A

2 D

3 **a** 1 **b** 1 **c** 3 **d** 1
e 4 **f** 1 **g** 6 **h** 2
i 1 **j** 24 **k** 1 **l** 1

4 **a** $7^0 = 1$
b Any number to the power of 0 is equal to 1.

5 **a** T **b** F **c** T **d** F
e T **f** F **g** T **h** F
i T **j** F **k** F

6 6

7 **a** 4^{11} **b** 6^4 **c** 2^{12} **d** 3^6
e 5^{12} **f** 7^8 **g** $3^0 = 1$ **h** $7^0 = 1$
i 4^9 **j** 5^4

Language activity

1 INDEX
2 DIVISIBLE
3 SQUARE ROOT
4 BASE
5 CUBE ROOT
6 PRIME
7 MULTIPLY
8 COMPOSITE
9 POWER
10 ZERO
11 DIVIDE
12 SUBTRACT
13 TERM
14 NUMBER
15 DIVISIBILITY

Practice test 2

Part A

1 2400
2 18 000
3 1, 2, 3, 4, 6, 12
4 $2x$
5 7

6 11.4
7 $160
8 $d = 5$
9 $21r$
10 12

Part B

11 B
12 D
13 B
14 13, 17, 19, 23
15 6^5
16 a 625 b 13 c –4
17 a 4^8 b 8^5 c 2^6
18 a 5^5 b 7^8 c 9^6
19 a 3^8 b 6^{15} c 4^{12}
20 a 1 b 4 c 1

CHAPTER 3

Exercise 3-01

1 B
2 A
3 a c b r c m
d z e p f s
4 a $c^2 = a^2 + b^2$ b $r^2 = p^2 + v^2$
c $m^2 = n^2 + a^2$ d $z^2 = x^2 + y^2$
e $p^2 = v^2 + w^2$ f $s^2 = q^2 + r^2$
5 a T b T c F
d T e F f T
6 a, b, d, f

Exercise 3-02

1 C
2 B
3 24, 900, $\sqrt{900}$, 30
4 a 17 b 26 c 25 d 50
5 a 10.8 b 12.1 c 15.2
d 10.6 e 19.2 f 14.8

Exercise 3-03

1 A
2 C
3 5, 5, 25, $\sqrt{144}$, 12
4 a 5 b 6 c 8
d $\sqrt{96}$ e $\sqrt{108}$ f $\sqrt{700}$
5 a 21.2 b 7.8 c 11.3 d 5.8

Exercise 3-04

1 A
2 D
3 a 4 cm b 16 m c 41 m
d 16 m e $\sqrt{260}$ m f $\sqrt{3471}$ cm
4 a 11.9 b 31.6 c 23.8
5 7.21 m
6 5.3 cm

Language activity

Across

1 Pythagoras
3 square
7 square root
9 add
10 triangle

Down

2 hypotenuse
4 theorem
5 three
6 subtract
8 right angled

Practice test 3

Part A

1 2:56 p.m.
2 458
3 220
4 56 m
5 $\frac{31}{40}$
6 –27
7 $42
8 6
9 6
10 $13.65

Part B

11 A
12 C
13 a 25 b $\sqrt{180}$
14 a 6.32 b 14.71
15 a 16.5 b 185.9
16 4.24 cm
17 12 m

CHAPTER 4

Exercise 4-01

1 B
2 B
3 D
4 a 22 b –5 c –12
d 28 e –50 f 100
5 a T b F c T
d T e F f F
g T h T i T
6
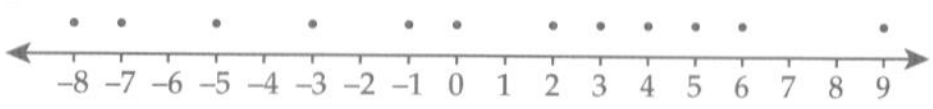

7 –15, –9, –6, –4, –2, 0, 1, 8, 9, 12, 18, 21
8 25, 23, 18, 8, 6, 0, –3, –4, –6, –12, –14, –17
9 a INTEGERS
b ARE
c INTRIGUING

ISBN 9780170350990

10 a > b < c > d >
e > f > g < h <
i < j < k < l >

Exercise 4-02

1 D
2 A
3 a F b T c T d F
4 a −4 b 4 c −12 d −2
e 2 f −12 g −3 h 3
i −15 j −4 k 4 l −18
m −7 n −11 o 15 p 4
q −16 r 19 s −18 t 10
u −16 v −27 w −18 x 22
5 On the 10th level
6 Three attempts

Exercise 4-03

1 B
2 D
3 a −9 b 9 c 1 d −9
e 9 f −3 g −9 h 9
i −5 j −9 k 9 l −7
m −14 n −3 o 16 p −15
q 14 r 35 s −30 t −15
u 53 v −33 w −21 x 65
4 a −71 b −25 c 41
d −146 e 127 f −145
g 132 h −150 i 28
5 $278
6 28 steps forward

Exercise 4-04

1 C
2 A
3 a −12 b −40 c 36 d −35
e 24 f −48 g −55 h −84
i 63 j −72 k 42 l −96
m −88 n −54 o 48

4

×	−2	5	−6	8	−4
3	−6	15	−18	24	−12
−7	14	−35	42	−56	28
−9	18	−45	54	−72	36
10	−20	50	−60	80	−40

5 a −81 b 228 c −416 d −333
e 248 f −616 g 390 h −258
i −3112 j −4255 k 11 070 l −15 004
m −5496 n −49 842 o 7168

Exercise 4-05

1 C
2 A
3 a −2 b −5 c 6 d −8
e 9 f −7 g 9 h 9
i 8 j −6 k 6 l −7
m 12 n −11 o 8
4 a −8 b −2 c 9 d −8
e 7 f −4 g −8 h −7
i −9 j 30 k −40 l −7
m 8 n −2 o 16 p 20

Exercise 4-06

1 B
2 D
3 a ×
b Whichever appears first from left to right
c ÷
d Whichever appears first from left to right
4 a × b ÷ c ÷
d ÷ e × f ÷
g ÷ h × i ×
5 a 43 b 28 c 69
d 29 e −40 f 66
g 24 h 104 i −120
6 a −36 b 156 c −5 d 93
e −31 f 279 g −236 h 2
i −4 j −10 k −23 l 35
m 32 n −7 o 40 p −202
q −2 r 7 s 6
7 a $33.55 b $16.45
8 $65

Language activity

			1 I								2 Z	
3 S			N				4 S				E	
U			T				U				R	
B			5 E	S	T	I	M	A	T	6 I	O	N
T			G							N		
R			E		7 A					F		
A		8 P	R	O	D	U	C	T		I		
C			S		D					N		
T					I					I		
I					T					T		9 O
O		10 D	I	V	I	D	E			E		R
N					O							D
					11 N	E	G	A	T	I	V	E
12 W	H	O	L	E								R

Practice test 4

Part A

1 171
2 3^8
3 56 400
4 $120
5 66 cm
6 200
7 $\frac{1}{8}$
8 1, 2, 3, 6, 9, 18
9 4
10 31 days

Part B

11 C
12 C
13 B
14 a 11 b 52
15 a –44 b 80
16 a –54 b 72
17 a 7 b –9
18 230

CHAPTER 5

Exercise 5-01

1 B
2 C

3	a	1	b	8	c	5	d	7
4	a	$\frac{9}{10}$	b	$\frac{9}{100}$	c	$\frac{9}{1000}$	d	$\frac{9}{10\,000}$
5	a	$\frac{5}{10}$	b	$\frac{6}{100}$	c	$\frac{7}{10\,000}$	d	$\frac{4}{1000}$
	e	$\frac{6}{100}$	f	$\frac{54}{100}$	g	$\frac{75}{1000}$	h	$\frac{386}{1000}$
	i	$\frac{24}{10\,000}$	j	$1\frac{8}{10}$	k	$2\frac{36}{100}$	l	$6\frac{82}{1000}$
	m	$5\frac{25}{100}$	n	$3\frac{4}{100}$	o	$7\frac{6}{1000}$	p	$2\frac{186}{10\,000}$
6	a	$\frac{1}{2}$	b	$\frac{3}{50}$	c	$\frac{7}{10\,000}$	d	$\frac{1}{250}$
	e	$\frac{3}{50}$	f	$\frac{27}{50}$	g	$\frac{3}{40}$	h	$\frac{193}{500}$
	i	$\frac{3}{1250}$	j	$1\frac{4}{5}$	k	$2\frac{9}{25}$	l	$6\frac{41}{500}$
	m	$5\frac{1}{4}$	n	$3\frac{1}{25}$	o	$7\frac{3}{500}$	p	$2\frac{93}{5000}$
7	a	0.6	b	0.03	c	0.004	d	0.9
	e	0.21	f	0.045	g	0.451	h	0.0079
	i	1.2	j	2.03	k	3.025	l	6.52
	m	7.8	n	0.524	o	0.178	p	6.0045
8	a	0.8	b	0.15	c	0.75	d	0.36
	e	0.2	f	0.14	g	0.2	h	0.44

Exercise 5-02

1 B
2 B
3 0.300, 0.030, 0.350, 0.003
4 0.003, 0.03, 0.3, 0.35
5 0.62, 0.61, 0.6, 0.06
6 a 0.003, 0.03, 0.3, 0.31, 0.312, 0.38
b 0.006, 0.05, 0.502, 0.516, 0.555, 0.56
c 0.009, 0.09, 0.119, 0.9, 0.911, 0.92
d 1.004, 1.014, 1.04, 1.114, 1.4, 1.41
7 a 0.666, 0.61, 0.601, 0.6, 0.06, 0.006
b 0.244, 0.242, 0.24, 0.024, 0.02, 0.002
c 0.853, 0.835, 0.8, 0.083, 0.08, 0.008
d 4.555, 4.55, 4.515, 4.5, 4.05, 4.005

8	a	T	b	T	c	T		
	d	F	e	T	f	T		
9	a	<	b	>	c	>	d	>

Exercise 5-03

1 D
2 A
3 128.685

4	a	6.8	b	3.9	c	4.2	d	12.5
	e	17.5	f	21.6	g	123.8	h	38.3
5	a	4.57	b	9.12	c	8.49	d	11.38
	e	183.65	f	34.53	g	78.89	h	982.48
6	a	45.8			b	45.829		
	c	45.8295			d	45.82945		
7	a	$4.57	b	$23.62	c	$60.11	d	$2.09

Exercise 5-04

1 C
2 A

3	a	22.5	b	5.97	c	32.65
	d	30.36	e	13.386	f	132.16
	g	57.333	h	44.55	i	182.252
4	a	28.1	b	31.9	c	47.25
	d	12.7	e	23.6	f	94.83
	g	82.7	h	318.9	i	744.65
5	a	269.54	b	63.85	c	1213.216
	d	16.245	e	213.06	f	0.696
	g	102.393	h	74.2	i	357.817

6 $11.55
7 a $479.20 b $20.80
8 $560.44

Exercise 5-05

1 C
2 B

3	a	2	b	3	c	3
	d	2	e	3	f	3
	g	3	h	5	i	4
4	a	0.16	b	0.063	c	0.024
	d	0.63	e	1.224	f	0.21
	g	0.21	h	0.05508	i	0.0234

ISBN 9780170350990

5 a 9.52 b 46.646 c 5.428
d 18.144 e 5.985 f 2.2815
g 0.0752 h 48.8824 i 2813.36
6 0.0208
7 0.504
8 0.001 96
9 a $310.10 b $89.90
10 $261.13

Exercise 5-06

1 D
2 B
3 C
4 a 14.3 b 21.3 c 12.8 d 2.17
e 43.1 f 8.53 g 31.2 h 4.4
i 46.4 j 5.65 k 18.0905 l 45.7908
5 a 486.8 ÷ 4 = 121.7 b 3755.84 ÷ 5 = 751.168
c 3465.3 ÷ 4 = 1155.1 d 5696.4 ÷ 4 = 1424.1
6 a 147.1 b 551.16 c 11 551 d 116.05
e 2540 f 121 000
7 a 1723 pieces b Yes c $49 105.50

Exercise 5-07

1 A
2 D
3 $5.30 for 4 L
4 $12.50 for 5 kg tray
5 a 5.5 kg for $8.50 b 1 kg for $18.20
c 5 L for $7.90 d 2.5 kg for $22.90
e 8 bread rolls for $3.60
6 Sam's salami, Bill's bacon, Pam's pastrami

Exercise 5-08

1 D
2 B
3 a $0.\dot{5}$ b $0.\dot{3}\dot{4}$ c $2.6\dot{8}$
d $6.4\dot{3}$ e $28.\dot{2}$ f $6.\dot{2}\dot{5}$
g $12.\dot{2}1\dot{3}$ h $1.0\dot{4}5\dot{2}$ i $72.745\dot{6}$
4 a $0.81\dot{6}$ recurring b 9.1 terminating
c 32.1 terminating d $35.7\dot{6}$ recurring
e 82.15 terminating f 2070 terminating
5 a 406 b $10.\dot{8}\dot{1}$ c 372.9
d $107.1\dot{6}$ e 13.65 f 4275.6
6 a F b T c F d T
7 a Divide 3 by 4 b 0.75
c terminating
8 a 0.8 b $0.\dot{3}$ c 0.375 d $0.8\dot{3}$
e 0.875 f $0.\dot{1}4285\dot{7}$ g $0.\dot{5}$ h 0.5
9 $0.\dot{8}$

Language activity

DECEPTIVE DECIMALS

Practice test 5

Part A

1 $90
2 22 000
3 6, 12, 18, 24
4 90
5 54.76
6 $\frac{1}{12}$
7 16
8 88 m^2
9 $\frac{1}{3}$
10 30

Part B

11 B
12 C
13 D
14 0.002, 0.02, 0.201, 0.211
15 B
16 a 545.58 b 65.75
17 1.081
18 D
19 Shop 1
20 $195.4\dot{6}$
21 Convert each fraction to a decimal.
a 0.375 b $0.\dot{2}$

CHAPTER 6

Exercise 6-01

1 C
2 D
3 a, e, k, u, b
4 a $b - b = 0$
b $m \times 1 = m$
c $n + n = 2 \times n$
5 a $3b$ b $5m$ c $5w$
d $a + b$ e $12n$ f $a + b$
g $10c^2$ h $3s + 4t$ i $2m + n$
j $24b^2$ k $2a + b$ l $5m - 3n + 4p$
6 a F b T c F d F
e T f F g F
h F i T j T
k F l F m T
7 a $5 + 2m$ b $3ab - 2$ c $20 - 3n$
d $4v - 3w$ e $8 + 5a$ f $d + 6$
g $20 - a^2$ h $14 + 2n$ i $2m + 5$

Exercise 6-02

1 C
2 A
3 a − b + c ×
d + e ÷ f ×2
g $(\,)^2$ h − i ×3
4 a $3n$ b $n - 8$ c $n + 7$ d $n \div 6$
e $n + 2$ f $n - 12$ g $2n - 5$ h n^2
i n^3 j $3n + 9$

5 **a** The sum of m and n
b The product of 6 and a
c Twice b plus 4
d The quotient of triple n and 4
e The difference of 8 and b
f Triple v less 2
g Four times m less n
h The quotient of two m and n
i Triple the difference of a and b

6 **a** $3w - 5$ **b** $(a + b) - 6$ **c** $2(d \div 9)$
d $mn + 8$ **e** $(r + s)^2$ **f** $(c - 4)^3$
g abc **h** $3(w + v + u)$

Exercise 6-03

1 B
2 D
3 **a** –81 **b** 21 **c** 29 **d** 24
e –48 **f** –99 **g** 97 **h** 105
i 54 **j** –354 **k** –123 **l** –648
4 **a** $w \times v$ **b** $3 \times w$ **c** $8 \times v$
d $3 \times v$ **e** $v \times w$ **f** $6 \times v$
g $8 \times w$ **h** $2 \times w \times v$
5 **a** –43 **b** –26 **c** 58 **d** –2
e 60 **f** 24 **g** –10 **h** 120
6 **a** –60 **b** 10.2 **c** –74 **d** 1.2
e –120 **f** 5 **g** –19.2 **h** 14.4
i –18 **j** –176 **k** 18 **l** 20
7 **a** T **b** F **c** T **d** F

Exercise 6-04

1 C
2 B
3 $5a, -3a, 12a, \frac{a}{3}$
Also $7n$ and $\frac{2n}{5}$
4 **a** $9b$ **b** a **c** $11m$ **d** $5a$
e $5x$ **f** $2x$ **g** $3b$ **h** $5w$
i 0 **j** $2m$ **k** $8w$ **l** $-3ab$
m $-y$ **n** $-5m$ **o** $5b$ **p** $2n$
q $6ab$ **r** $3r^2$ **s** $11mn$ **t** $8a - 3$
u $3v + 2w$ **v** $5s + 4st$
w $5p + 10p^2$ **x** $-4a - 2$
5 **a** T **b** F **c** T
d F **e** F **f** T
g T **h** F **i** F
6 $10x + 12$
7 **a** $5ab$ **b** $9mn - 8$ **c** $12 - 12uv$
d $5sr - 7$ **e** $-10n$ **f** $28a - 12b$
g $28 - 9y$ **h** $10ab$ **i** $24 + 6w$

Exercise 6-05

1 A
2 C
3 **a** $2a$ **b** $5m$ **c** $-4n$
d $7uv$ **e** $-16rs$ **f** $-48ab$
g $15bc$ **h** $11de$ **i** $20yz$
j $-30a$ **k** $-6w$ **l** $-24ac$
m $-12abc$ **n** $12n^2$ **o** $36m^2$
p $-60uw$ **q** $10a^2b^2$ **r** $24a^3$
4 **a** $6mn$ **b** $14b^2$
5 **a** $-60mn$ **b** $18abc$ **c** $-30mnq$
d $-12a^2b^2$ **e** $30bcd$ **f** $-24a^3$
g $-36t^3$ **h** $-15e^3$ **i** $-96c^2d^2e$
6 **a** \$600 **b** \$120m
c \$2400 **d** \$120mw

Exercise 6-06

1 B
2 A
3 **a** F **b** T **c** F **d** F
e T **f** F **g** T **h** F
i T **j** F **k** F **l** F
m F **n** F **o** T **p** T
q F **r** T
4 **a** $2m$ **b** $-3a$ **c** $-5mn$ **d** $-4b$
e $5mn$ **f** $\frac{-2w}{v}$ **g** $2c$ **h** $\frac{-2r}{s}$
i $2st$ **j** $6b$ **k** $-2m$ **l** $\frac{b}{2d}$
m $\frac{4r}{t}$ **n** $\frac{-2v}{w}$ **o** $\frac{-8}{w}$
p $-8e$ **q** $\frac{e}{-3g}$ **r** $\frac{6ab}{d}$

Language activity

ABSTRACT ALGEBRA ANTICS

Practice test 6

Part A

1 4
2 72
3 –27
4 $27a^3$
5 56
6 19.8
7 \$615
8 160 m^2
9 2.7
10 $\frac{3}{7}$

Part B

11 D
12 B
13 D
14 $2(a + b)$
15 **a** 55 **b** 43
16 **a** $10w$ **b** $3a - 2b$ **c** $-ab$ **d** $3bc - 12$
17 **a** $20ab$ **b** $48mn^2$
18 **a** $-\frac{6s}{r}$ **b** $\frac{6a}{c}$

ISBN 9780170350990

CHAPTER 7

Exercise 7-01

1 D
2 B
3 a $\angle PQR$ or $\angle RQP$, acute
b $\angle SRT$ or $\angle TRS$, reflex
c $\angle WUV$ or $\angle VUW$, obtuse
d $\angle CBA$ or $\angle ABC$, right
4 a reflex b revolution c acute
d obtuse e straight f right
5 a $\angle B$, acute b $\angle P$, obtuse c $\angle G$, reflex
6 D
7 Teacher to check.
8 a obtuse b reflex c acute d reflex

Exercise 7-02

1 B
2 C
3 a 48° b 140° c 310°
d 89° e 65° f 195°
4 Teacher to check.
5 Teacher to check.

Exercise 7-03

1 C
2 B
3 a They add to 90°. b They add to 360°.

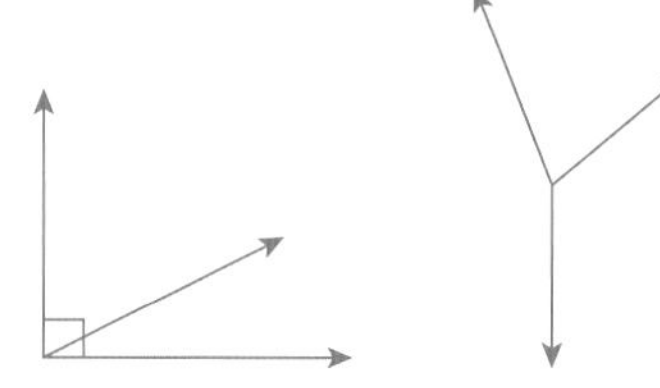

4 a F b T c T d F
5 a $w = 22$, angles in a right angle
b $n = 48$, angles in a straight angle
c $a = 54$, angles in a straight angle
d $c = 34$, angles in a right angle
e $b = 84$, vertically opposite angles
f $m = 108$, angles at a point
g $n = 48$, vertically opposite angles
h $r = 174$, angles at a point
i $b = 132$, angles at a point

Exercise 7-04

1 C
2 B
3 a

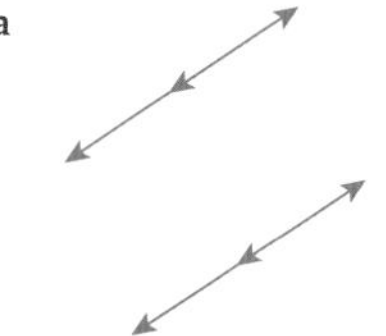

b

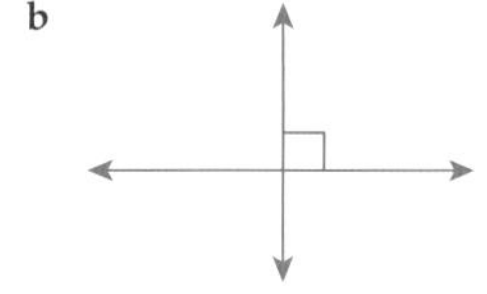

4 Teacher to check.
5 a ED b ED or BC
c CD d DE or BC

Exercise 7-05

1 C
2 B
3

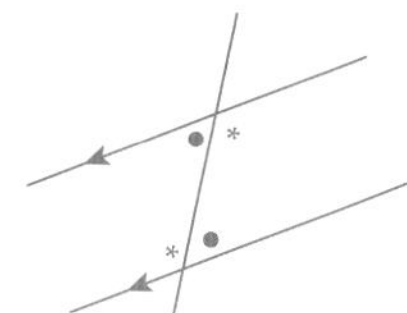

4 a $n = 116$ (corresponding angles on parallel lines)
b $m = 57$ (alternate angles on parallel lines)
c $v = 65$ (alternate angles on parallel lines)
d $c = 98$ (co-interior angles on parallel lines)
e $a = 106$ (co-interior angles on parallel lines)
f $b = 81$ (corresponding angles on parallel lines, vertically opposite angles)
g $d = 47$ (co-interior angles on parallel lines)
h $x = 120$ (corresponding angles on parallel lines), $y = 60$ (angles on a straight line), $z = 60$ (corresponding angles on parallel lines)
i $a = 88$ (alternate angles on parallel lines), $c = 92$ (angles on a straight line), $c = 92$ (co-interior or alternate angles on parallel lines)
5

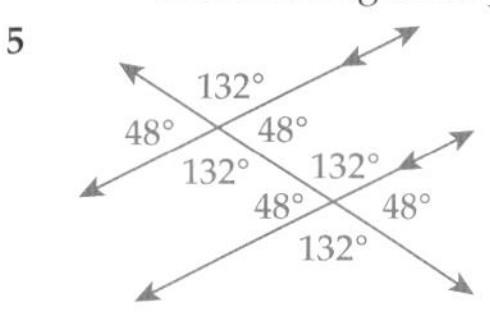

6 a T b T c F
d T e F f T
7 $a = 88$, $b = 92$, $c = 88$

Exercise 7-06

1 D
2 B
3 a AB and CD not parallel, alternate angles are not equal.
b EF and GH not parallel, co-interior angles are not supplementary.
c $IJ \parallel KL$, corresponding angles are equal.
d WX and YZ not parallel, alternate angles are not equal.
e $MN \parallel PQ$, co-interior angles are supplementary.
f $RS \parallel TU$, corresponding angles are equal.
4 a F b T c T d F
5 B
6 a Other answers are possible.
b Other answers are possible

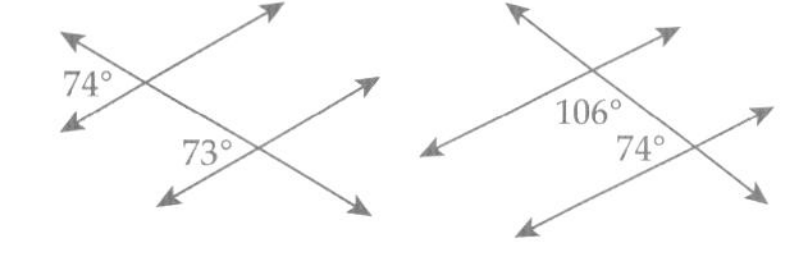

Exercise **7-07**

1 C

2 C

3 a

b

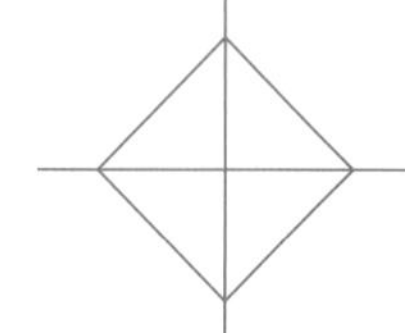

c

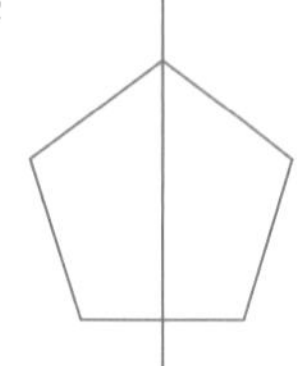

d

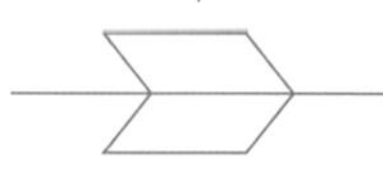

e

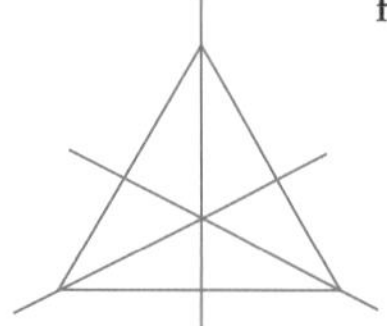

f

4 **a** Yes, 2 **b** No **c** Yes, 4
d Yes, 5 **e** Yes, 4 **f** No
g Yes, 4 **h** Yes, 6 **i** No

5 **a** 180° **c** 90° **d** 72°
e 90° **g** 90° **h** 60°

6 **a** F **b** F **c** T **d** T

Exercise **7-08**

1

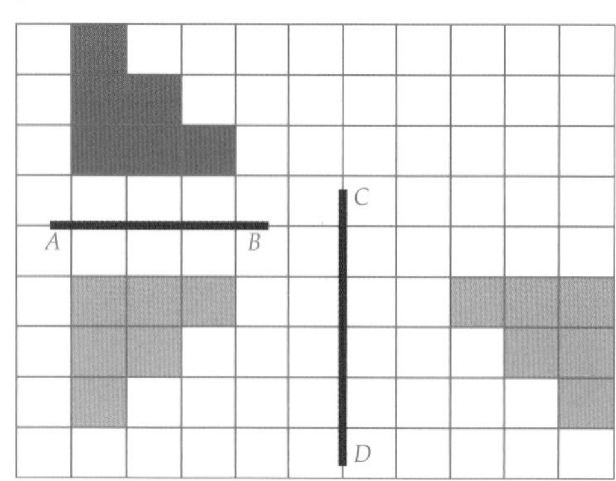

2

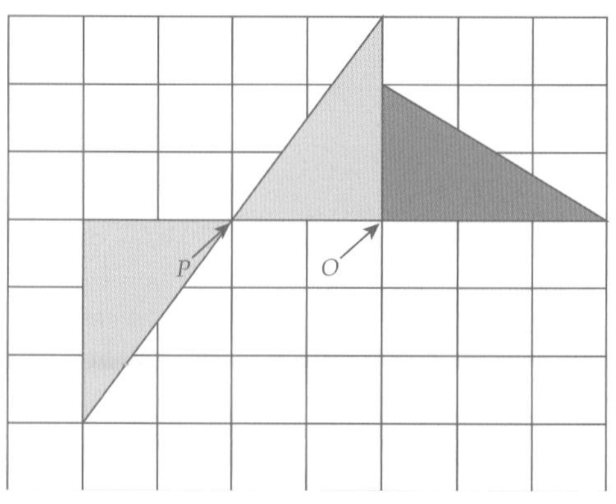

3 **a** Reflect in line *AB* and then translate 4 units right and 1 unit down.

b Rotate 90° clockwise about *O*, reflect in line *CD* and then translate 4 right and 2 units up.

4

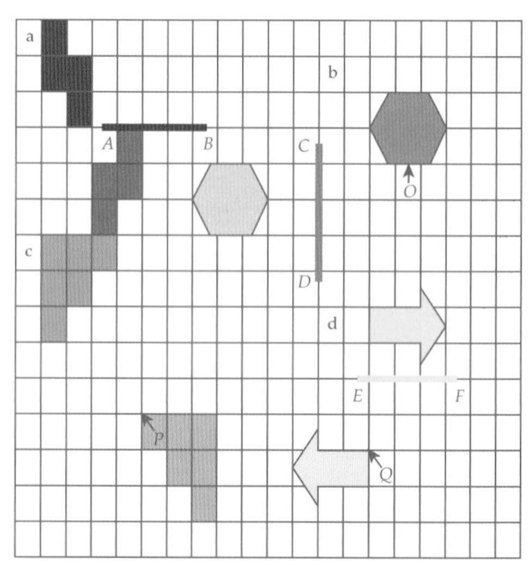

Exercise **7-09**

1 **a**

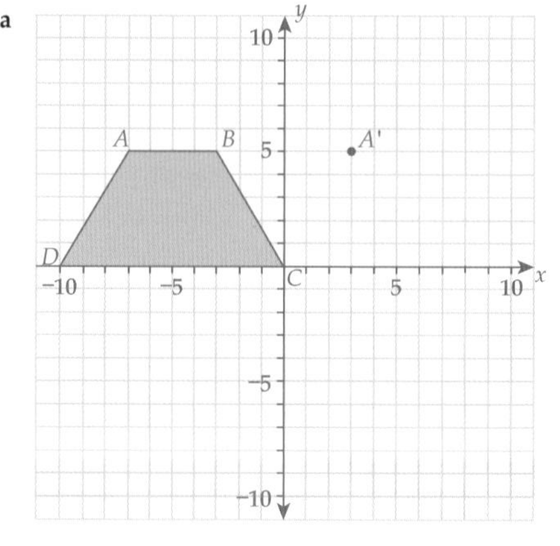

ISBN 9780170350990

b $A(-7, 5) \rightarrow A'(3, 5)$ $B(-3, 5) \rightarrow B'(7, 5)$
$C(0, 0) \rightarrow C'(10, 0)$ $D(-10, 0) \rightarrow D'(0, 0)$
The x-coordinate of each vertex increases by 10 whereas the y-coordinate stays the same.

2 $P(0, -6) \rightarrow P'(6, 0)$ $Q(3, -6) \rightarrow Q'(6, 3)$
$R(3, -4) \rightarrow R'(4, 3)$ $S(0, -4) \rightarrow S'(4, 0)$
The x-coordinate of each vertex becomes the y-coordinate of the image vertex, whereas the y-coordinate of each vertex changes sign and becomes the x-coordinate of the image vertex.

3 **a**

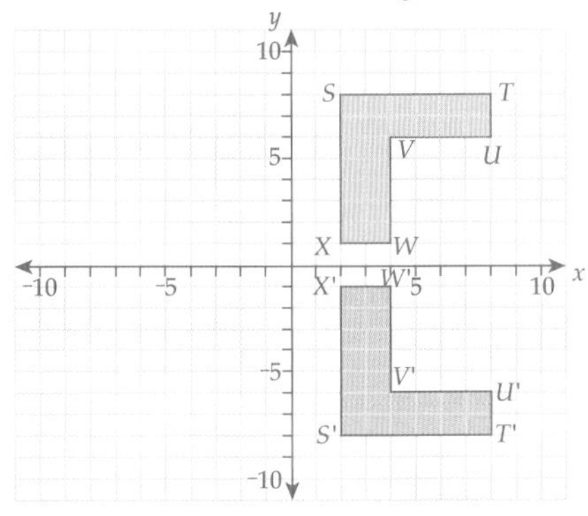

b $S(2, 8) \rightarrow S'(2, -8)$ $T(8, 8) \rightarrow T'(8, -8)$
$U(8, 6) \rightarrow U'(8, -6)$ $V(4, 6) \rightarrow V'(4, -6)$
The x-coordinate of each vertex stays the same whereas the y-coordinate changes sign (becomes negative).

4 **a** Translated 8 units right and 2 down.
b $W(-6, 8) \rightarrow W'(2, 6)$ $X(-2, 1) \rightarrow X'(6, -1)$
$Y(-6, 1) \rightarrow Y'(2, -1)$
The x-coordinate of each vertex increases by 8 whereas the y-coordinate decreases by 2.

5 **a**

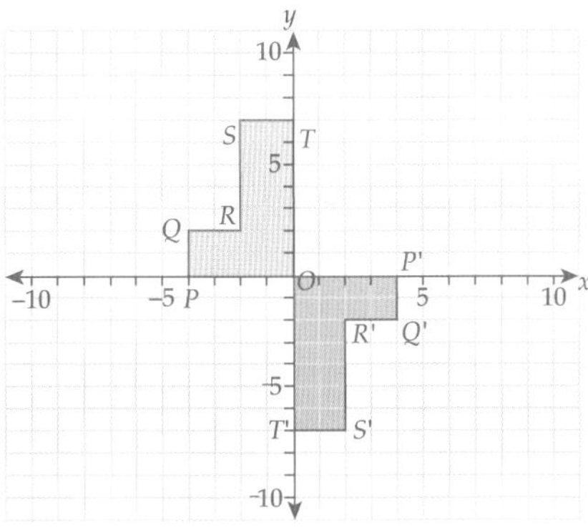

b $P(-4, 0) \rightarrow P'(4, 0)$ $Q(-4, 2) \rightarrow Q'(4, -2)$
$S(-2, 7) \rightarrow S'(2, -7)$
Both x- and y-coordinates of each vertex change sign.

Practice test 7

Part A

1 360
2 90
3 $-8b$
4 140 cm^3
5 \$15
6 \$43.15
7 6
8 4
9 8.19, 8.9, 8.909, 8.95
10 8, 32

Part B

11 $\angle FGE$ or $\angle EGF$
12 D
13 **a** $n = 46$, angles in a straight angle
b $k = 82$, vertically opposite angles
14 $AB \parallel CD$, $AB \perp AC$ (or $AC \perp CD$)
15 co-interior
16 $c = 72$
17 No, corresponding angles are not equal.
18 4
19 C
20 **a** Rotation of 90° clockwise about P.
b $M(-3, 6) \rightarrow M'(6, 3)$ $N(0, 6) \rightarrow N'(6, 0)$
$Q(-3, 0) \rightarrow Q'(0, 3)$
The x-coordinate of each vertex changes sign and becomes the y-coordinate of the image vertex, whereas the y-coordinate of each vertex becomes the x-coordinate of the image vertex.

CHAPTER 8

Exercise 8-01

1 A
2 C
3 **a** equilateral **b** isosceles
c scalene
4 **a** acute-angled **b** obtuse-angled
c acute-angled
5 **a**

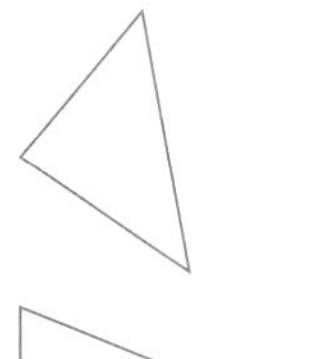

b

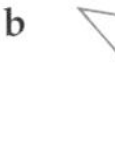

c

d

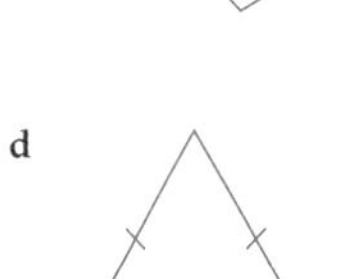

6 **a** Yes
b Because they are equal and add to 180°. Each angle = 180 ÷ 3 = 60
c No **d** 90° **e** No
7 **a** ΔRST **b** ΔDEF **c** ΔXYZ
8 **a** scalene right-angled
b isosceles acute-angled
c scalene obtuse-angled
9 **a** right-angled, obtuse-angled, acute-angled, scalene and isosceles
b 34

Exercise 8-02

1 D
2 D
3 a $n = 64$ b $b = 52$ c $m = 40$
d $w = 120$ e $c = 22$ f $a = 36$
4 a 55° b $p + q + r = 180$
5 a An equilateral triangle has all sides equal in length.
b A right triangle has one angle that is 90°.
c An isosceles triangle has two sides equal in length.
d A scalene triangle has all sides different lengths.
6 a They are all 60°. b Two angles are equal.
7 a $n = 60$ b $m = 58$ c $c = 78$
d $w = 45$ e $e = 41$ f $v = 144$

Exercise 8-03

1 D
2 B
3 a $\angle BAC$, $\angle ACB$, $\angle CBA$ b $\angle BCD$
c $\angle BCD = \angle BAC + \angle CBA$
4 a $x = 86$ b $y = 99$ c $q = 137$ d $n = 104$
5 a

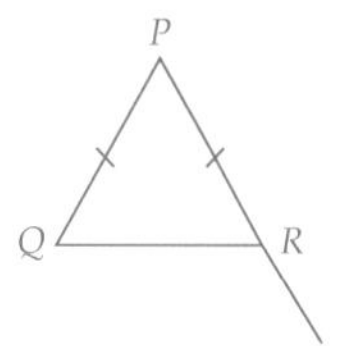

b $\angle RPQ$ and $\angle PQR$ c exterior angle = 116°

Exercise 8-04

1 B
2 D
3 a trapezium b square
c kite d irregular quadrilateral
e parallelogram f rectangle
4 a

b

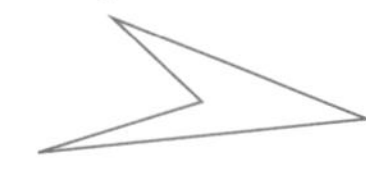

c

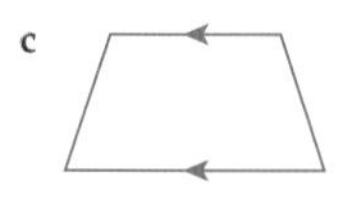

d

e

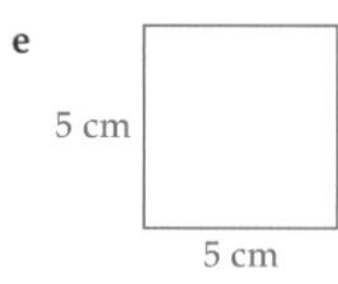

f

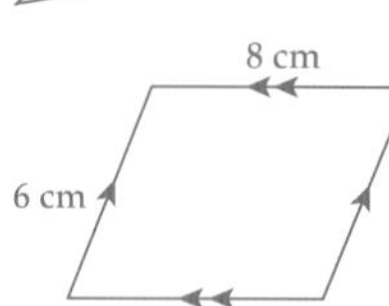

g

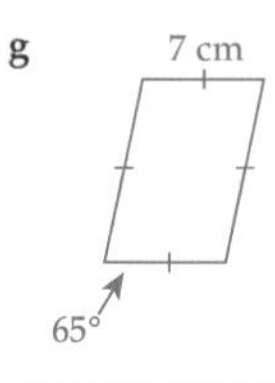

5 a F b F c T
d T e T f F

Exercise 8-05

1 B
2 A
3 a F b T c T
d F e F
4 a $a = 102$ b $b = 90$ c $c = 68$
d $m = 55$ e $n = 56$ f $w = 115$
5 a

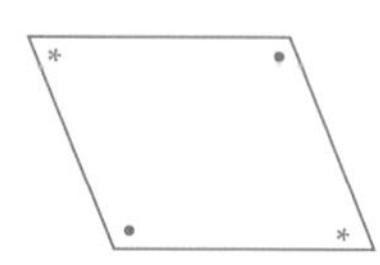

b 180° c Two d 360°
6 a a square b They are equal.
c Yes d Yes e 360°
f 90° g 90°

Exercise 8-06

1 B
2 A
3 a A square has all sides equal and a rectangle does not.
b A parallelogram has opposite sides parallel (or equal) and a quadrilateral does not.
c A rhombus has all sides equal and a parallelogram does not.
d A parallelogram has both pairs of opposite sides parallel and a trapezium has one pair of opposite sides parallel.
4 a T b T c T
d T e F f T
5 a rectangle and square
b parallelogram, rhombus, rectangle and square
c kite, rhombus and square
d rhombus and square
6

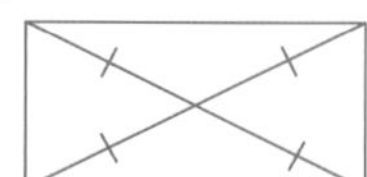

7

Quadrilateral	Angles 90°	Equal sides	Equal diagonals
Parallelogram	No	No	No
Rhombus	No	Yes	No
Rectangle	Yes	No	Yes
Kite	No	No	No
Square	Yes	Yes	Yes

ISBN 9780170350990

8

Quadrilateral	Sides	Angles	Diagonals
Parallelogram	Opposite sides are equal and parallel.	Opposite angles are equal.	Diagonals bisect each other.
Trapezium	One pair of opposite sides are parallel.	All angles are different.	Diagonals are not equal.
Rhombus	All sides are equal.	Opposite angles are equal.	Diagonals bisect each other at right angles and bisect the angles of the rhombus.
Rectangle	Opposite sides are equal.	All angles are 90°.	Diagonals are equal.
Square	All sides equal.	All angles are 90°.	Diagonals bisect each other at right angles and bisect the angles of the square. Diagonals are equal.

Language activity

Across

1 KITE
7 RIGHT
9 EQUILATERAL
11 PARALLELOGRAM
12 SCALENE
13 TRIANGLE
15 RECTANGLE

Down

2 ISOSCELES
3 IRREGULAR
4 QUADRILATERAL
5 CONVEX
6 SQUARE
8 TRAPEZIUM
10 RHOMBUS
14 NONCONVEX

Practice test **8**

Part A

1 An angle less than 90°.
2 142
3 7, 14, 21, 28, 35, 42
4 composite
5 7
6 Yes
7 $\frac{13}{24}$
8 16.8 m^2
9 $-32ab^2c$
10 28

Part B

11 D
12 B
13 60°
14 78°
15 $w = 115$
16 **a** $\angle PQR$ **b** 129°
17 **a** trapezium **b** rectangle
18 360°
19 **a** $m = 46$ **b** $w = 50$
20 **a** One diagonal bisects the other at right angles.
b Diagonals are equal and bisect each other.

CHAPTER 9

Exercise **9-01**

1 C
2 A
3 **a** cm **b** cm **c** mL **d** km
e s **f** kL **g** mL **h** h
4 **a** 2 **b** 4900 **c** 72 000 **d** 1.44
e 125 000 **f** 8.5 **g** 0.0864 **h** 7.25
i 1.25 **j** 45 **k** 0.82 **l** 4600
5 **a** F **b** F **c** T
d F **e** F **f** T
6 710.05 m
7 2350 mL, 4500 mL, 6.2 L, 16.4 L, 0.98 kL
8

Millimetres	Centimetres	Metres	Kilometres
3500	350	3.5	0.0035
6400	640	6.4	0.0064
28 000	2800	28	0.028
6 500 000	650 000	6500	6.5
52 000	5200	52	0.052
420 000	42 000	420	0.42

ISBN 9780170350990

Exercise 9-02

1 C
2 D
3 a 28 cm b 19.2 m c 20.4 cm
d 20.4 m e 25 m f 29 cm
g 25.2 m h 30.8 m i 26.2 m
4 a 42.2 m b 23 cm c 29.8 m
d 124 mm e 18 cm f 36 m
5 $l = 22$ m, $w = 20$ m; or $l = 24$ m, $w = 18$ m (other answers are possible)
6 a 34 m b 48.1 m c 36 cm

Exercise 9-03

1 D
2 C
3

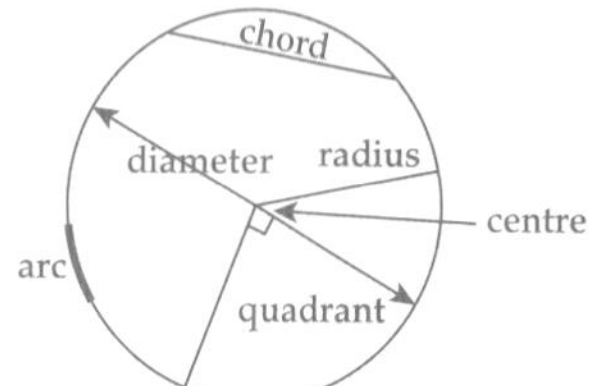

4 a 6 cm b diameter = 2 × radius
5 a diameter b arc
c centre d circumference
e chord f quadrant
6 D
7 semicircle

Exercise 9-04

1 A
2 B
3 b 9 cm
c diameter = 2 × radius
d 28.27 cm
4 a 28.27 cm b 35.19 m
c 28.90 m d 47.12 mm
e 57.81 m f 42.73 cm
5 a Halve the circumference, then add the diameter.
b 20.6 m
6 a 18.5 m b 16.9 m c 12.9 cm

Exercise 9-05

1 D
2 D
3 a at night b 7:20 a.m.
c 9:40 a.m. d 12:05 p.m., 1205
4 a 0300 b 1800 c 0420 d 1725
e 0742 f 2248 g 1200 h 2054
i 0400 j 1415 k 2130 l 0850
m 0320 n 2230 o 0545 p 2015
5 a 5:20 a.m. b 2:40 p.m. c 11:15 p.m.
d 6:05 a.m. e 10:56 p.m. f 3:38 a.m.
g 4:50 p.m. h 11:46 a.m. i 2:21 a.m.
j 12:45 a.m. k 12:12 p.m. l 1:48 a.m.
6 a 1220 b 11:40 a.m.
c 10:05 p.m., 2205

Exercise 9-06

1 D
2 C
3 a 5 h b 14 h c 11 h
d 7 h e 15 h f 23 h
4 a 23 min b 49 min c 53 min
5 a 95 b 384
6 a 4 h 25 min b 5 h 30 min c 15 h 20 min
d 7 h 45 min e 14 h 40 min f 46 min
g 3 h 55 min h 12 h 29 min i 9 h 15 min
j 12 h 24 min k 7 h 25 min l 11 h 15 min
7 a 3 h 20 min b 7 h 30 min c 6 h 25 min
8 a 1 min 38 s b 8 min 20 s c 2 min 34 s
9 a 1 h 52 min b 5:40 p.m. c 8:04 p.m.
10 17 h 22 min

Exercise 9-07

1 C
2 D
3 a 23 min b 10 min c 23 min d 22 min
4 9:18
5 5 min
6 8:58 Strathfield
7 24 min, no
8 Change trains at Redfern or Central.
9 a 26 min
b Bus 2 doesn't go to George St; Bus 3 takes the same time.
c Bus 2 d 9:27 e 13 min
f Less traffic for the 7:45 bus
10 Teacher to check

Exercise 9-08

1 B
2 D
3 a 11 a.m. b 3 a.m. c 1:30 p.m.
d 5 p.m. e 2 a.m. f 6 p.m.
4 a 11 a.m. b 12 midday c 6 a.m.
d 7 p.m. e 7 p.m. f 9 p.m.
5 3:20 a.m. Friday
6 2:30 p.m.
7 a 8:30 a.m. b 8:30 a.m.
c 6:30 a.m. d 8:00 a.m.
8 a 9:00 p.m. b 8:30 p.m.
c 9:00 p.m. d 9:00 p.m.
9 2:30 p.m.
10 4 a.m. to 7:20 a.m.
11 a 12 noon b 4 p.m.
12 10 p.m.

Language activity

1 LENGTH
2 METRIC
3 MASS

ISBN 9780170350990

4 TIME
5 PERIMETER
6 CIRCLE
7 RECTANGLE
8 KITE
9 TRIANGLE
10 PARALLELOGRAM
11 CIRCUMFERENCE
12 QUADRANT
13 RADIUS
14 DIAMETER
15 TIMETABLE
16 TIMEZONE

Practice Test 9

Part A

1 $40
2 34
3 6
4
5 $\frac{3}{4}$
6 trapezium
7 512
8 one
9 126.48
10 $\frac{1}{2}$

Part B

11 C
12 C
13 B
14 The distance from one edge of a circle to another going through the centre.
15 **a** arc **b** semicircle
16 **a** 25.13 cm **b** 16.34 m
17 17.1 m
18 **a** 1855 **b** 3:42 p.m.
19 **a** 1:54 p.m. **b** 43 min
20 9:12 a.m.
21 4 p.m.

CHAPTER 10

Exercise 10-01

1 C
2 B
3 **a** m² **b** cm² **c** km²
d ha or km² **e** mm² **f** m²
4 **a** 25 **b** 280 000 **c** 5.6
d 750 000 **e** 2 300 000 **f** 0.18
g 9 600 000 **h** 8.55 **i** 34
5 **a** 10 **b** 100 **c** 100
d 1000 **e** 1000 **f** 1 000 000
g 1000 **h** 1 000 000 **i** 5
j 0.5 **k** 0.08 **l** 0.0008
m 120 **n** 0.12 **o** 28 000
p 28 000 000 **q** 650 000 **r** 65 000 000
6 8 449 486 m²

Exercise 10-02

1 D
2 C
3 B
4 36 m² **b** 18 m² **c** 36 m²
5 **a** They have the same base length and height.
b The areas of the rectangle and parallelogram are the same, but the area of the triangle is half of these.
6 **a** 62.16 m² **b** 53.76 cm² **c** 90.1 m²
d 28.16 m² **e** 32.49 m² **f** 6.48 mm²
7 **a** 89.28 m² **b** 369.72 cm² **c** 23.22 m²
8 18.62 m²

Exercise 10-03

1 **a** 60 m² **b** 73.44 m² **c** 125.96 m²
d 17.9 cm² **e** 14.18 m² **f** 19.44 m²
3 **a** 174 m² **b** 316.98 m²
4 **a** **i** $12 \times 8 + 4 \times 9 + \frac{1}{2} \times 4 \times 3 = 138$ m²
ii $12 \times 12 - \frac{1}{2} \times 4 \times 3 = 138$ m²
b Subtracting areas
5 25.104 m²
6 Teacher to check.

Exercise 10-04

1 A
2 **a** 28 m² **b** 42 m² **c** 11.2 m²
d 23.8 m² **e** 27.88 m² **f** 18.6 m²
3 **a**

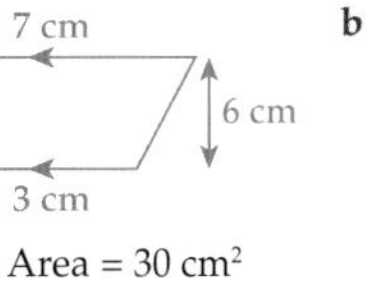

Area = 30 cm²

b

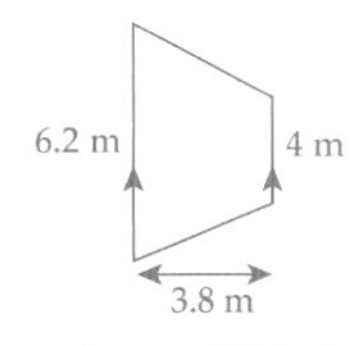

Area = 19.38 m²

4 **a** 72.66 m² **b** 34.645 m² **c** 361.34 m²
5 168.48 cm²

Exercise 10-05

1 B
2 C
3 **a** $A = \frac{1}{2} \times 3 \times 8 = 12$ m²
4 **a** 20 m² **b** 14 cm² **c** 48.72 m²
d 16 cm² **e** 3024 mm² **f** 31.02 m²
5 9.92 m²
6 **a** 1.26 m² Canvas costs $16.13 **b** $142.63

ISBN 9780170350990

Exercise 10-06

1 B
2 A
3 a $A = \pi \times 6.8^2 = 145.2672... = 145.27$ m²
4 a 153.9 m² b 113.1 m² c 55.4 cm²
d 3631.7 mm² e 8.6 m² f 84.9 cm²
5 a 12.57 m² b 100.91 cm² c 83.71 m²
d 166.99 mm² e 72.66 cm² f 108.52 m²

Exercise 10-07

1 A
2 D
3 a cm³ b m² c m d m²
e cm³ f mm³ g cm h m³
4 a 550 b 47 000 000
c 56 000 d 7 500 000 000
e 0.000 043 f 0.28
g 9 260 000 h 0.855
5 a mL b kL c mL
d mL e kL f mL
6 a 4 b 222 c 7.5 d 10 400
e 8.504 f 67 000 g 0.68 h 2.56
7 a 200 b 20 000 c 2 000 000
d 5000 e 500 000 f 50 000 000
g 80 000 h 8 000 000 i 800 000 000

Exercise 10-08

1 A, C, E, F, G, H, I
2 a I b F, H c A
d G e E f C
3 a b c

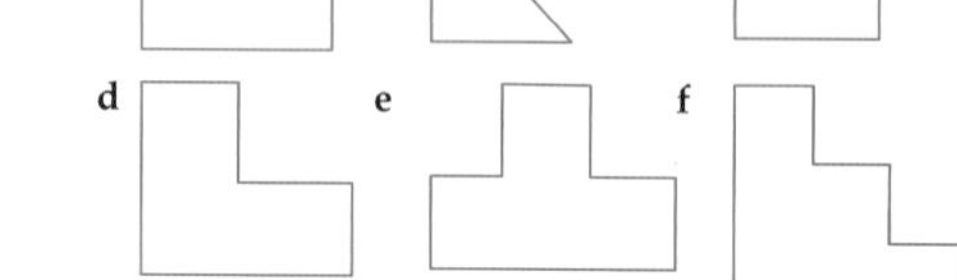

d e f

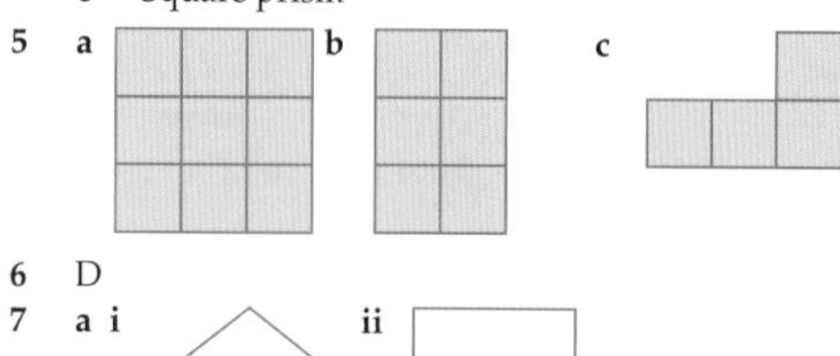

4 a Pentagonal prism b Triangular prism
c Square prism
5 a b c
6 D
7 a i ii
b i ii iii

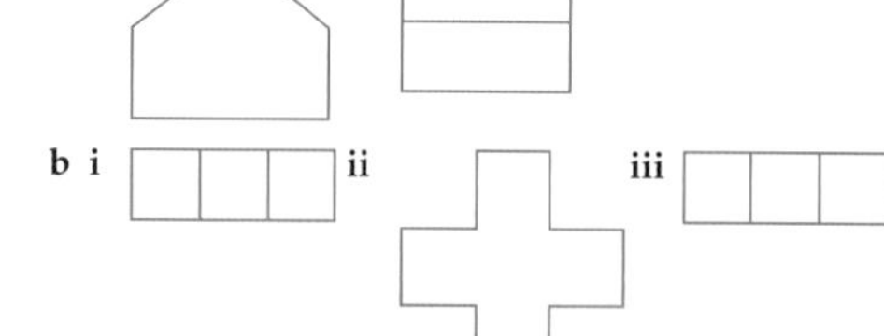

c i ii iii
d i ii iii

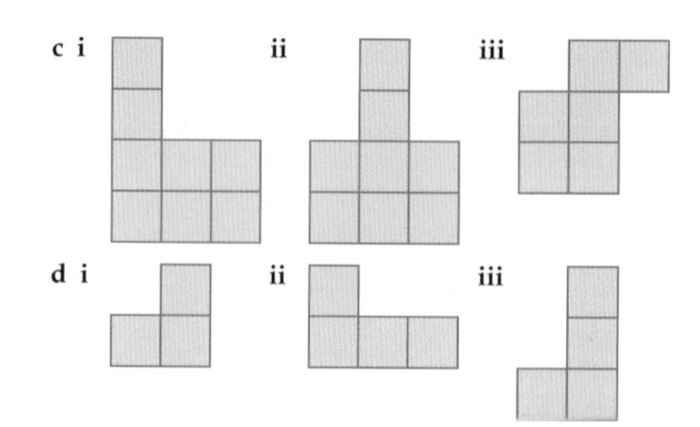

Exercise 10-09

1 C
2 B
3 a 189 m³ b 160 cm³ c 13.824 m³
d 196 cm³ e 94.5 cm³ f 42 m³
g 69 cm³ h 176 mm³ i 63 m³
j 900 mm³ k 99.846 m³ l 69.984 cm³

Exercise 10-10

1 A
2 D
3 $V = \pi \times 6^2 \times 8.4 = 950.02$ cm³
4 a 353.8 m³ b 268.7 m³
c 445.4 m³ d 1147.2 cm³
5 a Find the volume of the cylinder and then halve it.
b 231.2 m³
6 a 28.95 m³ b 28 953 L c 29 kL
7 1357 mm³
8 a 327.1 cm³ b 327 mL

Language activity

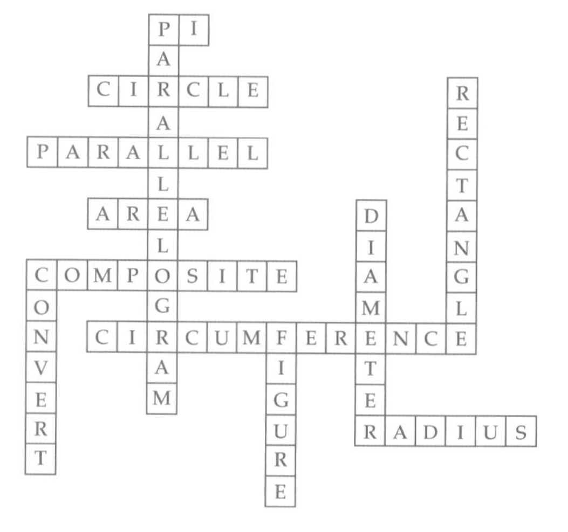

 ISBN 9780170350990

Practice test 10

Part A

1 56
2 1, 2, 3, 4, 5, 6
3 1
4 63
5 –7
6 180°
7 10:40 p.m.
8 2.68
9 $\frac{3}{5}$
10

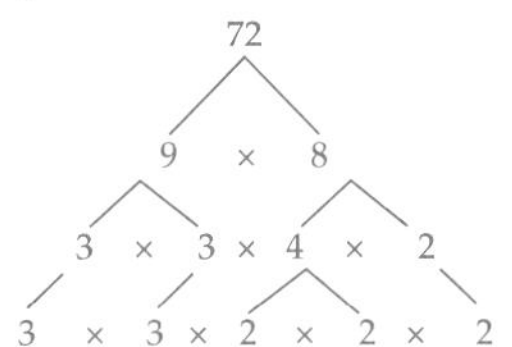

Part B

11 D
12 a 92 000 b 7.7
13 A
14 a 51.66 cm² b 17.92 m²
15 155.52 m²
16 71.02 m²
17 a 16.74 m² b 3168 mm²
18 83.7 cm²
19 a 0.042 356 b 56.8
20 a b c

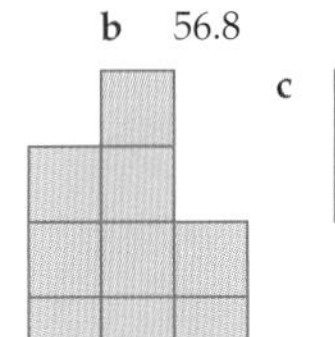

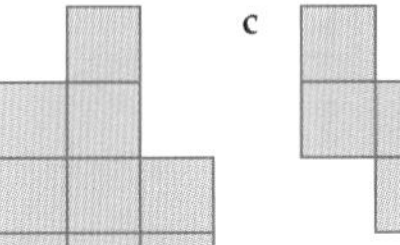

21 29.92 m³
22 a 147.03 m³ b 1240.25 cm³

CHAPTER 11

Exercise 11-01

1 C
2 B
3 numerator, denominator, numerator, denominator, divide, denominator, simplest
4 $\frac{12}{15}, \frac{4}{5}, \frac{24}{30}$
5 a 9 b 15 c 20
d 30 e 10 f 5
g 56 h 32 i 75
6 a $\frac{2}{3}$ b $\frac{1}{3}$ c $\frac{3}{5}$ d $\frac{8}{9}$
e $\frac{2}{3}$ f $\frac{1}{3}$ g $\frac{3}{4}$ h $\frac{4}{5}$
i $\frac{11}{17}$ j $\frac{3}{40}$ k $\frac{5}{28}$ l $\frac{4}{9}$
7 a F b T c F d F

Exercise 11-02

1 D
2 A
3 a M b P c M d P
e I f P g P h I
i I j P k M l I
4 a $1\frac{2}{5}$ b $1\frac{3}{5}$ c $2\frac{1}{4}$ d $1\frac{1}{6}$
e $2\frac{2}{5}$ f $1\frac{1}{5}$ g $1\frac{2}{9}$ h $4\frac{1}{4}$
5 a $\frac{7}{3}$ b $\frac{15}{4}$ c $\frac{23}{5}$ d $\frac{15}{8}$
e $\frac{11}{3}$ f $\frac{17}{6}$ g $\frac{24}{5}$ h $\frac{51}{8}$
6 a F b T c T
d F e T f F
7 a 34 pieces b 6 pieces

Exercise 11-03

1 B
2 A
3 a $\frac{7}{8}$ b $\frac{3}{5}$ c $\frac{7}{8}$
d $\frac{5}{6}$ e $2\frac{1}{3}$ f $3\frac{5}{7}$
4 a T b F c F
d F e T f T
5 a $-\frac{7}{12}, \frac{3}{12}, \frac{1}{2}, \frac{8}{12}, \frac{3}{4}$
b $-\frac{1}{2}, -\frac{3}{8}, \frac{1}{4}, \frac{5}{8}, \frac{9}{4}$
c $\frac{1}{6}, \frac{1}{3}, \frac{3}{4}, \frac{5}{6}, \frac{11}{12}$
6

$-\frac{4}{8}$, –1, $-\frac{3}{8}$, 0, $\frac{2}{8}$, $\frac{5}{8}$, 1, 2, $2\frac{2}{8}$

7 a $\frac{9}{6}, \frac{4}{6}, \frac{1}{2}, \frac{1}{3}, -\frac{3}{12}$
b $\frac{13}{9}, \frac{7}{9}, \frac{1}{3}, -\frac{5}{9}, -\frac{2}{3}$
c $\frac{7}{4}, \frac{9}{6}, \frac{5}{6}, \frac{2}{3}, \frac{7}{12}$
8

$-\frac{6}{9}$, –1, $-\frac{5}{9}$, 0, $\frac{3}{9}$, $\frac{7}{9}$, 1, $1\frac{4}{9}$

Exercise 11-04

1 C
2 B
3 a $\frac{4}{5}$ b $\frac{5}{8}$ c $\frac{6}{7}$ d $\frac{3}{5}$
4 a $\frac{13}{20}$ b $\frac{23}{30}$ c $1\frac{1}{24}$ d $1\frac{9}{40}$
e $\frac{7}{9}$ f $1\frac{1}{4}$ g $1\frac{1}{3}$ h $1\frac{17}{24}$
5 a $\frac{7}{20}$ b $\frac{7}{12}$ c $\frac{13}{24}$ d $\frac{3}{10}$
e $\frac{7}{18}$ f $\frac{1}{4}$ g $\frac{1}{3}$ h $\frac{1}{6}$
6 a F b T

7 a $3\frac{3}{4}$ b $3\frac{9}{20}$ c $2\frac{3}{7}$ d $2\frac{7}{15}$
e $\frac{3}{4}$ f $1\frac{1}{12}$ g $1\frac{8}{15}$ h $\frac{11}{12}$

8 $\frac{1}{4}$

Exercise 11-05

1 B
2 C
3 a 7 b 14 c 21
d 28 e 7 f 14
g 35 h 49 i 15 min
j $24 k 150 km l $150
m 18 m n 80 min o 56 pages
p $200 q 56 L r 280 mL
s 44 min t $90
4 a T b F c T
d F e T f F
5 14
6 a $1200 b $600 c $\frac{17}{20}$
7 a 100 b 64

Exercise 11-06

1 B
2 A
3 a T b F c T d F
4 a 3, 8, 6, $\frac{3}{20}$ b 3, 4, 36, $\frac{2}{3}$
5 a $\frac{1}{2}$ b $\frac{2}{5}$ c 1 d $\frac{5}{18}$
e $\frac{2}{7}$ f 1 g 1 h $\frac{2}{3}$
6 improper, numerators, denominators
7 a $3\frac{3}{10}$ b $5\frac{1}{2}$ c $2\frac{2}{3}$ d $\frac{7}{15}$
e $1\frac{3}{4}$ f $1\frac{13}{15}$ g $\frac{1}{3}$ h $3\frac{2}{3}$
i $3\frac{1}{8}$ j 2 k $4\frac{1}{3}$ l $15\frac{3}{5}$

Exercise 11-07

1 B
2 A
3 a $1\frac{1}{3}$ b 5 c $1\frac{4}{5}$ d 12
e $\frac{1}{5}$ f $1\frac{1}{4}$ g $1\frac{2}{5}$ h $\frac{1}{8}$
4 multiply, reciprocal
5 a F b T c T d F
6 a 3, 16, $1\frac{1}{15}$ b 3, 63, $\frac{8}{21}$
7 a $\frac{2}{5}$ b 4 c $\frac{3}{10}$ d $1\frac{1}{2}$
e 9 f 12 g $1\frac{1}{5}$ h $\frac{3}{4}$
8 improper, reciprocal
9 a $1\frac{1}{2}$ b $\frac{15}{28}$ c $\frac{8}{15}$ d $\frac{1}{3}$
e $\frac{1}{2}$ f $\frac{63}{80}$ g $3\frac{7}{16}$ h $\frac{2}{3}$
10 eight

Language activity

FRACTION FRENZY

Practice test 11

Part A

1 5:45 p.m.
2 $2y$
3 1
4 $x^2 = d^2 + e^2$
5 –8
6 $24d$
7 $x = 9$
8 54
9 $0.\dot{3}$
10 $\frac{1}{2}$

Part B

11 B
12 C
13 D
14 a $1\frac{5}{7}$ b $3\frac{1}{8}$
15 $\frac{3}{8}, \frac{1}{3}, \frac{1}{4}, \frac{1}{6}$
16 a $1\frac{5}{12}$ b $\frac{11}{40}$
17 a $510 b 1 h 20 min
18 a $\frac{3}{4}$ b $8\frac{1}{8}$
19 a $1\frac{1}{2}$ b $\frac{14}{19}$

CHAPTER 12

Exercise 12-01

1 D
2 a $\frac{1}{2}$ b $\frac{3}{4}$ c $\frac{1}{4}$ d 1
e $\frac{7}{10}$ f $\frac{11}{25}$ g $\frac{3}{10}$ h $\frac{13}{20}$
i $\frac{29}{50}$ j $1\frac{1}{10}$ k $\frac{7}{100}$ l $\frac{63}{100}$
m $\frac{1}{20}$ n $\frac{19}{20}$ o $\frac{4}{5}$ p $\frac{4}{25}$
3 a 100, 3, 100, 1 b $62\frac{1}{2}$, 2, 125, 5
4 a $\frac{5}{6}$ b $\frac{3}{8}$ c $\frac{2}{3}$ d $\frac{33}{200}$
5 A
6 a 10% b 75% c 20% d $22\frac{2}{9}$%
e 35% f $66\frac{2}{3}$% g $37\frac{1}{2}$% h $83\frac{1}{3}$%
7 $\frac{3}{5}$, 62.5%, $\frac{65}{100}$, 68%
8 88%, $\frac{85}{100}$, 81.5%, $\frac{4}{5}$

Exercise 12-02

1 D
2 A
3 a F b T c F
d T e T f F

ISBN 9780170350990

4 a 0.25 b 0.4 c 0.65 d 0.8
e 0.12 f 0.71 g 1 h 1.2

5 a 70% b 2% c 28% d 50%
e 47% f 55% g 90% h 85%
i 52.4% j 75% k 120% l 260%

6 50.2%, 0.505, 0.51, 53%

7 98%, 97.5%, 0.9, 0.099

8

Percentage	Fraction	Decimal
10%	$\frac{1}{10}$	0.1
15%	$\frac{3}{20}$	0.15
25%	$\frac{1}{4}$	0.25
30%	$\frac{3}{10}$	0.3
50%	$\frac{1}{2}$	0.5
60%	$\frac{3}{5}$	0.6
75%	$\frac{3}{4}$	0.75
80%	$\frac{4}{5}$	0.8
95%	$\frac{19}{20}$	0.95
100%	1	1.0

Exercise 12-03

1 B

2 A

3 a T b F c T
d T e F f T

4 a 0.25, 212 b 96, 72 c 24, 8

5 a 3 m b 6 cm c $900 d $72
e 30 L f 48 kg g $126 h 36 km
i $3900 j 12 min k 3 h l $4800
m 78 kg n 15 km o 12 mL p $237.50
q 66 L r 638 g

6 14

7 a 50 b 80% c 200 d $450

Exercise 12-04

1 C

2 A

3 a 500, 10 b 400, 100

4 a $\frac{1}{300}$ b $\frac{11}{300}$ c $\frac{1}{2}$
d $\frac{17}{1000}$ e $\frac{31}{40}$ f $\frac{3}{500}$
g $\frac{3}{50}$ h $\frac{3}{10}$ i $\frac{11}{200}$

5 a 500, 10 b 400, 9

6 a 5% b 2% c 8%
d 50% e 0.7% f 2.5%
g 0.6% h 0.05% i 0.15%

7 $\frac{5}{6}$

8 a 2.5% b 40

Exercise 12-05

1 D

2 C

3 a 0.1, 68, 68, 748 b 850, 170, 170, 680

4 a $60 b 480 kg
c 42.5 kg d 11 375 m
e $3.52 f 417.6 mL
g $3808 h 96 L
i 6.552 m j 176.64 kg

5 a No b $1920

6 $528

Exercise 12-06

1 B

2 A

3 a T b F c T

4 4800, 25, 192, 192, 19 200

5 a $800 b $4000 c $34 615.38
d $30.77 e $622.22 f 1578.9

6 $3000

7 $1566.67

8 6516

9 $550 000

Exercise 12-07

1 C

2 B

3

Cost price	Selling price	Profit or loss
$50	$80	Profit $30
$125	$105	Loss $20
$170	$220	Profit $50
$560	$545	Loss $15
$298	$380	Profit $82

4 a Profit $25 b 125%

5 a $4 b 25%

6 a $18 b $33\frac{1}{3}$%

7 $196

8 a $210 b $49 c $280 d $74.90

9 a $12, $132 b $5.50, $60.50
c $8.60, $94.60 d $11, $121
e $3.80, $41.80 f $2.40, $26.40
g $4.40, $48.40 h $2.50, $27.50

10 a $79.20 b $36.30 c $56.76 d $72.60
e $25.08 f $15.84 g $29.04 h $16.50

Language activity

Teacher to check.

Practice test 12

Part A

1 $\frac{2}{3}$

2 right angle

3 −6

4 7
5 $\frac{1}{2}$
6 11.24
7 125
8 4
9 $5a$
10 $C = 2\pi r$

Part B

11 D
12 D
13 B
14 **a** 2% **b** 65% **c** 12.5%
15 64.5%, 62%, 0.608, 0.06
16 **a** \$168 **b** 21 km
17 4.5%
18 $\frac{1}{250}$
19 **a** 254.6 kg **b** \$300
20 \$450
21 **a** \$60 **b** 37.5%

CHAPTER 13

Exercise 13-01

1 A
2 C
3 D
4 **a** a column graph
b James, Sarah, Tim and George
c Yes
5 **a** Sector angles: Sleep 120°, Work 90°, Chores 30°, Exercise 15°, Meals 30°, Relaxation 75°
b

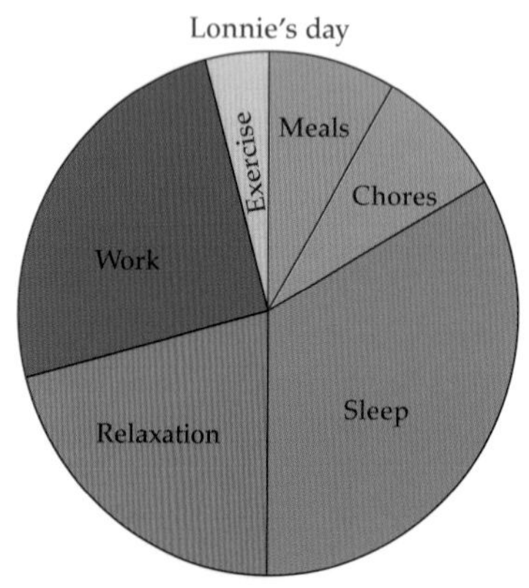

c $\frac{1}{3}$ **d** 20.8%
d Segment lengths: Sleep 4 cm, Work 3 cm, Chores 1 cm, Exercise 0.5 cm, Meals 1 cm, Relaxation 2.5 cm

Lonnie's day

Sleep | Work | Chores | Exercise | Meals | Relaxation

6 **a** 8:30 a.m. **b** 10 a.m.
c 25 km **d** 1 h
e 2 h **f** 12.5 km/h
7 **a** 15
b 25, at 3 p.m.
c Same number of messages sent each hour
d 100

Exercise 13-02

1 C
2 B
3 adding, dividing, number, often, common (or popular), no
4 **a** 6.5 **b** 13.9 **c** 28.3
d 56.1 **e** 85.4
5 **a** 7 **b** 14 **c** 28
d 54, 55 and 59 **e** None
b None
6 mean = 4.14, mode = 4
7 **a** 18.875 **b** 16 and 18

Exercise 13-03

1 C
2 B
3 highest, lowest, low, high, middle, scores, even, average, middle
4 **a** 7, 10 **b** 30, 13 **c** 41, 11
d 7, 8 **e** 60.5, 14 **f** 71, 11
5 **a** 712 **b** 407.5 **c** 390
d 441.85 **e** mode **f** median
6 mode = 71, range = 8, median = 73.5, mean = 73.75

Exercise 13-04

1 D
2 C
3 **a**

<table>
<tr><th>Score</th><th>Tally</th><th>Frequency</th></tr>
<tr><td>2</td><td>|</td><td>1</td></tr>
<tr><td>3</td><td>|||</td><td>3</td></tr>
<tr><td>4</td><td>|||</td><td>3</td></tr>
<tr><td>5</td><td>||</td><td>2</td></tr>
<tr><td>6</td><td>||||</td><td>4</td></tr>
<tr><td>7</td><td>𝍸 𝍸 ||</td><td>12</td></tr>
<tr><td>8</td><td>|||</td><td>3</td></tr>
<tr><td>9</td><td>||</td><td>2</td></tr>
</table>

b range = 7, mode = 7

ISBN 9780170350990

4 a

x	f	cf	fx
2	3	3	6
3	4	7	12
4	8	15	32
5	5	20	25
6	2	22	12

range = 4, mode = 4, median = 4, mean = 3.95

b

x	f	cf	fx
20	5	5	100
21	7	12	147
22	8	20	176
23	6	26	138
24	5	31	120

range = 4, mode = 22, median = 22, mean = 21.97

5 a

Score	Frequency	fx
0	4	0
1	8	8
2	18	36
3	12	36
4	6	24
5	2	10
Total	50	114

b 2 c 2.28

6

Score	Frequency	cf	fx
0	6	6	0
1	3	9	3
2	8	17	16
3	5	22	15
4	5	27	20
5	3	30	15
Total	30		69

a 5 b 2 c 2 d 2.3

Exercise 13-05

1 B
2 A
3 a T b F c T
d F e F

4 a

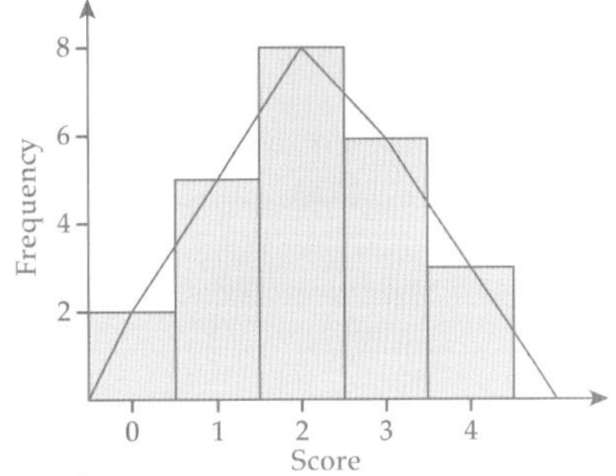

b

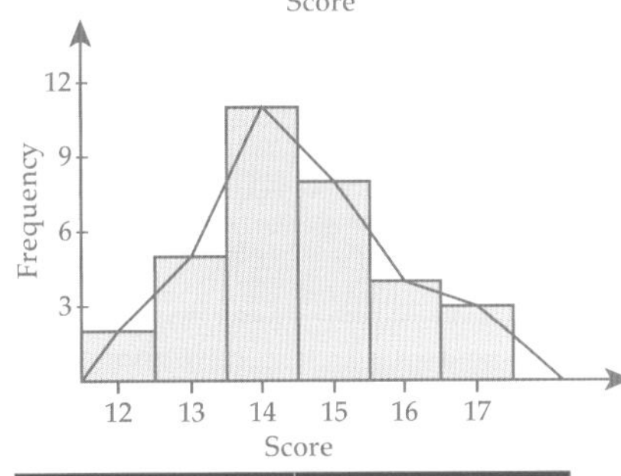

5 a

Score	Frequency
1	10
2	17
3	13
4	5
5	3
6	2

b

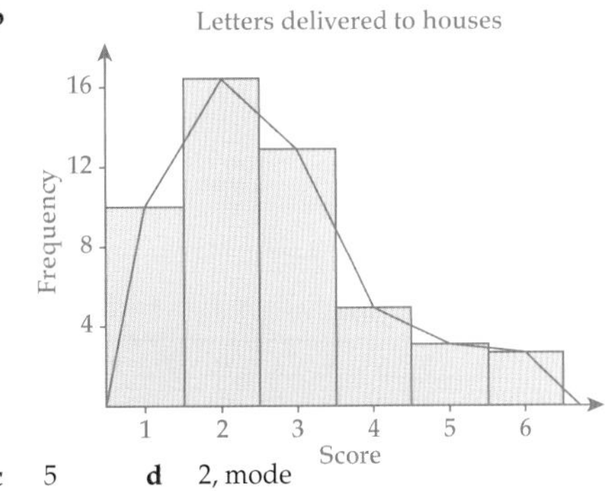

c 5 d 2, mode

6 a

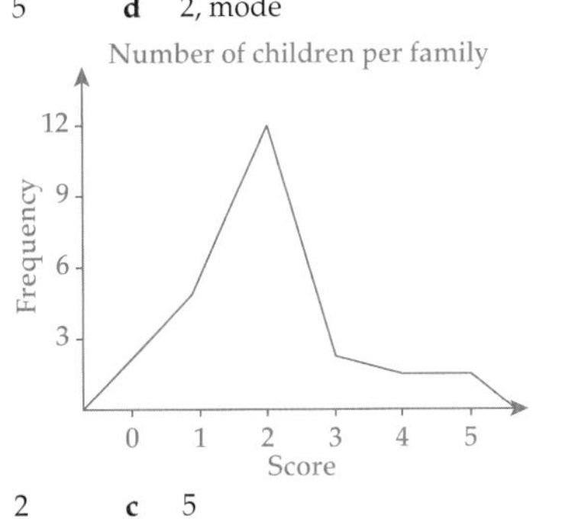

b 2 c 5

ISBN 9780170350990

Exercise 13-06

1 D

2 D

3 a $15 b $10 c 30

d $10 e $9.63

4 A

5 a

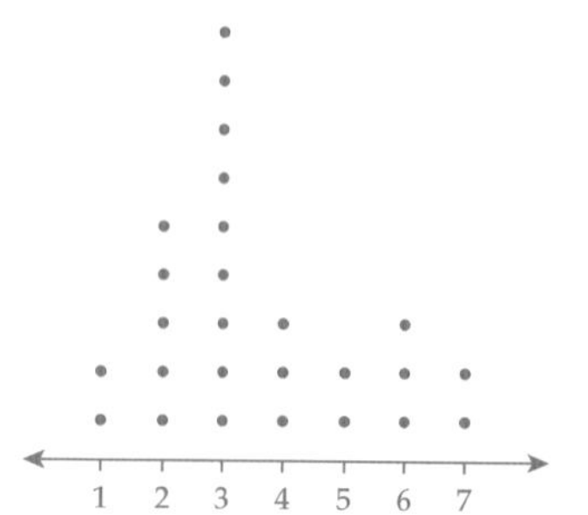

b 3 hours c 6 hours

d 3 hours e 3.6 hours

6 a $350 b $800 c $885.71

d $875 e $1100 f From $800 to $950

7 a

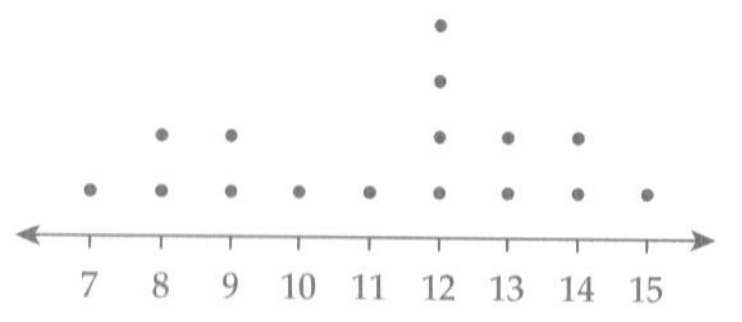

b Range = 8, Mode = 12 c 12

d 11.2 e 12

Exercise 13-07

1 C

2 A

3 a 30 b 14

c 14.5 d 4.9

e 29 f Cluster in the 10s, outlier 34

4 a

Stem	Leaf
3	5 6 6 6 7 8 9
4	2 4 5 7
5	2 4 5 7 8 8
6	2 3 3 3 3 3 5

b 53 c 50.46 d 30 e 63

5 a 39 b 124

c 124 d mean = 122.95

e There are no clusters or outliers.

6 a **Oburgo**

Stem	Leaf
28	4 4
29	0 5
30	2 5 7 8
31	6 7 7

b **Hungry Jill's**

Stem	Leaf
29	8
30	6 6 8
31	2 6 7 8
32	5 5 8

b Hungry Jill's, 328

c Oburgo, 284

d Hungry Jill's

e Oburgo, 305 and Hungry Jill's, 316, so Hungry Jill's median is higher

Language activity

1 PLOT
2 GRAPH
3 SECTOR
4 COLUMN
5 RANGE
6 MEAN
7 MODE
8 LEAF
9 STEM
10 DOT
11 MEDIAN
12 FREQUENCY
13 TABLE
14 AVERAGE
15 HISTOGRAM
16 POLYGON

Practice test 13

Part A

1 An angle between 180° and 360°

2 64°

3 1, 2, 3, 6, 9, 18

4 composite

5 7

6 1258 is not divisible by 6 evenly

7 $\frac{3}{8}$

8 12.8 m²

9 $-12ab^2c$

10 280 000

Part B

11 C

12 a

Elle's weekly spending

Savings
Rent
Clothes
Food

b 108°

ISBN 9780170350990

13 24.3

14 D

15 a 8 b 9

16 a 3 b 7 c 6.67

17 Number of toothpicks per packet

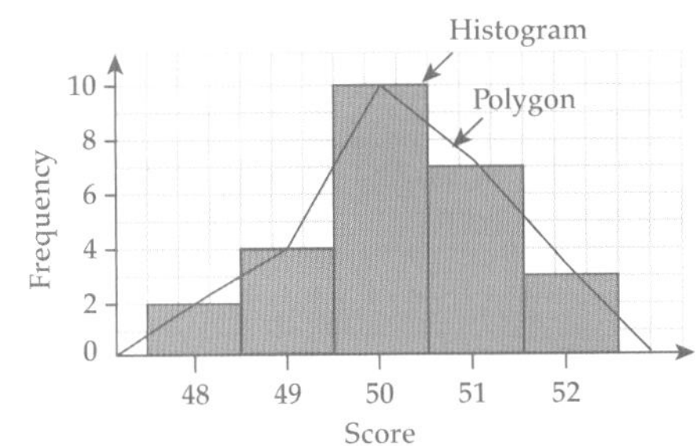

18 a 6 b 5 c 5.5 d 9

19 a 21 b 134
c 133 d in the 130s (13 stem)

CHAPTER 14

Exercise 14-01

1 B

2 C

3 a likely b even chance
c impossible d even chance
e impossible f unlikely
g unlikely h even chance
i certain j likely

4 0 — c e — f g — b d h — a j — i — 1

5 a T b T c F
d T e F f T

6 Teacher to check

Exercise 14-02

1 C

2 D

3 a 1, 2, 3, 4, 5, 6
b A, B, C, D, E, F, G, H, I, J, K, L, M, N, O, P, Q, R, S, T, U, V, W, X, Y, Z
c 5, 6, 7, 8, 9, 10, 11, 12, 13, 14, 15
d HH, HT, TH, TT
e 5c, 10c, 20c, 50c, $1, $2
f Monday, Tuesday, Wednesday, Thursday, Friday, Saturday, Sunday
g H1, H2, H3, H4, H5, H6, T1, T2, T3, T4, T5, T6
h 11, 13, 15, 17, 19, 21, 23
i $5, $10, $20, $50, $100
j 1, 2, 3, 4, 6, 12

4 a 6 b 26 c 11 d 4
e 6 f 5 g 12 h 7
i 5 j 6

5 a red, blue and green $n(S) = 3$
b no

6 a red, blue, yellow and green
b yes c no d yes

7 a purple, orange and green b no
c yes d unlikely
e no f likely

8 a b

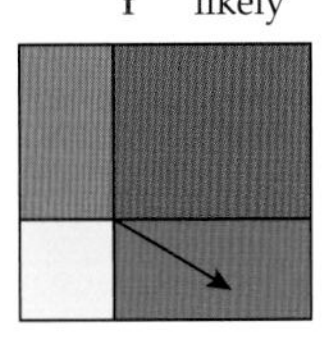

Exercise 14-03

1 B

2 A

3 a $\frac{1}{2}$ b $\frac{1}{6}$ c $\frac{8}{11}$ d $\frac{2}{5}$
e $\frac{7}{15}$ f 1 g 0
h $\frac{1}{26}$ i 1 j $\frac{3}{5}$

4 a No
b i $\frac{3}{14}$ ii $\frac{5}{14}$ iii $\frac{3}{7}$ iv 0
c $\frac{5}{27}$

5 a $\frac{1}{13}$ b $\frac{1}{4}$ c $\frac{1}{13}$
d $\frac{4}{13}$ e $\frac{1}{52}$ f $\frac{1}{4}$
g $\frac{1}{52}$ h $\frac{2}{13}$ i $\frac{2}{13}$

Exercise 14-04

1 C

2 C

3 a tossing a tail b Not rolling a 5
c Not selecting an 'e' d Not choosing a 5
e Selecting a blue marble
f Selecting 't' or a letter before 't'
g Tossing an even number

4 a $\frac{4}{9}$ b $\frac{5}{9}$ c $\frac{1}{3}$ d $\frac{2}{3}$
e 0 f 1 g $\frac{2}{3}$ h $\frac{1}{3}$

5 75%

6 a 0.3 b 0.7 c 0.9 d 0.6

7 a $\frac{2}{3}$ b train

Exercise 14-05

1 D

2 A

3 a 10 b 10 c–e Teacher to check

4 a 10 b–d i–iv Teacher to check

5 a $\frac{7}{10}, \frac{3}{10}$ b $\frac{113}{200}, \frac{87}{200}$ c The whole table
d The more trials, the closer the result is to the theoretical probability.

6 a 10, No b $\frac{1}{6}$ c–d Teacher to check

ISBN 9780170350990

Exercise 14–06

1 C

2 D

3 a $\frac{15}{19}$ b $\frac{6}{19}$ c $\frac{4}{19}$

4 a $\frac{1}{5}$ b $\frac{11}{35}$ c $\frac{16}{35}$ d $\frac{24}{35}$

5

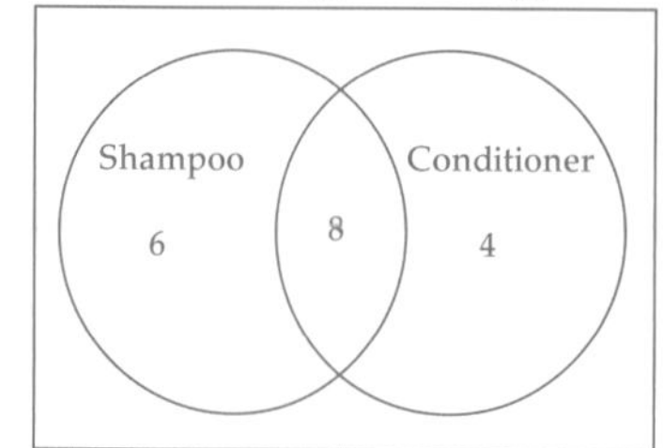

a $\frac{1}{3}$ b $\frac{2}{9}$

6

Italy 20, both 25, France 27, neither 8

a $\frac{5}{16}$ b $\frac{27}{80}$

7 a i $\frac{13}{20}$ ii $\frac{4}{15}$ iii 0 iv $\frac{1}{12}$

b Yes

8

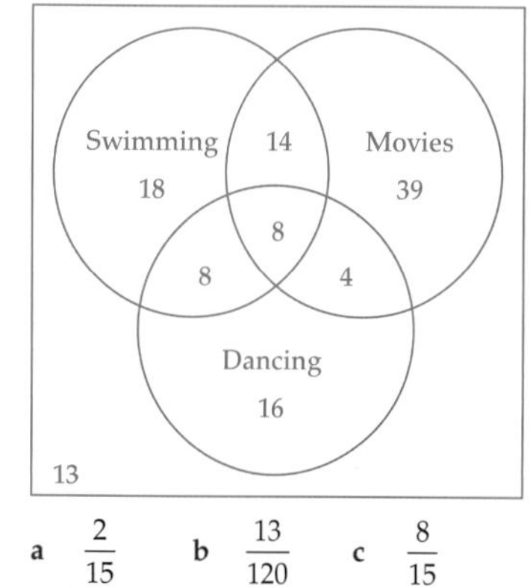

a $\frac{2}{15}$ b $\frac{13}{120}$ c $\frac{8}{15}$

Exercise 14–07

1

	Works full time	Works part time	Total
Male	75	15	90
Female	48	26	74
Total	123	41	164

a 48 b 15

c i $\frac{45}{82}$ ii $\frac{1}{4}$ iii $\frac{13}{82}$ iv $\frac{29}{41}$

2

	Married	Single	Total
Northside	65	27	92
Southside	75	19	94
Total	140	46	186

a $\frac{25}{62}$ b $\frac{9}{62}$ c $\frac{23}{93}$ d $\frac{17}{31}$

3

	Liked	Disliked	Total
Adult	78	34	112
Child	65	47	112
Total	143	81	224

a i 65 ii 34 b i $\frac{47}{224}$ ii $\frac{81}{224}$

4

	Hip-hop	Not hip-hop	Total
Rock	7	9	16
Not rock	8	11	19
Total	15	20	35

a $\frac{8}{35}$ b $\frac{24}{35}$ c $\frac{1}{5}$ d $\frac{11}{35}$

Exercise 14–08

1 a $\frac{1}{12}$ b $\frac{8}{9}$ c $\frac{1}{12}$ d $\frac{5}{18}$

2 a i $\frac{2}{9}$ ii $\frac{2}{3}$ iii $\frac{2}{3}$ b $\frac{4}{13}$

3 a

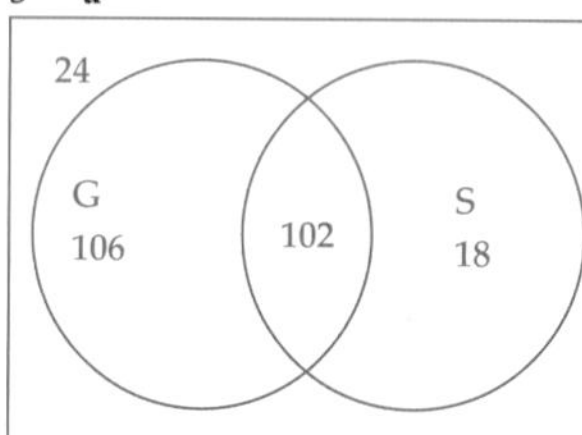

G: Houses with a garage
S: Houses with a swimming pool

b i $\frac{51}{125}$ ii $\frac{53}{125}$

4 a

	Beach	Pool	Total
Male	46	18	64
Female	28	42	70
Total	74	60	134

b i $\frac{9}{67}$ ii $\frac{14}{67}$ c 71.9%

5 a Because there are six equal sections, with number taking up two of the six sections.

b $\frac{1}{6}$ c 1, 2, 3 d $\frac{1}{3}$

6 a i $\frac{16}{48}=\frac{1}{3}$ ii $\frac{24}{48}=\frac{1}{2}$

iii $\frac{32}{48}=\frac{2}{3}$ iv $\frac{40}{48}=\frac{5}{6}$

b $\frac{24}{46}=\frac{12}{23}$

ISBN 9780170350990

Language activity

Across

3 CERTAIN
4 VENN DIAGRAM
7 TWO WAY
8 UNLIKELY
9 EVEN

Down

1 SAMPLE SPACE
2 PROBABILITY
5 LIKELY
6 IMPOSSIBLE

Practice test **14**

Part A

1 \$210
2 $2x - 10y$
3 7.5
4

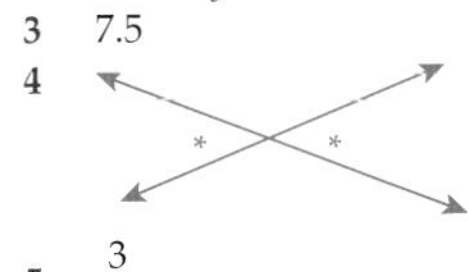

5 $\frac{3}{10}$
6 5
7 11:35 p.m.
8 $8ab(7b - c)$
9 2
10 3

Part B

11 D
12 A
13 A
14 $\frac{1}{2}$
15 $\frac{1}{3}$
16 $\frac{4}{7}$
17 $\frac{2}{5}$
18 a $\frac{3}{10}$ b $\frac{23}{30}$
19

	Soft Centred	Hard Centred	Total
Milk Chocolate	12	18	30
Dark Chocolate	16	22	38
Total	28	40	68

a 68 b $\frac{4}{17}$

20 a $\frac{3}{10}$ b $\frac{7}{10}$ c $\frac{1}{2}$

CHAPTER 15

Exercise **15-01**

1 B
2 D
3 a $w, 3, 6w$ b $5, 4, 20$
c $3, 3, 27$ d $8, 8, 6, 16a$
e $-4, w, 6, -4w$ f $r, -7, 42$
4 a $5a + 30$ b $4w - 24$ c $6a + 14$
d $-4c - 20$ e $-6m + 18$ f $-15a + 3$
g $21n + 14$ h $12a - 16$ i $-7m - 42$
j $12a - 16$ k $-15n - 30$ l $-16w - 48$
5 a $a^2 + 4a$ b $2v^2 - 3v$ c $6w^2 + 21w$
d $2m^2 - 8m$ e $-3a^2 - 5a$ f $-8b^2 + 6b$
g $b^2 + 5b$ h $n^2 - 4n$ i $2m^2 - 12m$
j $-2a^2 - 3a$ k $3v^2 - 21v$ l $-4r^2 - 20r$
m $12w^2 - 24w$ n $-18n^2 - 63n$ o $12m^2 - 24m$
6 a T b T c F d F
7 a Yes, both sides of the equation equal 35 when $x = 3$
b Yes

Exercise **15-02**

1 D
2 B
3 $4, xy, 4, xy, 4xy$
4 a $4b$ b $6a$ c m
d 5 e $2m$ f 4
g $2b$ h $4n$ i $2a$
5 a $a, 2$ b $m, 3$
c $w + 4$ d $n - 2$
e $3 - 5s$ f $3u + 4w$
g $a - 3$ h $3c + 4b$
i $4b$ j $8a$
6 a $6(2a - b)$ b $3(m + 3n)$ c $4b(2a + 3c)$
d $2a(a - 3)$ e $8b(3a + 2c)$ f $8w(1 - 3v)$
g $4n(4m + 5p)$ h $3m(m - 5n)$ i $6c(3b + 4d^2)$
j $4w(3w - 4v)$ k $2bc(9c - 4)$ l $7uv(4 - v)$
7 a $3cd$ is not the HCF b $6cd(5d - 3c)$
8 a $a - 2c + 3ac$ b $1 - 3a + 4v$
c $3a, 4b$ d $2m - n + 3mn$

Exercise **15-03**

1 C
2 A
3 a $m - 2$ b $w + 6$
c $3 + 4m$ d $3u - 2w$
e $a - 3$ f $2n + 3m$
g $-4b$ h $-2m$
4 a $-4(2a + 3)$ b $-6(m - 3)$
c $-5(n + 6)$ d $-4(3w - 2)$
e $-5(3r + 4)$ f $-6(4m - 3)$
g $-7n(1 + 3m)$ h $-9y(1 - 5a)$
i $-4b(3a + 2c)$ j $-8s(3r - 2t)$
k $-9g(2f + 3h)$ l $-10v(2u - 3w)$

ISBN 9780170350990

5 a $-4a$ b $-4a$

6 a $-n(n + 3)$ b $-5b(b + 4)$ c $-8c(c - 4)$
d $-7m(m - 8n)$ e $-8r(r + 3s)$ f $-4w(w - 3a)$
g $-3n(5m + n)$ h $-9v(2u - v)$ i $-5d(5bd + 4e)$

7 a $-3n$ b $-3n$

8 a $-4b(a - 3c + 2ac)$ b $-3s(2r + 3t - 6rt)$
c $-3v(4u - 5w + 6uw)$ d $-3b(2ab + 3ac - 8c^2)$
e $-5h(4g - 3hf + 5gf)$ f $-9st(3r - 2s + 4rt)$

Exercise 15-04

1 D
2 B
3 inverse, same, balanced, underneath
4 a subtraction b division
c multiplication d addition
5 a $x = 21$ b $m = 4$ c $x = 6$
d $n = 2$ e $w = 24$ f $m = 32$
g $a = -3$ h $b = -15$ i $x = -30$
6 a subtracting 6 b dividing by 8
c adding 7 d multiplying by 4
7 a 7, 1 b 11, 13 c 9, 4 d 4, 4, 16
8 a $n = 11$ b $b = 7$ c $m = 12$ d $x = 48$
e $a = -7$ f $n = -36$ g $v = 4$ h $m = -6$
i $y = -18$ j $r = -7$ k $m = -56$ l $b = -5$
9 a 42 b 8
10 a $m = -29$ b $n = -4$ c $b = 1\frac{3}{4}$
d $m = -5\frac{2}{3}$ e $v = -6$ f $a = -17$
g $t = 3\frac{4}{5}$ h $m = -1\frac{1}{2}$

Exercise 15-05

1 A
2 B
3 a 6, 6, 20, 10 b 4, 4, 15, 3 c 3, 3, 6, 24
4 a -4 b $+3$ c -7 d $+3$
e -5 f $+4$ g -6 h $+6$
i -8 j -20 k -12 l -19
5 a $a = 2$ b $g = 9$ c $m = 2$ d $v = 7$
e $x = 3$ f $c = -2$ g $a = 4$ h $e = -2$
i $v = 9$ j $y = 3$ k $m = -5$ l $n = 7$
6 a $-3, x = 4$ b $+5, m = 24$ c $-5, n = 12$
d $+4, s = 72$ e $-7, w = 48$ f $-5, x = -60$
g $+6, x = 21$ h $+9, s = 63$ i $-6, m = -60$
7 a $2n - 8 = 28$, 18 b $3n + 7 = 52$, 15
8 a true b false c true d false

Exercise 15-06

1 C
2 A
3 a $3x - 5$ b $4a + 2$ c $8 - 3x$
4 a $2x + 8$ b $3a - 4$ c $x + 4$
5 a m, 4, 12 b n, 2, 15, 5
6 a $w = 14$ b $m = -9$ c $n = 3$ d $a = 4$
e $m = 7$ f $b = -7$ g $m = -7$ h $v = 5$
i $w = 13$ j $a = 11$ k $b = -6$ l $n = 11$
7 a $6m$, 6, −12, 6 b $6v$, 6, −18, 6
8 a $w = 8$ b $m = -15$
c $n = 18$ d $m = -14$

Exercise 15-07

1 C
2 D
3 a F b F c T d T
4 a 10, 10, 10, 10, 8, 4 b $8m$, 12, 12, 12, 24, 3
5 a $a = 11$ b $n = 4$ c $b = 1$
d $a = -7$ e $m = -9$ f $v = 2$
g $v = 11$ h $a = 3$ i $a = 2$
j $x = 4$ k $m = -7\frac{1}{2}$ l $c = -7$
6 a Correct b Incorrect
c Incorrect d Correct

Language activity

1 ALGEBRA
2 SOLVE
3 PATTERN
4 EQUATION
5 VARIABLE
6 SOLUTION
7 PRONUMERAL
8 VALUES
9 EXPRESSION
10 EXPAND
11 FACTORISE
12 BRACKETS

Practice test 15

Part A

1 0.054
2 120
3 $12a$
4 $64a^3$
5 11
6 31.2
7 \$11 520
8 $b = 15$
9 -48
10 $\frac{2}{9}$

Part B

11 C
12 B
13 C
14 $8b(a - 3c^2)$
15 a $-5y(x + 4)$ b $-6ab(b + 3a)$
16 a $w = 7$ b $d = 12$ c $a = 7$
17 a $m = 9$ b $x = 6$
18 a $a = 15$ b $x = 6$
19 a $m = 3$ b $x = -2$

ISBN 9780170350990

CHAPTER 16

Exercise 16-01

1 C
2 B
3 a 45 b 9 c 20 d 22
e 12 f 4 g 6 h 5
i 8 j 4 k 9, 21 l 11, 12
4 a 7:5 b 1:2 c 10:7 d 5:8
e 8:7 f 7:10 g 1:6 h 3:1
5 a 7:8 b 5:7 c 3:5 d 2:3
e 1:3 f 3:4 g 2:3 h 3:5
i 2:3 j 3:4 k 2:5 l 1:3
m 1:3 n 7:9 o 4:3 p 4:3
q 4:5:6 r 8:12:15 s 3:4:9
6 D
7 a 1:2 b 2:5 c 8:1 d 1:20
e 3:2 f 1:3 g 3:10 h 90:1
i 5:1 j 6:5 k 63:31 l 7:10
8 a 1:4 b 5:24 c 1:6 d 5:4
e 1:200 f 3:2 g 3:10 h 7:8
i 49 : 60 j 3 : 1 k 1 : 4 l 1 : 12

Exercise 16-02

1 A
2 a 87 m b 203 m
3 1152
4 18 cm, 24 cm, perimeter 72 cm
5 a 18 b 20
6 84 km/h
7 35
8 $210
9 4 kg
10 200 m
11 a 14 b 5
12 360 kg
13 $80
14 150 mL pink, 450 mL purple
15 75

Exercise 16-03

1 C
2 D
3 a 1.64 m b 1 m c 2.96 m
4 a 2.55 m b 28.5 m c 1.35 m d 22.5 cm
5 a 3 cm b 4.1 cm
6 a 2.6 cm by 2.6 cm b 312 cm by 312 cm
c 2.1 cm by 2.6 cm d 252 cm by 312 cm
e $987.91
7 a 1.5 km b 1.9 km c 1.1 km
d 2.0 km e 3.0 km f 2.2 km

Exercise 16-04

1 B
2

Ratio	Total parts	Total amount	One part	New ratio
3 : 5	3 + 5 = 8	$640	$640 ÷ 8 = $80	$240 : $400
4 : 3	4 + 3 = 7	$5600	$800	$3200 : $2400
2 : 7	2 + 7 = 9	$720	$80	$160 : $560
5 : 2	5 + 2 = 7	$7700	$1100	$5500 : $2200

3 3, 7, 7, 100, 100, 100, 400, 300
4 $2560
5 a 27 wins b No
6 Lee gets $20 000 and Nathan gets $15 000
7 a 4 : 3 : 1 b $675 c $135
8 Ante 16, Josh 12
9 $52
10 a 3 : 5 b Sophie $56 250, Claire $93 750

Exercise 16-05

1 A
2 D
3 a km/h b $/kg
c words/min d c/L
e m/s f $/kg
g runs/wicket h $/h
4 a 16 b 2.25 c 80 d 22
e 120 f 140 g 1.21 h 54
5 a 6 goals/match b $90/day
c $7/kg d 28 runs/wicket
e 96 m/s f 35 students/teacher
g $900/ha h 1500 rev/min
6 $1.14/L
7 3.14 persons/km^2
8 $18.90/h
9 $57/h
10 97.8 km/h
11 520 words in 5 min

Exercise 16-06

1 a 1380 b 70 minutes
2 a $\frac{\$}{\text{h}}$ b $624 c 15 h
3 a 162 min b 27 min/car
c 4 h 30 min d 13 cars
4 $1724.80
5 84 goals
6 40 min
7 $842.40
8 a 5 m/s b 15 m c 35 s
9 a 1840 words b 45 min
10 a $\frac{\text{c}}{\$}$ b $19 968 c $54 500
11 8.5 km/L
12 a $5.73 b 1.97 kg

Exercise 16-07

1 C
2 D
3 a 20 km/h b 50 km/h c 40 km/h
d 110 km/h e 3 km/h f 210 km/h
4 a A b F c D
d B e C f E
5 a 6 h b 425 km c 60 m/s
d 5 s e 2016 m
6 a 225 km b 90 km/h
c 216 km d 2 h 42 min
7 a 5 min 33 s b 3.6 km
8 a 165 km b 165 km/h
9 a 16 km b 2 h c 8 km/h

Language activity

1 BRAIN WAVE 5 : 4
2 OVER ALL 4 : 3
3 HIGH CHAIR 4 : 5
4 CHAIR MAN 5 : 3
5 TOMB STONE 4 : 5
6 SWIM WEAR 1 : 1
7 PREMIER SHIP 7 : 4
8 FRY PAN 1 : 1
9 DASH BOARD 4 : 5
10 COB WEB 1 : 1
11 SCHOLAR SHIP 7 : 4
12 GRAVE YARD 5 : 4
13 MILE STONE 4 : 5
14 BROAD BAND 5 : 4
15 CLOCK WORK 5 : 4
16 CYCLE WAY 5 : 3
17 PINE APPLE 4 : 5
18 COMMON SENSE 6 : 5
19 MOCKING BIRD 7 : 4
20 MASTER PIECE 6 : 5

Practice test 16

Part A

1 37
2 28
3 11
4 60 m
5 $1\frac{5}{12}$
6 25%
7 10:18 p.m.
8 $13x - 5y$
9 $1126.55
10 $\frac{1}{2}$

Part B

11 B
12 C
13 35
14 45
15 75 cm
16 1.5 cm
17 Tanya $3084, Mikayla $4112
18 a 7 goals/match b $125/day
19 a 24 min/car b 60 cars
20 80 km/h

CHAPTER 17

Exercise 17-01

1 B
2

Point	Move left or right?	Move up or down?
(–2, 6)	Left	Up
(–4, –5)	Left	Down
(3, –4)	Right	Down
(1, 6)	Right	Up

3 anticlockwise
4 a $A(-6, 8)$, $B(-4, 8)$, $C(0, 6)$, $D(3, 6)$, $E(-7, 0)$, $F(-5, -4)$, $G(7, -4)$, $H(1, -8)$
b i 2nd ii 3rd iii 1st iv 4th
5

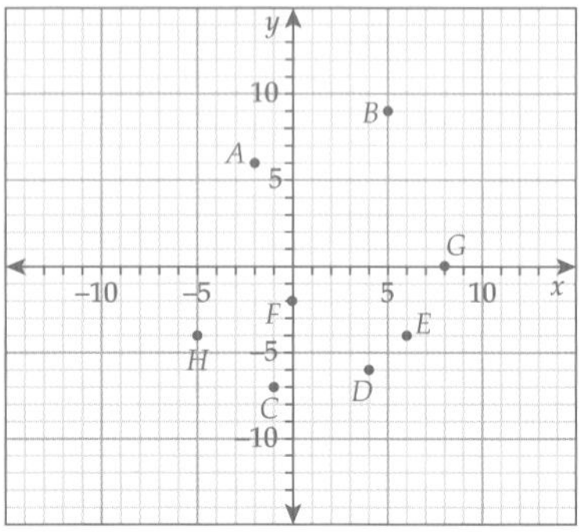

6 D and E
7 2nd quadrant
8 A truck

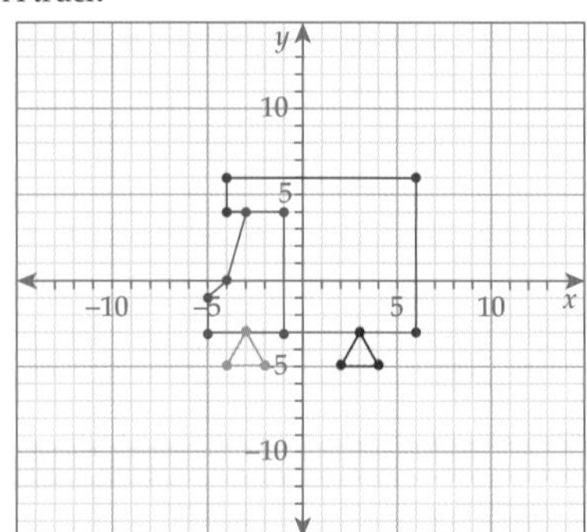

 ISBN 9780170350990

Exercise 17-02

1 C

2 B

3 a −6 b −4 c −2 d 0
e −9 f −10 g −12 h −8

4 a 4 b 0 c 6 d −8
e 14 f 18 g 24 h 16

5 a $y = 2x$

x	0	1	2	3
y	0	2	4	6

b $y = x - 1$

x	4	3	2	1
y	3	2	1	0

c $y = x \div 2$

x	10	8	6	4
y	5	4	3	2

d $y = x + 3$

x	0	1	2	3	4
y	3	4	5	6	7

e $p = 3n$

n	1	2	3	4
p	3	6	9	12

f $b = 5 - a$

a	−3	−2	−1	0	1	2
b	8	7	6	5	4	3

g $e = 2 - d$

d	0	1	2	3	4
e	2	1	0	−1	−2

h $z = \frac{y}{3}$

y	12	9	6	3	0	−3	−6
z	4	3	2	1	0	−1	−2

i $v = 2u + 5$

u	1	2	3	4
v	7	9	11	13

j $h = 10 - 3f$

f	−3	−2	−1	0	1	2
h	19	16	13	10	7	4

Exercise 17-03

1 C

2 A

3 a

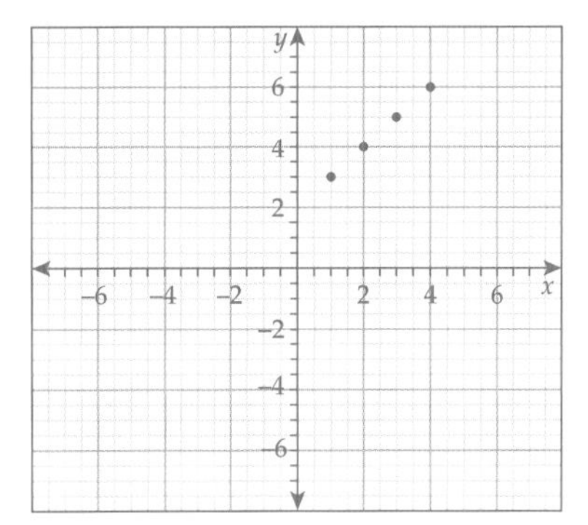

b

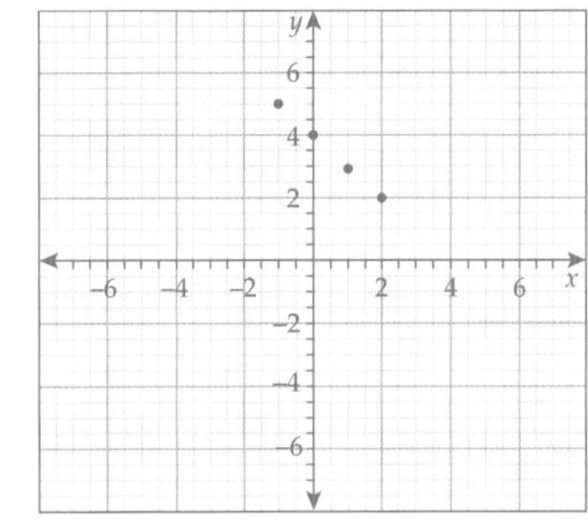

c

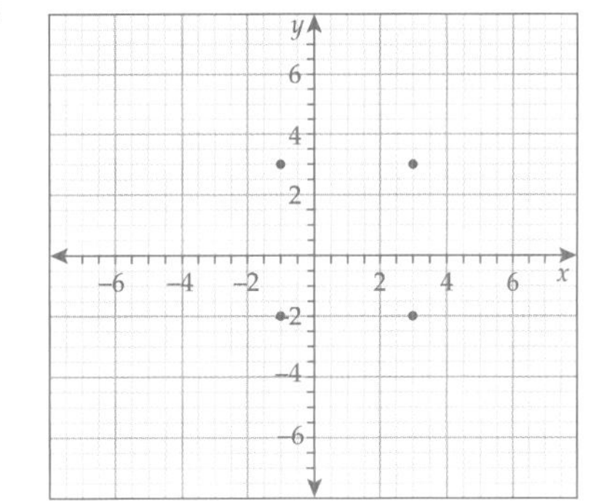

d

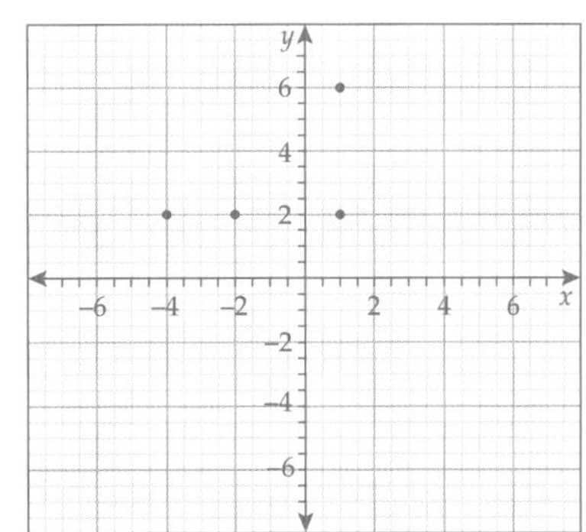

4 a a line b a line
c a rectangle d a triangle

ISBN 9780170350990

5 **a**

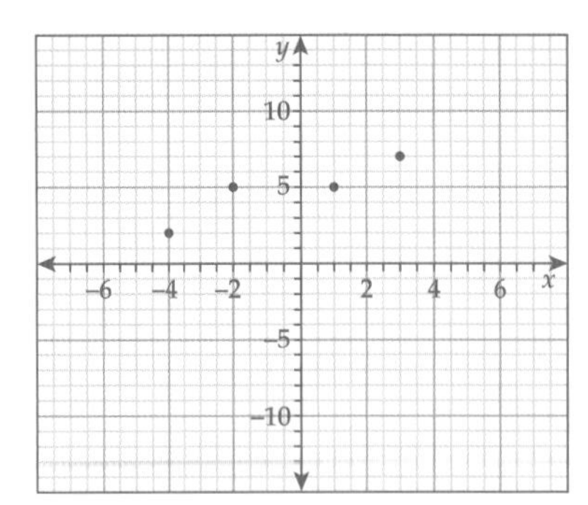

b

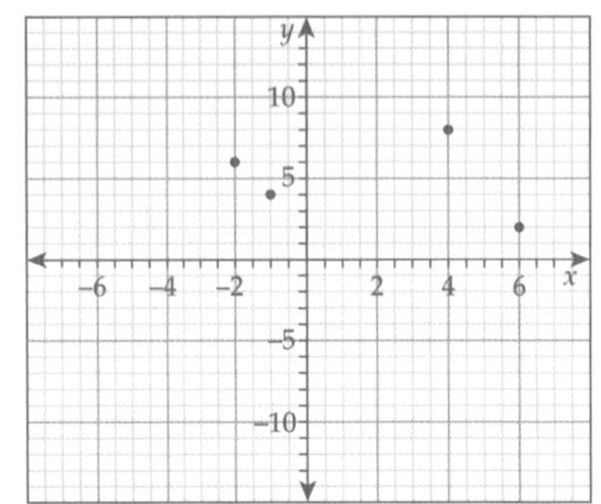

c straight line

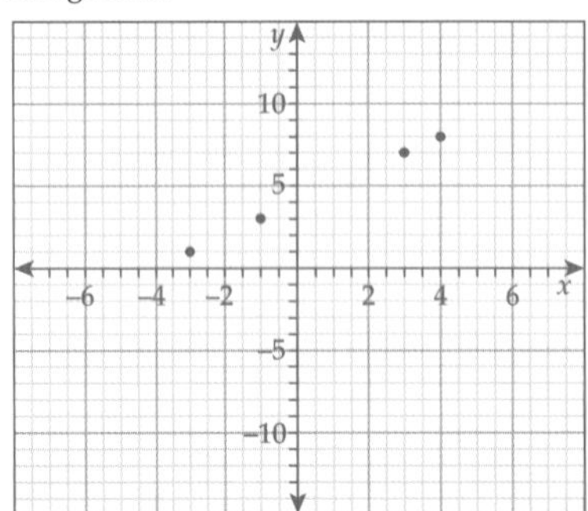

d

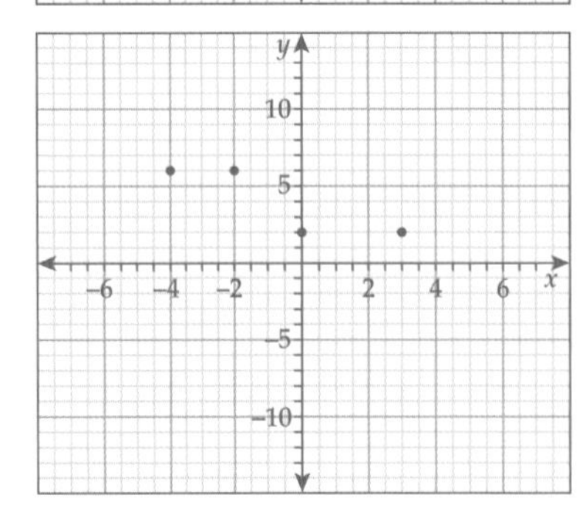

6 **a** $y = x + 2$

x	0	1	2
y	2	3	4

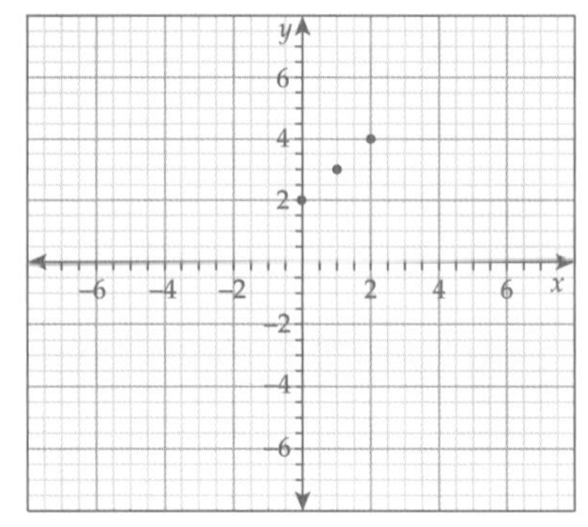

b $y = x - 1$

x	0	1	2
y	−1	0	1

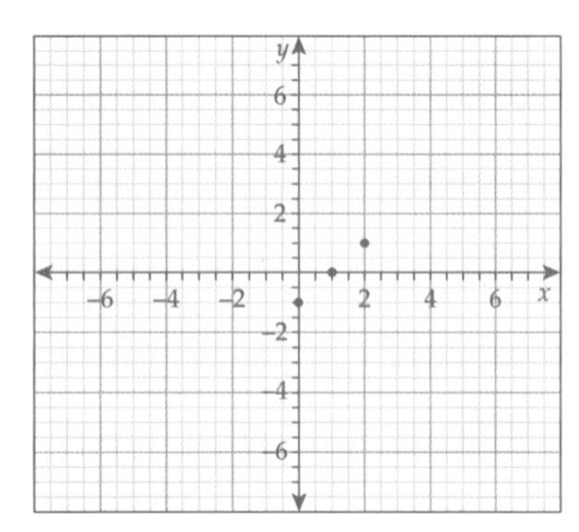

c $y = 2x + 3$

x	0	1	2
y	3	5	7

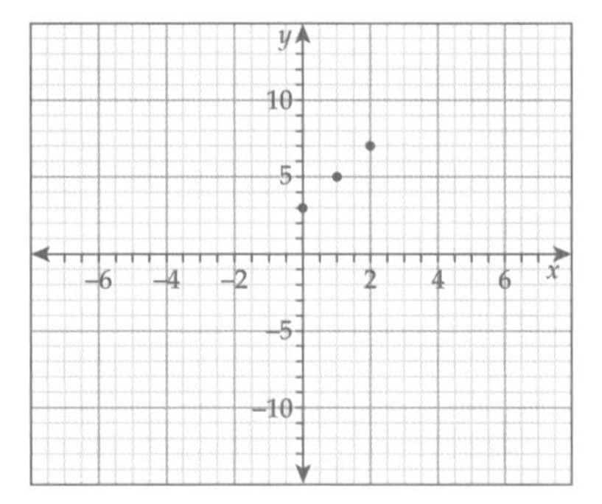

 ISBN 9780170350990

d $y = 3x - 1$

x	0	1	2
y	–1	2	5

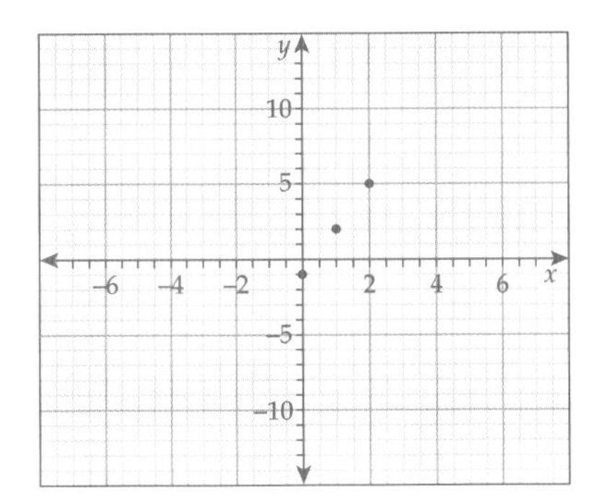

Exercise **17–04**

1 C
2 values, table, points, line
3 **a** –2 **b** 4
4 **a** $y = x + 5$

x	0	1	2
y	5	6	7

y-intercept = 5

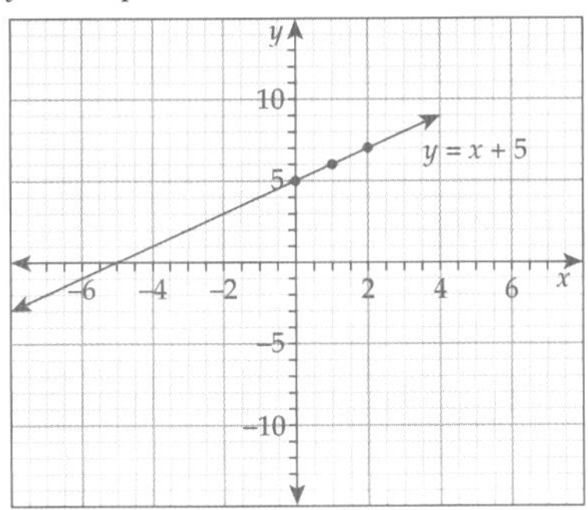

b $y = x - 3$

x	0	1	2
y	–3	–2	–1

y-intercept = –3

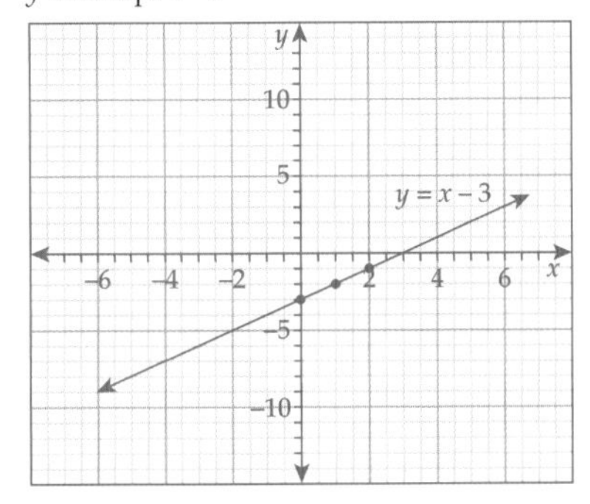

c $y = 2x + 1$

x	0	1	2
y	1	3	5

y-intercept = 1

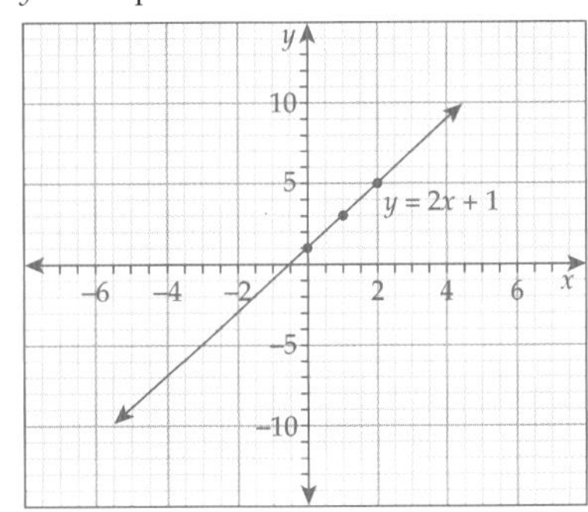

d $y = 4x - 2$

x	0	1	2
y	–2	2	6

y-intercept = –2

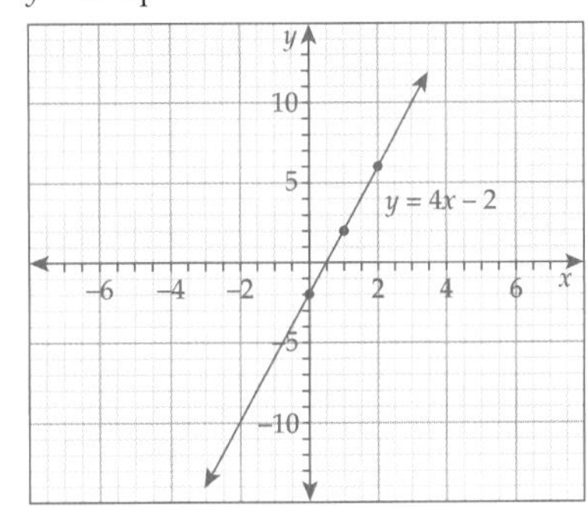

5 **a** $y = 2x + 4$

x	0	1	2
y	4	6	8

y-intercept = 4

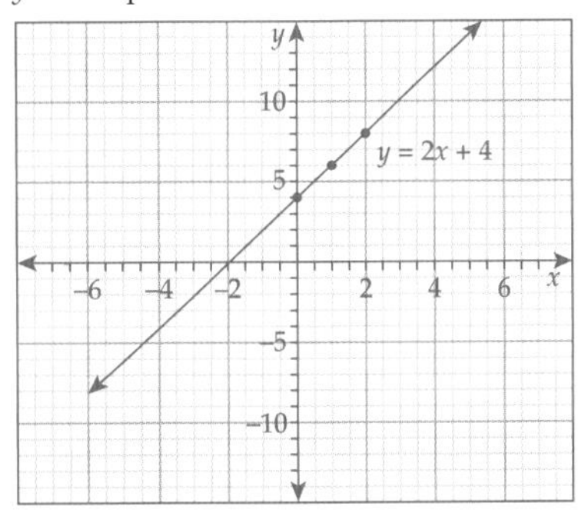

ISBN 9780170350990

b $y = 4x - 2$

x	0	1	2
y	–2	2	6

y-intercept = –2

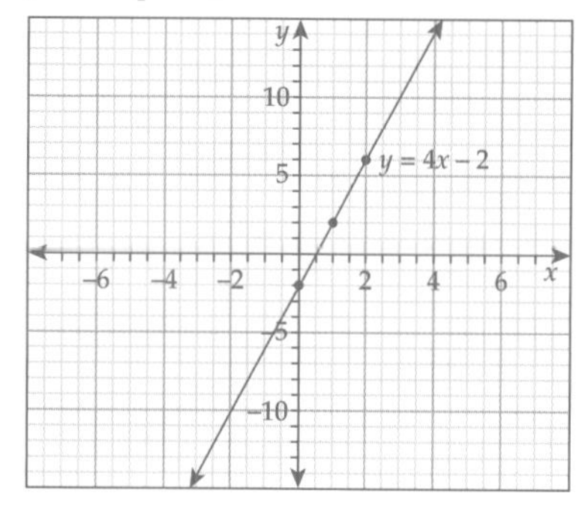

c $y = 6 - x$

x	0	1	2
y	6	5	4

y-intercept = 6

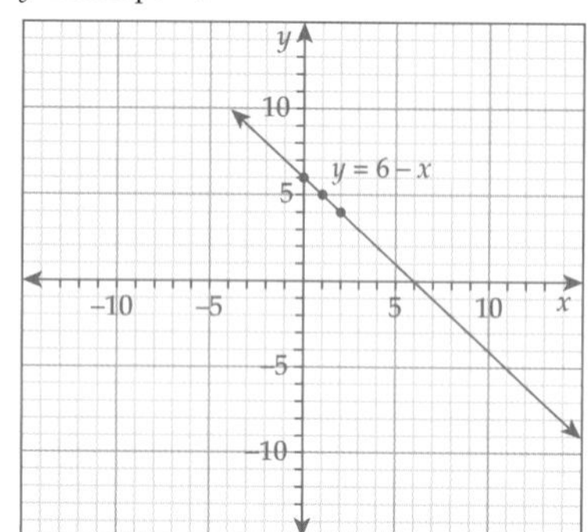

d $y = 5 - 2x$

x	0	1	2
y	5	3	1

y-intercept = 5

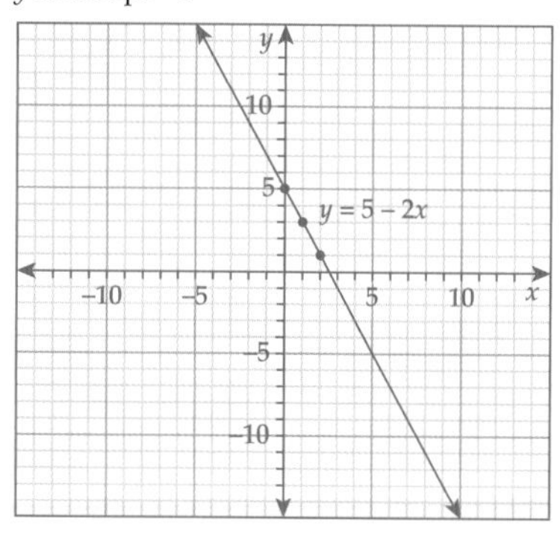

e $x + y = 5$

x	0	1	2
y	5	4	3

y-intercept = 5

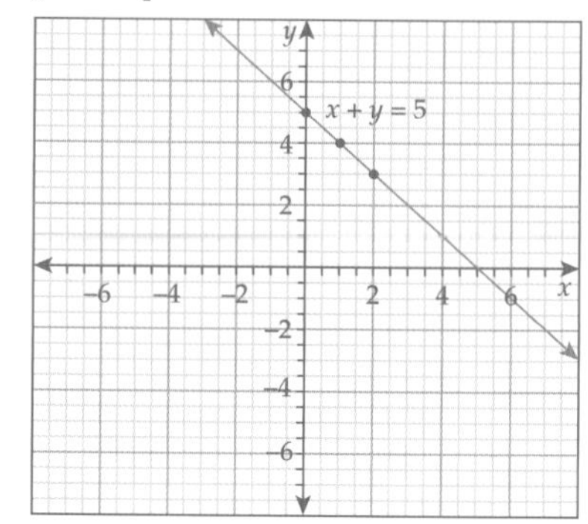

f $y = -2x + 1$

x	0	1	2
y	1	–1	–3

y-intercept = 1

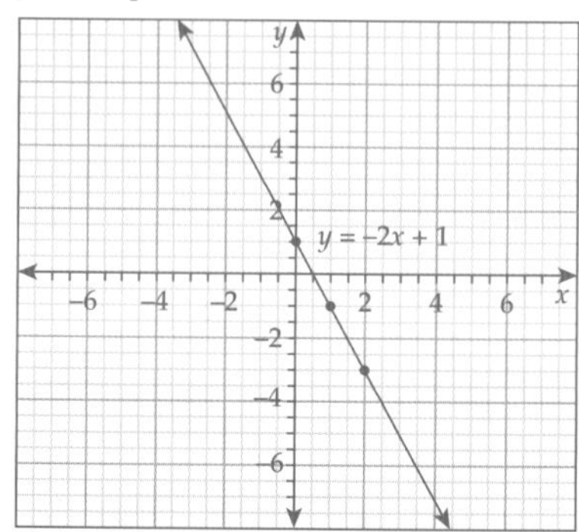

Exercise 17–05

1 C

2 **a** $x = -1$ **b** $x = 4$ **c** $y = 2$ **d** $y = -3$

3 **a** $x = 1$

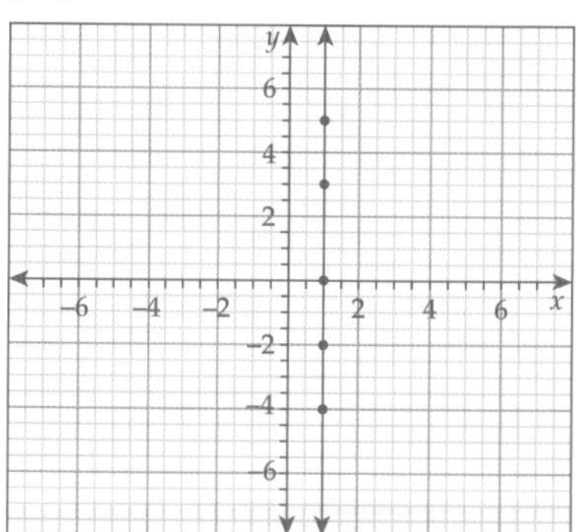

ISBN 9780170350990

b $y = 2$

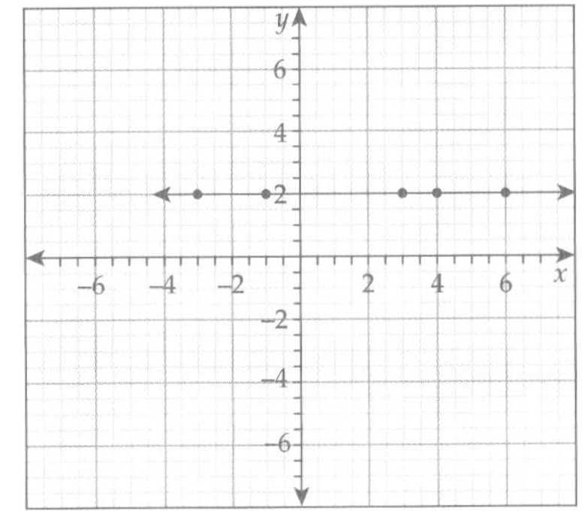

c $y = -3$

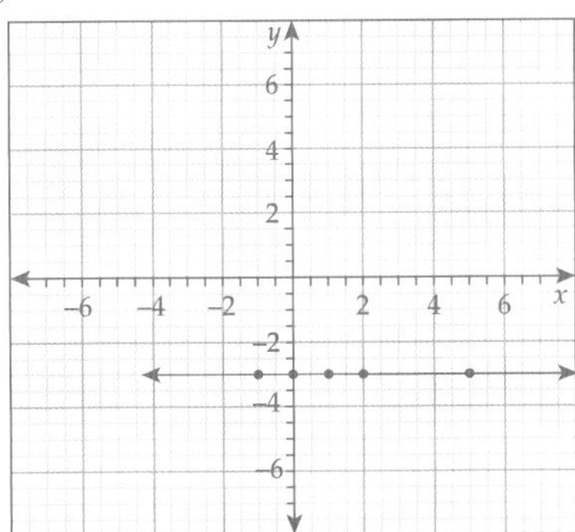

d $x = -2$

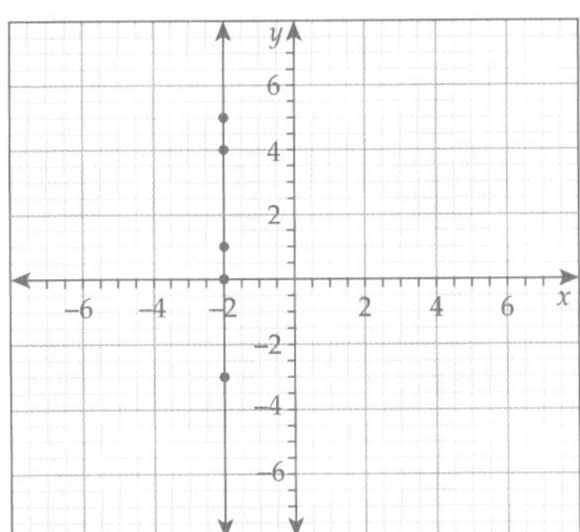

4

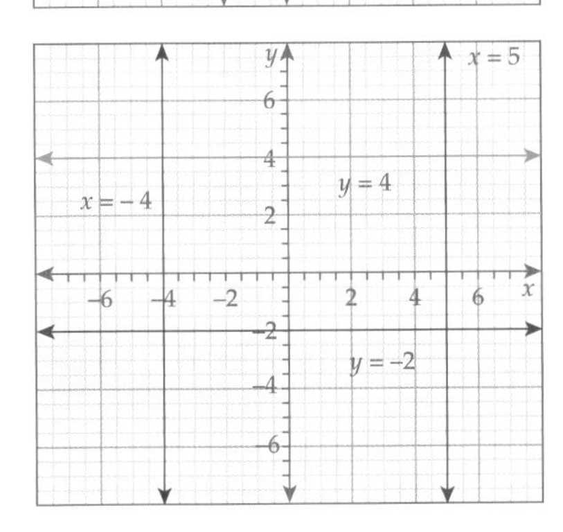

5 B

6 **a** y-axis **b** x-axis

7 **a** $y = 5$ **b** $x = -3$ **c** $y = 3$ **d** $x = 4$

8

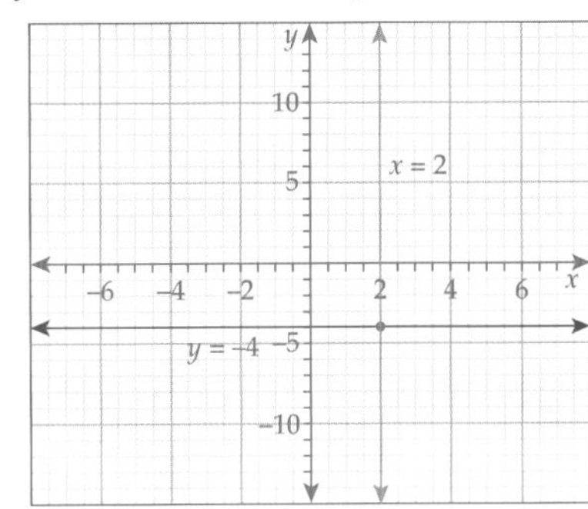

Point of intersection (2, –4)

Language activity

POINT PLOTTING PANDEMONIUM

Practice test **17**

Part A

1. 0.00032
2. 330
3. $\frac{2}{a}$
4. $4mn$
5. 1, 2, 3, 5, 6, 10, 15, 30
6. 36 cm
7. $540
8. $b = 3$
9. $765
10. $\frac{1}{26}$

Part B

11 C

12

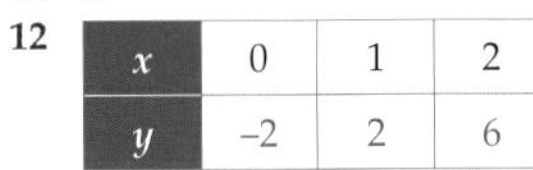

x	0	1	2
y	–2	2	6

13

14 y-intercept = –1

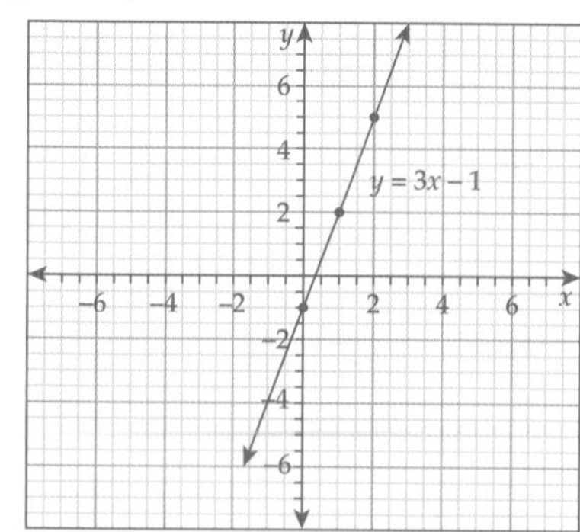

15

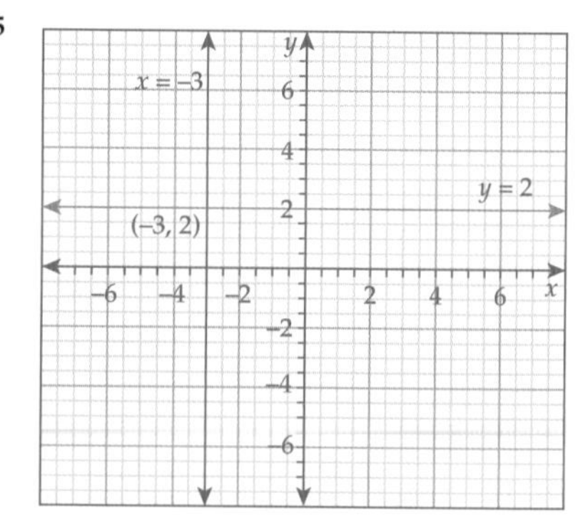

Point of intersection (–3, 2).

ISBN 9780170350990

INDEX

ISBN 9780170350990

ISBN 9780170350990

INDEX

 ISBN 9780170350990